Biocontrol of Plant Diseases by *Bacillus subtilis*

Basic and Practical Applications

T0139278

New Directions in Organic and Biological Chemistry

Series Editor

Philip Page

For more information about this series, please visit: https://www.crcpress.com/New-Direct
ions-in-Organic--Biological-Chemistry/book-series/CRCNDOBCHE

Biocontrol of Plant Diseases by *Bacillus subtilis*

Basic and Practical Applications

Makoto Shoda

CRC Press
Taylor & Francis Group
Boca Raton London New York

CRC Press is an imprint of the
Taylor & Francis Group, an **informa** business

CRC Press
Taylor & Francis Group
6000 Broken Sound Parkway NW, Suite 300
Boca Raton, FL 33487-2742

First issued in paperback 2021

ISBN-13: 978-0-367-13610-9 (hbk)
ISBN-13: 978-1-03-208939-3 (pbk)

Library of Congress Cataloging-in-Publication Data

Names: Shoda, Makoto, author.
Title: Biocontrol of plant diseases by bacillus subtilis : basic and
practical applications / Makoto Shoda.
Description: Boca Raton, Florida : CRC Press, 2019. | Series: New
directions in organic & biological chemistry | Includes bibliographical
references and index. | Summary: "Plant diseases are a serious threat to
food production. This unique volume provides the fundamental knowledge
and practical use of B.subtilis as a promising biocontrol agent. In
order to replace chemical pesticides, one possibility is microbial
pesticides using safe microbes. Bacillus subtilis is one of several
candidates. Screening of the bacterium, the application of plant tests,
clarification of its suppressive mechanism to plant pathogens and
engineering aspects of suppressive peptides production are presented
here. The author illustrates how B. subtilis is far more advantageous
than, for example, Pseudomonas in biocontrol and can be considered as an
useful candidate"-- Provided by publisher.
Identifiers: LCCN 2019030557 (print) | LCCN 2019030558 (ebook) | ISBN
9780367136109 (hardback) | ISBN 9780429027635 (ebook)
Subjects: LCSH: Microbial pesticides. | Bacillus subtilis--Biotechnology. |
Phytopathogenic microorganisms--Biological control.
Classification: LCC SB976.M55 S56 2019 (print) | LCC SB976.M55 (ebook) |
DDC 632/.96--dc23
LC record available at https://lccn.loc.gov/2019030557
LC ebook record available at https://lccn.loc.gov/2019030558

Visit the Taylor & Francis Web site at
http://www.taylorandfrancis.com

and the CRC Press Web site at
http://www.crcpress.com

CV 06.09.2021 2024

Contents

Preface

The aim of this book is to summarize the results of more than 30 years of research on *Bacillus subtilis*, which was isolated in my laboratory, as a candidate biocontrol agent. Although isolated *B. subtilis* has antifungal and antibacterial activities *in vitro*, the application conditions for an *in vivo* test, the suppressive spectrum of plant pathogens, the persistency of *B. subtilis* in soil, its suppressive mechanisms, the stability of the suppressive effect, and the productivity of products remain to be determined.

It is obvious that chemical pesticides have been contributing to the stable production of agricultural products and are necessary for modern agriculture. However, the overuse and risk of chemical pesticides have been noted, and safer chemicals are needed even now. One possible alternative to chemical pesticides is to utilize microbial activities for biocontrol.

Various studies have attempted to improve or change environments using biological interaction, but practical examples are rare. This is mainly because of the lack of basic information and their high cost and instability. In this book, one possibility for using a single microorganism to solve environmental issues is demonstrated.

This book is composed of 12 chapters, and various aspects of isolated *B. subtilis* are presented, from scientific analysis to the engineering optimization of products to the practical usage of the bacteria in fruit gardens.

The features of this book on *B. subtilis* are as follows:

1. Several strains of *B. subtilis* were isolated from compost, and the characteristics of each strain were determined.
2. A suppressive test of isolated *B. subtilis* against various plant pathogens was carried out *in vitro*, and the suppressive spectrum of this bacterium was significantly larger than those of other well-known microorganisms.
3. Using several plant seedlings, the suppression of plant diseases by *B. subtilis* was confirmed via *in vivo* tests.
4. The mechanism of the wide suppressive spectrum demonstrated by *B. subtilis* was investigated according to a scientific analysis and the production of three lipopeptide antibiotics and an iron-chelating chemical, siderophore, was confirmed.
5. One lipopeptide antibiotic, iturin, was targeted to determine its synthetic mechanism by both biochemical and genetic techniques.
6. Two engineering methods to produce high concentrations of lipopeptide antibiotics were investigated, namely, submerged fermentation (SmF) and solid-state fermentation (SSF). As a solid waste substrate for SSF, soybean curd residue (okara) was selected, and the productivity of lipopeptide antibiotics in SSF was compared with that in SmF.
7. Optimization of lipopeptide antibiotics in SSF and spores of *B. subtilis* was conducted.

8. The product derived from treating okara with *B. subtilis* contained both *B. subtilis* cells and lipopeptide antibiotics, which can be reutilized in agriculture as a bifunctional product of an organic fertilizer and a biological pesticide.

9. The rapid death of *B. subtilis* in soil is well accepted. This information is based on the fact that vegetative cells were introduced into soil. In this study, it is shown that when spores of *B. subtilis* are introduced into soil, *B. subtilis* survival is guaranteed for more than 2 months, suggesting that spore treatment of *B. subtilis* in soil is a key factor for the long survival and suppressive activities of *B. subtilis*.

10. Co-utilization of *B. subtilis* and a chemical pesticide was confirmed to be possible.

11. In a fruit garden, more than 15 years of practical treatment with *B. subtilis* made it clear that *B. subtilis* is effective for growing fruits without the use of chemical pesticides.

I believe that this book will provide researchers in this field with a useful resource that supports current knowledge on biocontrol and will provide students with a logical and practical scheme for approaching biocontrol agents.

During my time as director of the Resources Recycling Process Laboratory, at the Chemical Resources Laboratory, Tokyo Institute of Technology, many professors, graduate and undergraduate students, and researchers from outside the campus and from private companies helped to accomplish these results. I am indebted to the following professors and researchers for giving me the opportunity to study biocontrol and for offering many valuable suggestions: Drs. H. Kubota, T. Ano, M. Hirai, K. Okuno, T. Imanaka, M. M. Szczech, S. Fujiwara, and E. Ogata. I am grateful to the graduate and undergraduate students: C. G. Phae, C.-C.Huang, K. Tsuge, O. Asaka, N. Tokuda, K. Nakasaki, S. Mizumoto, A. Ohno, M. Sasaki, S. Nakayama, M. Nakano, Y. Matsuno, M. Kondoh, T. Akiyama, S. Inoue, T. Ueda, and H. Hiraoka.

I sincerely thank Ms. K. Shimada for refining the tables and figures. I would like to express deep appreciation to Mr. S. Hirose, the owner of Hirose-en Tourist Fruit Garden, for applying *B. subtilis* in his garden and for accumulating valuable data.

I would like to acknowledge with gratitude the encouragement I have received from my wife during the preparation of this book.

Finally, special appreciation is given to Professor N. Kita, at Nihon University, the former director of the Kanagawa Institute of Agricultural Science, for providing help and support from the start of this research.

Author

Makoto Shoda started his academic career with undergraduate and graduate courses in biochemical engineering at the University of Tokyo, Japan. His PhD thesis was on the modeling and computer simulation of product inhibition in microbial reactions under the supervision of Professor S. Aiba, who was a pioneer in biochemical engineering. This work on the introduction of computers to modeling was the first trial in the microbial world, and his scientific paper received the 40th Anniversary Award for Papers with Special Impact on Biotechnology (John Wiley & Sons, Inc.)

As assistant professor at Nagoya University, Shoda gained experience in basic biochemistry and microbiology under the supervision of the late Professor S. Udaka, who was famous as the pioneer of producing amino acids using microbial reactions. Shoda then moved to the Tokyo Institute of Technology, where he concentrated on environmental biotechnology based on basic genetics and biochemistry, as well as computer control. As director of the Research Laboratory of Resources Recycling Process, he supervised numerous graduate students who contributed to this book. He has published more than 250 scientific papers, regarding isolation of several new microorganisms, including a new fungus that decomposes many toxic and colored chemicals by producing different unique enzymes and a bacterium that converts ammonia directly to N_2 gas without using conventional nitrification and denitrification processes at a 100-fold higher rate than the conventional method. He is active in research and application of microbial reactions as professor emeritus at the Tokyo Institute of Technology.

1 Introduction

The overuse of chemical pesticides to cure or prevent plant diseases has caused soil pollution and had harmful effects on human beings. Accordingly, to reduce the use of these chemicals, one possibility is to utilize the activity of microorganisms. It is desirable to replace chemical pesticides with materials that possess the following three criteria: (i) high specificity against the targeted plant pathogens, (ii) easy degradability after effective usage, and (iii) low cost of mass production. Products produced biologically or the microbial cells themselves are called biological control (biocontrol) agents or biological pesticides if they fulfill these criteria. Under these circumstances, the use of bacteria has been investigated mainly because genetic and biochemical analyses and the mass production of bacteria or bacterial products are much easier than those of fungi, and thus the advance of bacterial control is expected to have great potential. As bacterial control agents, *Agrobacterium*, *Pseudomonas*, *Bacillus*, *Alcaligenes*, *Streptomyces*, and others have been reported. Different mechanisms are involved in the actions of these bacteria against plant pathogens, such as parasitism, cross protection, antibiosis, and competition (1). Among them, the antibiotic mechanism is to operate when the metabolic products (antibiotics) produced by one species inhibit or suppress the growth of another species. This is the main mechanism that *B. subtilis* isolated in this study plays as a biocontrol agent.

It is critical to note in relation to biocontrol that while inhibition of the growth of plant pathogens is observed *in vitro*, often no correlation is observed between inhibition *in vitro* and reduction of diseases in a test in a greenhouse or in a field. Therefore, a plant test is essential to confirm the effectiveness of biocontrol agents.

As pseudomonads already have a long history as a candidate of bacterial control agents, and some reviews and reports of intensive research have been published (1), brief comments on pseudomonads are given and mainly the use of *Bacillus* spp. are emphasized.

1.1 BIOLOGICAL CONTROL OF PLANT PATHOGENS BY *PSEUDOMONAS* SPECIES

Pseudomonas strains are some of the most active and dominant bacteria in the rhizosphere and have been intensively investigated as biocontrol agents. *P. fluorescens*, *P. putida*, and *P. cepacia* were the predominant focus of research for practical applications. The strains of *Pseudomonas* produce several kinds of antibiotics, such as pyrolnitrin, pyoluteorin, and phenezine-1-carboxylate, all of which are closely related to the suppression of plant diseases.

Some *Pseudomonas* strains were confirmed to suppress the activities of plant pathogens by the production of siderophore or pseudobactin. Siderophores are extracellular, low-molecular-weight compounds that have a high affinity for ferric ions and play a role in biocontrol.

A siderophore-producing *P. fluorescens* induced an increase in the emergence of cotton seedlings in a soil infested with *Pythium ultimum*, while siderophore-deficient mutants of *P. fluorescens* derived by Tn*5* mutagenesis induced a decrease in the emergence of seedlings.

1.2   BIOLOGICAL CONTROL OF PLANT PATHOGENS BY *BACILLUS* SPECIES

The use of the gram-positive *Bacillus* species as a biocontrol agent has been relatively rare and has been studied less intensively than that of the gram-negative bacteria. *Bacillus subtilis* is mainly studied and occasionally *B. megaterium*, *B. cereus*, *B. pumilus*, and *B. polymyxa*. As *Bacillus* spp. have the characteristics of omnipresence in soils, high thermal tolerance, rapid growth in liquid culture, ready formation of resistant spores, and are considered to be a safe biological agent, their potential as biocontrol agents is considered to be high. However, evaluation of the bacteria has focused primarily on the degree of disease suppression (2–6). The population dynamics and mechanism of suppression to plant pathogens in soil by *Bacillus* spp. have not been extensively investigated.

1.3   PRODUCTION OF ANTIFUNGAL SUBSTANCES

B. subtilis produces several kinds of antibiotics, e.g., bacillomycin, iturin, mycosubtilin, bacilysin, fengycin, plipastatin, and mycobacillin. However, the bacteria require a suitable substrate for production of these materials in soil. Investigation of the control of *Streptomyces scabies* in potatoes by *B. subtilis* revealed that more antibiotic was produced when the bacteria were grown in a water extract of soybean (7). Furthermore, when this bacterium was added to soil, buildup of potato scab was prevented, presumably because the soybean substrate supported antibiotic production in soil by *B. subtilis*, thus enhancing its ability to control the specific disease. A root system includes a wide variety of different substances whose compositions vary with changes in the metabolic state of different parts of the root system. The changes can be influenced by outside effects on the plants. Once a root is invaded by a pathogen, significant changes in exudation occur. Such changes could convert the environment into one hostile to the biocontrol agent. Because antibiotic activity is so dependent on outside events, biocontrol through its effects is likely to be unsatisfactory and certainly unpredictable, unless strains of microorganisms can be manipulated to make the synthesis of antagonistic compounds less susceptible to changes in nutrient sources. In this book, Chapters 2 and 3 explain the production of antifungal substances—iturin, surfactin, and plipastatin—and some examples of plant test related with these substances in soil.

1.4   PRODUCTION OF SIDEROPHORES

2,3-Dihydroxybenzoyl-glycine (2,3-DHBG) is known as only one siderophore produced by the gram-positive *B. subtilis* (8, 9). Bacterial siderophores are also known to have functions to help plant growth such as supplying iron via an iron-chelating function called the plant-growth-promoting rhizobacteria (PGPR) effect, which eventually leads to a decrease in disease occurrence. The strain of *B. subtilis* isolated in this study showed a wide suppressive spectrum on plant pathogens by producing antibiotics and 2,3-DHBG (Chapter 5). Gram-negative *Escherichia coli* produces a siderophore, enterobactin; and the *E. coli* mutant of *entD*-encoding enterobactin was complemented for enterobactin production by the *sfp*^{0} gene of *B. subtilis*, which differs from *sfp* (surfactin production gene) by five base substitutions and one base insertion that truncated a 224-amino-acid residue in *sfp* to a 165-amino-acid residue. Similar complementation was found in the regulatory gene *lpa-14* for the antibiotics iturin A and surfactin derived from a potential bacterial control agent of *B. subtilis*. The finding that *B. subtilis* produced not only antibiotics that suppress plant pathogens but also siderophores, and the regulation of these products by the gene *lpa-14* indicates the possibility of enhanced effectiveness of biocontrol by the manipulation of the gene (Chapter 5).

1.5   GENETIC ANALYSIS

Some antifungal peptides produced by *Bacillus* species are synthesized by multienzyme complexes via nonribosomal mechanisms, and the molecular organization of these peptide synthetases encoded by the bacterial operons has been well analyzed. However, no relation between the genetic information and the suppressive effect on plant pathogens in soil has been reported. One regulatory gene *lpa-14*, which is responsible for the production of antibiotics, was found to be closely involved in the bacterial suppressive function against a plant disease in soil (Chapter 5).

B. *subtilis* RB14 is a producer of the antifungal lipopeptide iturin A. Using a transposon, the iturin A synthetase operon of RB14 was identified and cloned, and the sequence of this operon was determined. The iturin A operon is composed of four open reading frames: *ituD*, *ituA*, *ituB*, and *ituC*. Each frame was characterized, and the homology with each frame of mycosubtilin was clarified. A 42 kb region of the *B. subtilis* RB14 genome, which contains a complete 38 kb iturin A operon, was transferred to the genome of a non-iturin A producer, *B. subtilis* 168, and the recombinant was converted into an iturin A producer (Chapter 6).

B. *subtilis* YB8 produces the lipopeptide antibiotic plipastatin. *B. subtilis* MIll3, which is a derivative of strain 168, was converted into a new plipastatin producer by competence transformation with the chromosomal DNA of strain YB8 (Chapter 6).

1.6   ECOLOGICAL FITNESS IN SOIL

A few investigations have been done from the ecological point of view of *Bacillus* spp. The environmental factors for the production of the antifungal compounds by *Bacillus* spp. are important. The *B. subtilis* strain showed the optimal temperature for the production of antifungal substances at 30°C in liquid cultivation, but at below 25°C in solid-state cultivation (Chapter 7). The temperature difference is reflected in the ecological behavior in soil, mostly in relation to the suppressive capacity of the bacterium, as the average soil temperature of mainland Japan is below 25°C. It is often said that *Bacillus* strains are not normally stimulated in the root region.

However, when introduced massively into soil, they multiply and colonize parts of the roots. They are good competitors and attach themselves firmly to root surfaces, especially if an inoculum of spores is used. A spore inoculum for *Bacillus* strains is advantageous to use in biocontrol preparations, because it is resistant to adverse environmental conditions and remains viable for several years under any storage condition. Environmental conditions must be appropriate for the spores to germinate to vegetative cells in the soil. As the generation time of *Bacillus* species is slower than that of *Pseudomonas* species, *Bacillus* species are less likely to lead to modified rhizosphere microflora.

Although *B. subtilis* is not considered a representative rhizosphere strain as is *Pseudomonas* spp., the rhizosphere population density as well as the persistence of the bacterium in soil is an important factor in the suppression of plant diseases in soil. *B. subtilis* is uniformly distributed throughout the soil and is not concentrated in the plant rhizosphere, and the population of *B. subtilis* is considered to be 10^3–10^5 cells/g of soil, indicating that it is a minor species in soil. The threshold population density of bacteria to show significant suppression of plant pathogens was reported as 10^5 cells/g of root in the case of *Pseudomonas* spp. (10). It was found that most of the cells of *B. subtilis* recovered from soil were in the form of spores that had no ability to produce antibiotics. However, the obvious reduction of plant disease suggests that the spores germinate in the rhizosphere and the antibiotics are produced.

As well as the survival of microorganisms, the persistence of antibiotics produced by bacteria in soil is crucial for the maintenance of their stable efficacy on plant pathogens. An example of the stability of the antibiotics iturin A and surfactin in soil into which the solutions of these substances were introduced is shown in Figure 3.8 (see Chapter 3). Surfactin was more persistent than iturin A in the soil. The decrease in the amount of these antibiotics was caused by leaching from soil during watering, biodegradation by microorganisms, or irreversible binding to soil materials or humic acids. Since the concentrations of iturin A and surfactin in a centrifuged culture supernatant were stably maintained for more than 1 month, such instability was related to the soil environment. When vegetative cells of *B. subtilis* were inoculated into the soil, the survival rate of the cells was very low. However, once the spores of the bacterium were introduced or sporulation from vegetative cells occurred, mainly because of the oligotrophic condition of the soil, the population was stabilized in accordance with the spore number on the order of 10^8 cells/g of dry soil, and almost no change was observed for 50 days (Chapter 4). Therefore, if the initial populations are well maintained in soil and most of the cells exist as spores, *B. subtilis* is stable in the soil as spores; the spore's stability and ease of handling are advantageous for the use of this bacterium as a biocontrol agent.

1.7   PRODUCTION OF ANTIBIOTICS IN LIQUID CULTIVATION

For the mass production of antagonistic bacteria or antibiotics produced by bacteria, liquid cultivation is the conventional method. With the advances in bioreactor-based biotechnological processes, it has become important to develop economical methods for the isolation and purification of products from cultures or cells that are suitable for large-scale application in biological control.

When *B. subtilis* isolated in this study was cultivated in conventional liquid cultivation, the lipopeptide was condensed only in the foam formed during cultivation. This indicates that continuous separation and condensation of the product is possible only by collecting the foam under properly controlled foaming conditions (Chapter 7). Some sophisticated separation methods, such as membrane separation or chromatographic procedures, are successful as laboratory methods but have limitations as large-scale processes. Foam separation is a possible and simpler separation method. Foam separation has been reported for the removal of several microorganisms from aqueous suspensions (11–13) or for the separation of extracellular proteins produced by *Saccharomyces cerevisiae* (14). However, foam separation and the concentrations of peptidolipidic substances in a jar fermenter have not been reported. Foaming in a jar fermenter is generally cumbersome, and the use of an antifoam agent is in most cases unavoidable. The feature of this method is closely related to the chemical nature of the product, which contains both hydrophobic and hydrophilic parts in its structure. The advantage of this method also lies in the minimal use of an antifoam agent, which sometimes causes physiological and economical disadvantage in fermentation (Chapter 7).

1.8   OPERATION OF SOLID-STATE CULTIVATION

In order to produce bacterial control agents, solid-state cultivation of bacteria using solid organic wastes is one possibility. More than 2 million tons of organic waste available for microbial treatment are disposed off annually from the food industry in Japan. However, direct application of these organic matters to soil causes severe damage to plants due to the low ratio of carbon to nitrogen in them. Therefore, some pretreatment before introduction into soil is needed. One of the materials is a soybean curd residue, which is a by-product of tofu manufacturing. Approximately 700,000 tons of the residue are produced annually; most of it is incinerated as an industrial waste. In the cultivation of soybean curd residue for the growth of *B. subtilis*, the accumulation of the antibiotics associated with suppression of plant pathogens was confirmed when proper temperature control and air supply were maintained (Chapter 7).

As iturin A has five homologues, the production of fractions with a longer side chain, which show stronger antibiotic activity, was greater in solid-state cultivation, indicating that solid-state cultivation has an advantage over liquid cultivation at least in regard to this point.

B. subtilis is a good candidate for solid-state fermentation and the product treated by the bacterium can be utilized as an organic material that functions both as an organic fertilizer and a microbial pesticide. As these organic matters are vulnerable to decay, a new oven-drying process was developed, and the processed product exhibited improved transportability and preservability (Chapter 7).

Solid-state cultivation of *B. subtilis* has the following features: (i) Traditional oriental foods are produced by solid-state cultivation (e.g., enzymes, organic acids, koji-making). (ii) Most of the microorganisms used are fungi. Little bacterial usage is encountered. (iii) Solid substrates require only addition of water or other nutrients. (iv) Vessels in solid-state cultivation are small relative to the product yield.

(v) Low-moisture content reduces the problem of contamination. (vi) Extraction processes are simpler. (vii) The amount of solvents used for extraction is much smaller. (viii) Spore inoculation does not need a seed tank. (ix) Products can be used for direct applications such as animal feed and fertilizer.

The extraction procedure of the solid material to yield antibiotic is much simpler in solid-state cultivation than in liquid cultivation, and smaller amounts of solvent are needed in solid-state cultivation. This is mainly because the water content of solid material is much lower than that of the liquid medium, and because the accumulated antibiotic concentration in solid-state cultivation is higher than that in liquid cultivation. In these respects, solid-state cultivation is more promising for the mass production of *B. subtilis* as a biocontrol agent (Chapter 7).

1.9 PLANT TEST

As a means of application of *B. subtilis* as an antagonist to plant pathogens in soil, the methods of dipping plant roots in bacterial solution, seed coating (15–21), and direct mixing with soil (22–27) are possible. However, these reports were rather phenomenological and there is a lack of scientific analysis for the mechanisms that are involved in the interaction with soil. When the culture samples obtained from liquid cultivation of *B. subtilis* were applied to the biological control of crown and root rot of tomato caused by *Fusarium oxysporum*, the tomato yield was three to four times greater as compared with the control in a greenhouse test (26). The sample was also effective to reduce occurrence of bacterial wilt of tomato caused by the bacterial pathogen *Pseudomonas solanacearum* (Chapter 3). So far, *B. subtilis* was considered to be not effective against *P. solanacearum* (28). In the soil, where occurrence of damping-off of tomato caused by *Rhizoctonia solani* was suppressed by the *B. subtilis* culture, cells of *B. subtilis* were recovered and antibiotics were detected (29). The product from solid-state cultivation was also applied to a plant test of tomato for the damping-off and the occurrence of the disease was reduced to one-third of that of only the control plant pathogen (30). Cells of the bacterium and the antibiotics were also recovered (Chapter 7).

As considerable difficulty was envisaged in large-scale inoculation of the soil of a field with bacterium, the seed bacterization or dipping plant roots to enhance the effectiveness of bacterial control agents will be alternative methods.

Liberation of organic nutrients by the germinating seeds and the roots has a profound effect on the development of specific microbial populations on seed and root surfaces as in the soil around them. The soilborne pathogens have to compete with the antagonistic microbial population of the rhizosphere in order to invade the host. The rhizosphere soils are normally known to contain higher concentrations of sugars and amino acids and thus would help in the increased production of the antibiotic once the bacterium introduced in this region is established. The bacteria on seed would create an inhibitory zone at these loci by production of antibiotics against invasion of roots of the seedlings by the pathogens. Moreover, additional nutrients will be effective if seed bacterization is to be attempted. Seed inoculation with *B. subtilis* has decreased wilt of pigeon peas grown in wilt-sick soil, especially when the bacterial suspension was prepared with molasses and groundnut cake materials (18).

The production of the antibiotic lasted for more than 10 days in soil and the amount of the produced antibiotic was higher when the seeds were bacterized in molasses and groundnut cake.

B. subtilis was used successfully to control Pythium and Rhizoctonia diseases of ornamental plants grown in soil treated by aerated steams (19). However, tests in field soils by inoculating several different strains of B. subtilis did not always increase the yield. The fact that the antibiotic is produced in unsterilized soil in appreciable quantities is important from the practical point of view, because most microorganisms that produce antibiotics in sterilized soil fail to do so under unsterilized conditions. Therefore, when inconsistency and low efficiency of biocontrol occur in the field, one possible solution is the use of an integrated biocontrol system, using a mixture of several bacteria, or a mixture of chemical and biological agents. B. subtilis with added chemicals yielded more effective control of onion white rot caused by Sclerotium cepivorum (31), root rot of field pea caused by R. solani (32), and potato scab caused by Streptomyces scabies (33) than chemicals alone. In the greenhouse, the occurrence of these diseases was reduced by 60–70%, and in the field a similar effect was observed. The synergistic phenomenon involved in integrated control using fungicides and biocontrol agents may be more efficient and longer lasting than the control achieved through biocontrol agents or fungicides alone (Chapter 9).

1.10 COMPOSTS

One of the several benefits of treating soil with composted organic waste is to stimulate microbial activity in the soil. Besides this, some microorganisms inhabiting composts have the potential to suppress plant diseases. With increased public awareness of the need to protect our environment, such biological control of plant diseases has received attention mainly as an alternative to the use of fungicides or bacteriocides. Some reviews on biological control using organic matter have been published (34, 35). Hoitink and his colleagues (36, 37) demonstrated suppression by matured hardwood compost mixed with a soilless container medium, and some bacterial antagonists to phytopathogenic fungi have been isolated and their effects compared (38). However, the occurrence of plant diseases is influenced by plant species, the types of diseases, cultivation methods, the time between soil amendment with compost and planting (39), and environmental factors that depend on local or regional conditions. In Japan, some composts are known to have a suppressive effect on the occurrence of plant disease, but little is known about the mechanism of these effects. In Chapter 2, effective composts and B. subtilis isolated from these composts are demonstrated.

REFERENCES

1. Shoda, M. Bacterial control of plant disease. J. Biosci. Bioeng., 89, 515–521 (2000).
2. Aldrich, J., and Baker, R. Biological control of Fusarium roseum f. sp. dianthi by Bacillus subtilis. Plant Dis. Rep., 54, 446–448 (1970).
3. Utkhede, R. S. Antagonism of isolates of Bacillus subtilis to Phytophthora cactorum. Can. J. Bot., 2, 1032–1035 (1984).

4. Silo-suh, L. A., Lethbridge, B. J., Raffel, S. J., He, H., Clardy, J., and Handelsman, J. Biological activities of two fungistatic antibiotics produced by *Bacillus cereus* UW85. *Appl. Environ. Microbiol.*, 60, 2023–2030 (1994).

5. Siala, A., and Gray, T. R. Growth of *Bacillus subtilis* and spore germination in soil observed by a fluorescent-antibody technique. *J. Gen. Microbiol.*, 81, 191–198 (1974).

6. van Elsas, J. D., Dijkstra, A. F., Govaert, J. M., and van Veen, J. A. Survival of *Pseudomonas fluorescens* and *Bacillus subtilis* introduced into two soils of different texture in field microplots. *FEMS Microbiol. Ecol.*, 38, 151–160 (1986).

7. Weinhold, A. R., and Bowman, T. Selective inhibition of potato scab pathogen by antagonistic bacteria and substrate influence on antibiotic production. *Plant Soil*, 27, 12–24 (1968).

8. Leong, J. Siderophores: Their biochemistry and possible role in the biocontrol of plant pathogen. *Annu. Rev. Phytopathol.*, 24, 187–209 (1986).

9. Ito, T., and Neilands, J. B. Products of "low iron fermentation" with *Bacillus subtilis*: Isolation, characterization and synthesis of 2,3-dihydroxybenzoylglycine. *J. Am. Chem. Soc.*, 80, 4645–4647 (1958).

10. Raaijmakers, J. M., Leeman, M., van Oorschot, M. M. P., van der Sluis, I., Schippers, B., and Bakker, P. A. H. M. Dose-response relationships in biological control of fusarium wilt of radish by *Pseudomonas* spp. *Phytopathology*, 85, 1075–1081 (1995).

11. Rubin, A. J., Cassel, E. A., Henderson, O., Johnson, J. D., and Lamb III, J. C. Microfloatation: New low gas-flow rate foam separation technique for bacteria and algae. *Biotechnol. Bioeng.*, 8, 135–151 (1966).

12. Grieves, R. B., and Wang, S. L. Foam separation of *Escherichia coli* with a cationic surfactant. *Biotechnol. Bioeng.*, 8, 323–336 (1966).

13. Wang, D. I. C., and Sinskey, A. Collection of microbial cells, pp. 121–152. In Perlman, D. (ed.), *Advances in Applied Microbiology*, vol. 12. Academic Press, New York (1970).

14. Effler, W. T., Tanner, R. D., and Malaney, G. W. Dynamic *in situ* fractionation of extracellular proteins produced in a baker's yeast cultivation process, pp. 235–255. In Fiechter, A., Okada, H., and Tanner, R. D. (eds.), *Bioproducts and Bioprocesses*. Springer-Verlag, Berlin (1989).

15. Singh, P., Vasudeva, R. S., and Bajaj, B. S. Seed bacterization and biological activity of bulbiformin. *Ann. Appl. Biol.*, 55, 89–97 (1965).

16. Chang, I., and Kommedahl, T. Biological control of seedling blight of corn by coating kernels with antagonistic microorganisms. *Phytopathology*, 58, 1395–1401 (1968).

17. Merriman, P. R., Price, R. D., Kollmorgen, J. F., Piggott, T., and Ridge, E. H. Effect of seed inoculation with *Bacillus subtilis* and *Streptomyces griseus* on the growth of cereals and carrots. *Aust. J. Agric. Res.*, 25, 219–226 (1974).

18. Kommedahl, T., and Mew, I. C. Biocontrol of corn root infection in the field by seed treatment with antagonists. *Phytopathology*, 65, 296–300 (1975).

19. Broadbent, P., Baker, K. F., Franks, N., and Holland, J. Effect of *Bacillus* spp. on increased growth of seedlings in steamed and in nontreated soil. *Phytopathology*, 67, 1027–1034 (1977).

20. Fillippi, C., Bagnoli, G., and Picci, G. Antagonistic effects of soil bacteria on *Fusarium oxysporum* f. sp. *dianthi* (Prill and Del.) Synd and Hans. IV: Studies on controlled *Solanum lyecopersicum* L. and *Dianthus caryophyllus* L. rhizosphere. *Agr. Med.*, 119, 327–336 (1989).

21. Turner, J. T., and Backman, P. A. Factors relating to peanut yield increases after seed treatment with *Bacillus subtilis*. *Plant Dis.*, 75, 347–353 (1991).

22. Olsen, C. M., and Baker, K. F. Selective heat treatment of soil and its effect on the inhibition of *Rhizoctonia solani* by *Bacillus subtilis*. *Phytopathology*, 58, 79–87 (1968).

23. Broadbent, P., Baker, K. F., Franks, N., and Holland, J. Effect of *Bacillus* spp. on increased growth of seedlings in steamed and nontreated soil. *Phytopathology*, 67, 1027–1034 (1977).
24. Yuen, G. Y., Schroth, M. N., and McCain, A. H. Reduction of *Fusarium* wilt of carnation with suppressive soil and antagonistic bacteria. *Plant Dis.*, 69, 1071–1075 (1985).
25. Gupta, V. K., and Utkhede, R. S. Factors affecting the production of antifungal compounds by *Enterobacter aerogenes* and *Bacillus subtilis*, antagonists of *Phytophthora cactorum*. *J. Phytopathol.*, 120, 143–153 (1986).
26. Bochow, H. Use of microbial antagonists to control soil borne pathogens in greenhouse crops. *Acta Hortic.*, 255, 271–280 (1989).
27. Phae, C. G., Shoda, M., Kita, N., Nakano, M., and Ushiyama, K. Biological control of crown and root rot and bacterial wilt of tomato by *Bacillus subtilis* NB22. *Ann. Phytopath. Soc. Japan*, 58, 329–339 (1992).
28. Podile, A. R., Prasad, G. S., and Dube, H. C. *Bacillus subtilis* as antagonist to vascular wilt pathogens. *Curr. Sci.*, 54, 864–865 (1985).
29. Asaka, O., and Shoda, M. Biocontrol of *Rhizoctonia solani* damping-off of tomato with *Bacillus subtilis* RB14. *Appl. Environ. Microbiol.*, 62, 4081–4085 (1996).
30. Mizumoto, S., Hirai, M., and Shoda, M. Production of lipopeptide antibiotic iturin A using soybean curd residue cultivated with *Bacillus subtilis* in solid-state fermentation. *Appl. Microbiol. Biotechnol.*, 72, 869–875 (2006).
31. Utkhede, R. S., and Rahe, J. E. Chemical and biological control of onion white rot in muck and mineral soil. *Plant Dis.*, 67, 153–155 (1983).
32. Hwang, S. F., and Chakravarty, P. Potential for the integrated control of *Rhizoctonia* root-rot of *Pisum sativum* using *Bacillus subtilis* and a fungicide. *Z. PflKrankh. PflSchutz.*, 99, 626–636 (1992).
33. Schmiedeknecht, G., Bocbow, H., and Junge, H. Use of *Bacillus subtilis* as biocontrol agent. II. Biological control of potato diseases. *Z. PflKrankh. PflSchutz.*, 105, 376–386 (1998).
34. Hoitink, H. A. J., and Fahy, P. C. Basis for the control of soilborne plant pathogens with composts. *Annu. Rev. Phytopathol.*, 24, 93–114 (1986).
35. Lumsden, R. D., Lewis, J. A., and Papavizas, G. C. Effect of organic amendments on soilborne plant diseases and pathogen antagonists, pp. 51–70. In Lockeretz, W. (ed.), *Environmentally Sound Agriculture*. Praeger Press, New York (1983).
36. Chef, D. G., Hoitink, H. A. J., and Madden, W. Effects of organic components in container media on suppression of *Fusarium* wilt of chrysanthemum and flax. *Phytopathology*, 73, 279–281 (1983).
37. Kuter, G. A., Nelson, E. B., Hoitink, H. A. J., and Madden, L. V. Fungal population in container media amended with composted hardwood bark suppressive and conducive to *Rhizoctonia* damping-off. *Phytopathology*, 73, 1450–1456 (1983).
38. Kwok, O. C. H., Fahy, P. C., Hoitink, H. A. J., and Kuter, G. A. Interactions between bacteria and *Trichoderma hamatum* in suppression of *Rhizoctonia* damping-off in bark composted media. *Phytopathology*, 77, 1206–1212 (1987).
39. Millner, P. D., Lumsden, R. D., and Lewis, J. A. Controlling plant disease with sludge compost. *Biocycle*, 23, 50–52 (1982).

2 *In Vitro* Selection of *B. subtilis* as a Candidate of Biocontrol Agent and Characterization of Suppressive Products

2.1 SUPPRESSIVE EFFECT OF COMPOSTS ON PHYTOPATHOGENIC FUNGI (1)

2.1.1 MATERIALS AND METHODS

2.1.1.1 Compost Used

Fifty-three well-matured compost products produced commercially from various materials were collected. Thirty of them were bark composts, six were made from sewage or night soil sludge, four were garbage composts, and thirteen were cattle manure composts. Some immature samples were also collected at each stage during the composting operation in two composting facilities in addition to the matured products and the suppressive effect of each sample was investigated.

2.1.1.2 Preparation of Suspensions of Phytopathogenic Fungi

As phytopathogenic fungi, *F. oxysporum* f. sp. *cucumerinum*, *Pythium ultimum*, *Verticillium dahliae*, and *Rhizoctonia solani* were used in *in vitro* tests to find suppressive composts. Cells grown on potato dextrose agar (PDA) medium slants (potato dextrose agar, 39 g; distilled water 1 liter [pH 5.6]) were suspended in sterile distilled water. Then 0.1 ml of each suspension was spread on potato-dextrose agar plates. The plates were incubated at 27°C for 4 to 7 days until the hyphae of the fungi covered their surface. Next, 30 ml of sterile water was poured into the plates and fully grown fungi were suspended. The suspensions were filtered with four sheets of gauze, and then the filtrate containing spores and/or mycelia was diluted with sterile water to adjust the concentration to about 1.0×10^7 colony-forming units (cfu/ml).

2.1.1.3 Suppressiveness Testing of Composts

For the suppressiveness testing, 0.1 ml of each diluted filtrate of the fungi prepared as mentioned earlier was spread onto agar plates containing either no. 3 medium (glucose, 10 g; Polypepton, 10 g; KH_2PO_4, 1 g; $MgSO_4 \cdot 7H_2O$, 0.5 g; agar, 18 g; distilled water 1 liter [pH 6.8]) or potato-dextrose agar medium. Then, about 0.5 g of the composts was placed onto the center of the plates and they were incubated at 27°C. As a reference, composts sterilized either by autoclave treatment (123°C, 20 min) or by gamma-irradiation with ^{60}Co for 3 h at a dose of 1 Mrad/h were also placed on agar plates. Changes of the surface were observed for 20 to 30 days by the naked eye or using an observatory microscope.

2.1.2 RESULTS

Among the 53 compost samples collected, 10 showed clear inhibitory zones to the growth of at least one fungus among the four phytopathogenic fungi tested on the agar plates. Among 10 compost samples, four composts from the N, U, Y, and S plants showed a clear inhibition of the growth of the fungi on agar plates as shown in Figure 2.1A. These composts were initially covered with hyphae of the fungi, but after 5–15 days, clear inhibitory zones appeared around the composts, and the zone area expanded later. The materials of the composts and conditions of composting plants N, U, Y, and S are listed in Table 2.1. None of the sterilized composts showed a suppressive effect as indicated in Figure 2.1B. There was no difference in results between the two sterilization methods (2).

Several samples were collected from different stages of the two composting plants and the suppressive effect on the four phytopathogenic fungi was evaluated. In the N plant, this composter is a vertical multistage apparatus with forced aeration where the primary fermentation period is 7 days. No suppressive effect was observed for the composting material at the start, namely, a mixture of excess sludge and digested sludge in a night soil treatment plant. Only samples from the 7 days composting and

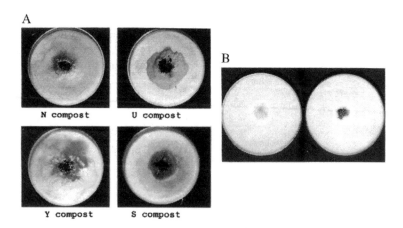

FIGURE 2.1 (A) Suppressive effect *in vitro* of N, U, Y, and S composts on *Fusarium oxysporum* f. sp. *cucumerinum* after a 26-day incubation period at 28°C. (B) Right: Suppressive effect of a composting material at the start (a mixture of excess and digested sludges) at N composting plant on *F. oxysporum* f. sp. *cucumerinum*. Left: Suppressive effect of sterilized N compost on *F. oxysporum* f. sp. *cucumerinum*.

TABLE 2.1

Compost Samples Collected and Operational Conditions

		Primary Stage		Curing Stage	
Plant Name	Main Material	Operation Period (Days)	Maximum Temperature (°C)	Operation Period (Days)	Maximum temperature (°C)
N	Night soil sludge	7	70	14	50
Y	Barks	85	75	—a	—a
U	Garbage	7	80	60	80
S	Garbage	4	50	21	60

a Data are not available.

final product after curing stage showed the same degree of suppressiveness. In the Y composting plant, the materials were piled with occasional turning without any forced air supply. As mites and nematodes present in raw bark preyed on fungi, no clear suppressive effect was observed until the final compost product after 80 days operation.

2.2 ISOLATION OF BACTERIA SUPPRESSIVE TO FUNGI FROM COMPOSTS

2.2.1 MATERIALS AND METHODS

2.2.1.1 Isolation of Suppressive Bacteria

Isolation of suppressive bacteria was tried from each compost that showed a significant suppressive effect on the agar plates against any of the phytopathogenic fungi mentioned earlier. The isolating medium for bacteria was Trypticase soy agar medium (BBL Microbiology Systems, Trypticase peptone, 17 g; phytone peptone, 3 g; NaCl, 5 g; K_2HPO_4, 2.5g; glucose, 2.5 g; agar, 20 g; distilled water 1 liter, pH 7.3). As preliminary experiments, several media for bacterial isolation from composts were tried and the largest number of isolates appeared on the Trypticase soy agar medium (3). From the outer vicinity of the suppressive composts placed on the agar plates, a small amount of compost was picked up by a sterilized platinum string and suspended into the isolating medium. The colonies that appeared on the plates were purified.

2.2.1.2 Suppressiveness Testing of Isolated Bacteria

Cells of purified bacteria grown on agar slants were spotted by platinum string onto the center of potato dextrose agar plates where 0.1 ml of the filtrate suspension of each fungus had been spread beforehand. During incubation of the plates at 27°C, bacteria that showed zones inhibitory to the growth of fungi on the plates were selected as suppressive strains.

2.2.1.3   Cultivation of Phytopathogenic Fungi and Bacteria

In addition to the four fungi used in the *in vitro* tests with the composts, eleven phytopathogenic fungi and two phytopathogenic bacteria were selected to verify the suppressive spectrum of the isolated bacterium, *B. subtilis*. The preparation of these fungi was the same as the procedure mentioned earlier. The phytopathogenic bacteria were grown in a shaking flask containing no. 3 liquid medium at 27°C for 2 days. The cell suspensions were diluted and spread onto no. 3 agar plates.

2.2.2   RESULTS

Two hundred four bacteria were isolated from suppressive composts as suppressive bacteria. Among them, four bacteria that were significantly inhibitory to all four fungi used on the agar plates were identified as *B. subtilis* and named as NB22, UB24, YB8, and SB4, respectively. Basic characteristics for identification of the bacteria are listed in Table 2.2. As the appearance of the colonies of these bacteria which formed on the no. 3 agar medium plates was obviously different from that of other mesophilic bacteria inhabiting each compost product, the number of suppressive bacteria in the final composts was counted by using no. 3 agar plates as shown in Table 2.3. The number of suppressive bacteria at each stage of the composting process was almost constant, indicating that suppressive bacteria were in a minority in the composts. These bacteria showed similar characteristics in the production of suppressive substances and then each isolate was used arbitrarily in the subsequent study.

TABLE 2.2
Some Characteristics of Isolated Bacteria

Gram staining	+
Spores	+
Spore shape	Elliptical or cylindrical
Spore position	Central
Sporangium distended distinctly	−
Catalase	+
Oxidase	+
Anaerobic growth in glucose agar	+
Growth in 7% NaCl	+
Acid from glucose	+
Gas from glucose	+
Casein decomposition	+
Gelatin decomposition	+
Citrate utilization	+
Starch hydrolysis	+
NO_3^- to NO_2^-	+
Arginine dihydrolase	−

TABLE 2.3
Number of Suppressive *B. subtilis* in the Compost Products

Compost Material (Extra Additives)	Total Cell Number[a] (cells/g dry solid)	Number of Suppressive Bacteria (cells/g dry solid)	Designation of *B. subtilis* Isolate[b]
Night soil sludge	4.6×10^9	7.4×10^4	NB22
Bark (fermented hen dropping and wood chips)	2.4×10^9	1.3×10^6	YB8
Garbage	6.3×10^9	6.0×10^4	UB24
Garbage	9.6×10^9	5.9×10^5	SB4

[a] The number was counted on the no. 3 agar medium.
[b] *B. subtilis* strain isolated from suppressive composts.

We also used another *B. subtilis* RB14 that showed a higher productivity of suppressive substances than those isolates. The characteristics of RB14 are explained later.

2.3   SUPPRESSIVENESS BY *B. SUBTILIS* AND BY CULTURE BROTH OF THE BACTERIUM TO PHYTOPATHOGENIC MICROORGANISMS

2.3.1   MATERIALS AND METHODS

B. subtilis NB22, which showed relatively larger zones inhibitory to phytopathogenic fungi than other isolates, was selected for further investigation. This bacterium was cultivated in no. 3 medium with shaking flasks at 27°C at a shaking speed of 120 strokes per minute. After 5 days incubation, 7 ml of the medium was sampled and centrifuged at $24,000 \times g$ for 10 min. The supernatant was filtered with a 0.45 µm membrane filter and the filtrate was used for suppressiveness testing.

Then, 0.1 ml of the suspensions of the 11 phytopathogenic fungi was spread onto potato dextrose agar medium plates in the same way as described earlier. Then, a sterile stainless cup (8 mm in diameter × 10 mm) was placed on the center of the plate. Next, 0.2 ml of culture broth of *B. subtilis* was prepared as described earlier was poured into the cup. The diameter of the inhibitory zones, which formed around the cup after 7 days incubation, was used as a measure of the degree of suppressiveness. The two phytopathogenic bacterial suspensions were diluted with sterile water and spread onto no. 3 agar medium plates. A stainless cup was put in place, and the effect of the culture broth was tested in a similar way as in the case of the fungi.

2.3.2   RESULTS

Table 2.4 shows the results of the tests for suppressiveness to various phytopathogenic microorganisms of the bacterium *B. subtilis* NB22. Suppressiveness was judged in

TABLE 2.4
Suppressiveness of *B. subtilis* NB22 and the Culture Broth of This Bacteria *In Vitro*

Plant Pathogens	Inhibitory Zone by the Cells (mm)	Inhibitory Zone by the Broth (mm)
Fungi		
Alternaria mali IFO 8984	21	33
Cercospora kikuchii NIAES 5039	28	40
Phytophthora infestans IFO 5547	25	30
Botrytis cinerea Bot1	40	34
Rhizoctonia solani NIAES 5219	40	38
Pyricularia oryzae NIAES 5001	40	35
Cochliobolus miyabeanus NIAES 5425	32	33
Fusarium oxysporum f. sp. *cucumerinum* NIAES 5117	10	18
Fusarium oxysporum f. sp. *lycopersici* race JISUF 119	12	20
Pythium ultimum Trow H-1	18	N[a]
Verticillium dahliae Klebahn V-3	25	25
Bacteria		
Xanthomonas oryzae IFO 3998	35	25
Pseudomonas syringae pv. *lachrymans* P1-7415	25	25

[a] No inhibitory zone.

3 days for *F. oxysporum* f. sp. *cucumerinum, V. dahliae, Pseudomonas lachrymans, Xanthomonas oryzae, Alternaria mali, Cercospora kikuchii, Phytophthora infestans, Cochliobolus miyabeanus,* and in 6 days for the others. Inhibitory zones of more than 40 mm were seen for *Botrytis cinerea, Pyricularia oryzae,* and *R. solani,* and of more than 30 mm for *C. miyabeanus* and *X. oryzae.* Figure 2.2 shows examples of the suppressive effect of *B. subtilis* NB22 on the four phytopathogens *in vitro.*

Table 2.4 also shows the suppression result of the culture broth of *B. subtilis* NB22 obtained in 5 days incubation. All fungi except *P. ultimum* were suppressed. *R. solani, P. oryzae,* and *C. miyabeanus,* whose inhibition zones by the cells were relatively larger, were also significantly suppressed by the broth. The growth of *P. ultimum* was suppressed for 2 days, but the growth of hyphae was so fast that hyphae came to cover the surface of plates and no apparent effect was observed later. Figure 2.3 shows examples of the suppressive effect of the culture broth of *B. subtilis* NB22 on the four phytopathogens *in vitro.*

Effect of culture broth of *B. subtilis* NB22 on the suppressive effect by exposure to various temperatures and pHs was investigated. Table 2.5 shows that the treatment of the culture broth exposed to different temperatures (30°C–100°C) and pHs (2–9) did not adversely affect the suppressive effect on the growth of *P. oryzae.* The stability of the culture broth is important for the control of plant pathogens in soil.

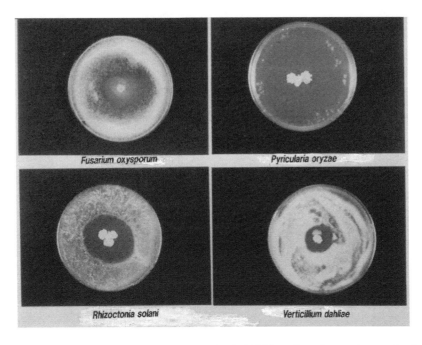

FIGURE 2.2   The suppressive effect of *B. subtilis* NB22 on four plant pathogens *in vitro*. The cells of NB22 were placed in the center of each plate.

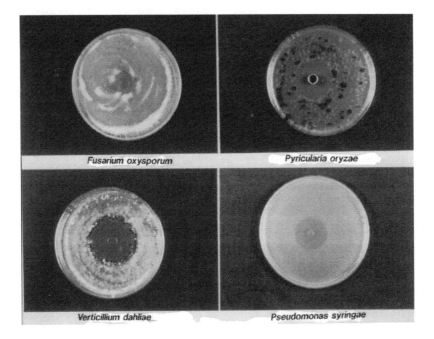

FIGURE 2.3   The suppressive effect of the culture broth of *B. subtilis* NB22 on four plant pathogens *in vitro*. The cell-free culture broth was placed in the center of each plate.

TABLE 2.5

Effects of Heat Treatment and pH on the Suppressive Effect of Culture Broth of *B. subtilis* **NB22**

Heat Treatment	pH			
	2	7	9	12
No treatment	n.d.	22	n.d.	n.d.
30°C, 180 min	20	21	23	0
60°C, 30 min	23	22	23	0
100°C, 5 min	22	24	24	0

Notes: Values are expressed as the diameter (mm) of the inhibitory zone to the growth of *P. oryzae* IFO 5279. n.d., not determined.

2.4   IDENTIFICATION OF SUPPRESSIVE SUBSTANCE, ITURIN

2.4.1   MATERIALS AND METHODS

The culture liquor was acidified to pH 2 with HCl. The resulting solution was centrifuged and the precipitate was extracted with methanol for 3 h. After the methanol was evaporated under reduced pressure, the residual solid was dissolved again in methanol. Then, the solution was applied to a silica gel column (Kiesel gel 60, Merck) using methanol–chloroform–water (65:30:5, v/v/v) followed by thin-layer chromatography on silica gel (Kiesel gel 60 F254, Merck). The individual components of the suppressive spot were further purified by high-performance liquid chromatography (HPLC) (Shim-pack, Pre-ODS, 2 $\varphi \times$ 15 cm) using 75% methanol or acetonitrile–ammonium acetate (10 mM) (2:3, v/v). The antifungal effect of the separated liquid or spot at each purification step was checked by an *in vitro* test using *F. oxysporum.*

Each purified component collected by HPLC was subjected to analyses by a fast atom bombardment mass spectrometer (FAB-MS) (JMS-DX303/JMA-DA5100), ^{13}C and ^{1}H nuclear magnetic resonance (NMR) (Jeol-GX-500), a Fourier transform infrared spectrometer (FTIR) (Hitachi 260-50), and an amino acid analyzer (Hitachi 835).

2.4.2   RESULTS

The HPLC separation pattern of the purified suppressive substance is shown in Figure 2.4, where the components are denoted as peaks 1 to 6 according to the elution time order. The structure of each component was analyzed by the following methods. The FTIR spectrum as shown in Figure 2.5 is dominated by the amide I and amide II bands (1658 and 1517 cm^{-1}), a broad O–H stretching band (3320 cm^{-1}), and two C–H stretching bands (2923 and 2851 cm^{-1}), which were identical to the previously reported spectrum (4).

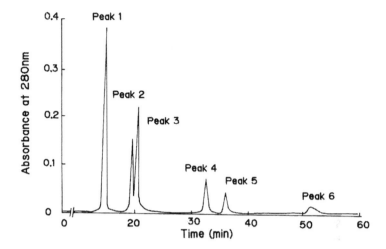

FIGURE 2.4 HPLC separation pattern of iturin in acetonitrile–ammonium acetate (10 mM) 2:3 (v/v) on an ODS column (2 cm in diameter × 25 cm) at a flow rate of 3 ml/min.

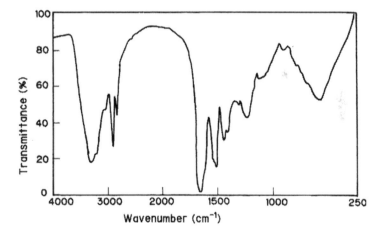

FIGURE 2.5 Fourier transform infrared (FTIR) spectrum of peak 1 shown in Figure 2.4. Solvent is pure methanol.

The hydrolysate of iturin (6N HCl, 150°C, 8 h) was analyzed by an amino acid analyzer. The result is shown in Table 2.6, indicating Asp, Glu, Pro, Ser, and Tyr in a molar ratio of 3:1:1:1:1. From the data determined by FAB-MS and NMR, Asp and Glu determined by the amino acid analyzer were judged as Asn and Gln, respectively. The configurations of amino acids and the ultraviolet (UV) spectrum were also compared with the results by Peypoux et al. (5).

The ^{13}C-NMR spectrum of peak 1 as shown in Figure 2.6 suggests that it contains β-amino acid and only one aliphatic methyl signal at 14.5 ppm, indicating that this β-amino acid is of *n* structure. The aliphatic region (15–65 ppm) containing the α-, β-, γ-, δ-carbon signals of the α- and β-amino acids was assigned with the previously

TABLE 2.6

Amino Acid Composition of Peak 1 and Peak 3 Isolated from HPLC

Amino Acid	Peak 1[a]	Peak 3[a]	Peak 1/Ser[b]	Peak 3/Ser[b]
Aspartic acid	0.223	0.262	3.338	3.403
Threonine	0.000	0.000		
Serine	0.066	0.077	1.000	1.000
Glutamic acid	0.063	0.075	0.995	0.974
Proline	0.072	0.086	1.091	1.116
Glycine	0.003	0.000		
Alanine	0.000	0.000		
Cystine	0.001	0.000		
Valine	0.000	0.000		
Methionine	0.000	0.000		
Isoleucine	0.000	0.000		
Leucine	0.001	0.000		
Tyrosine	0.066	0.078	1.000	1.013
Phenylalanine	0.000	0.000		
Lysine	0.000	0.000		
Arginine	0.000	0.000		
Histidine	0.000	0.000		

[a] Unit: amino acid μmol/ml.

[b] Ratio of each amino acid to serine.

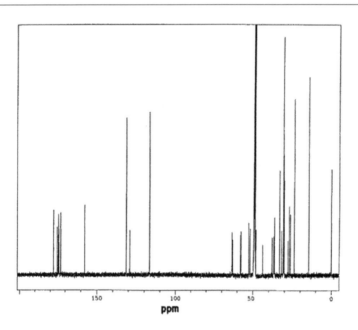

FIGURE 2.6   ¹³C-NMR spectrum of peak 1 shown in Figure 2.4. Chemical shifts are in ppm downfield from TMS.

reported data (4). The ^{1}H-NMR spectra of peaks 1 to 6 were distinguishable from each other in their methyl regions: a 3H distorted triplet for a *n*-type compound, a 6H doublet isotype, and a 6H multiplet due to triplet plus doublet for an *anteiso*-type compound (data not shown).

The mass spectrum of each component of iturin obtained with FAB-MS represented homologous (M+H)$^+$ ion peaks at *m/z* 1043 for peak 1, 1057 for peaks 2 and 3, 1071 for peaks 4 and 5, and 1085 for peak 6 as shown in Figure 2.7. It is obvious that these correspond to the homogeneous compounds with an equivalent chain length. By comparing the aforementioned data of the instrumental analysis described and the melting point tests of the purified substances with those according to previously published papers (4–6), the substances were judged to be of the iturin group, cyclic peptidolipidic antifungal antibiotics. The structure is as follows:

$$\text{RCHCH}_2\text{CO} \rightarrow \text{L-Asn} \rightarrow \text{D-Tyr} \rightarrow \text{D-Asn}$$
$$| \qquad\qquad\qquad\qquad\qquad\qquad \downarrow$$
$$\text{HN} \leftarrow \text{L-Ser} \leftarrow \text{D-Asn} \leftarrow \text{L-Pro} \leftarrow \text{L-Gln}$$

These contain seven residues of α-amino acid and a collection of β-amino acids that comprise *n*-C_{14}-β-amino acid, *anteiso*-C_{15}-β-amino acid, *iso*-C_{15}-β-amino acid, *n*-C_{16}-β-amino acid, *iso*-C_{16}-β-amino acid, and *n*-C_{17}-β-amino acid. Peaks 1 to 6 in

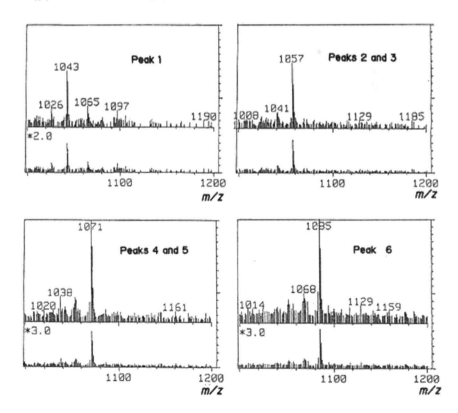

FIGURE 2.7 FAB-MS spectrum of peak 1 shown in Figure 2.4.

Figure 2.4 correspond to the component with n-C_{14}-β-amino acid to n-C_{17}-β-amino acid in this order.

2.5 *IN VIVO* TEST BY CULTURE BROTH

2.5.1 MATERIALS AND METHODS

The suppressiveness of the culture broths of two suppressive bacteria was investigated *in vivo* according to a standard method (7). Leaf and pot tests were conducted in a greenhouse. As a reference, commercially available chemicals were used. For the phytopathogenic bacteria, *in vitro* tests by paper disks were carried out to compare the broths with available antibiotics.

2.5.2 RESULTS

So far, the iturin group has been known to show an antifungal effect (10), but no data concerning the effect of iturin or iturin-producing bacteria on phytopathogenic fungi either *in vivo* or *in vitro* have been reported. Table 2.7 shows the list of plant

TABLE 2.7
Suppressive Effect of Culture Broth of *B. subtilis* NB22 and UB24 in Greenhouse or *In Vitro* Tests

Phytopathogenic Fungi or Bacteria	Name of Plant Disease	Name of Chemical Used	Test Method	Efficacy of Broth	
				NB22	**UB24**
Alternaria mali	Alternaria leaf spot of apple	Polyoxin	Leaf test	C	n.t.
Cercospora kikuchii	Purple seed stain of soybean	Benlate	Pot test	A	n.t.
Phytophthora infestans	Late blight of tomato	Ridomil	Pot test	n.t.	C
Botrytis cinerea	Gray mold of cucumber	Sumilex	Pot test	C	n.t.
Puccinia coronate	Crown rust of oat	Bayleton	Pot test	C	C
Rhizoctonia solani	Sheath blight of rice	Moncut	Pot test	A	A
Pyricularia oryzae	Blast of rice	Fuji-One	Pot test	A	B
Cochliobolus miyabeanus	Brown spot of rice	Sumilex	Pot test	A	B
Xanthomonas oryzae	Bacteria leaf blight of rice	Streptomycin	Paper disk	A	A
Psuedomonas lachrymans	Angular leaf spot of cucumber	Streptomycin	Paper disk	A	A

Notes: Efficacy A, 100%–95%; B, 80%–95%; C, 60%–80% (relative suppressive value to that by each chemical used). n.t., not tested.

pathogens used, the disease name, the chemicals used for reference, and the test methods *in vivo*, and represents the suppressive results *in vivo* of the culture broths of the bacteria *B. subtilis* NB22 and UB24. The broths were harvested after 5 days cultivation. The broth of NB22 showed a stronger and broader suppression than any of the other bacteria. The efficacy of the broth to some fungi was equivalent to that of commercial chemicals, especially for rice diseases. The reason why *Puccinia coronata* was used only *in vivo* is that no method has been established for the growth of this fungus *in vitro* due to its obligate parasitism. *B. subtilis* NB22 was intensively used for further investigation.

2.6 BIOLOGICAL PROPERTIES OF EACH COMPONENT OF ITURIN

2.6.1 MATERIALS AND METHODS

The minimum inhibitory concentration (MIC) values of purified components of iturin to seven fungi, two yeasts, and four bacteria were determined in a potato dextrose agar medium or in a no. 3 medium containing increasing amounts of each component.

2.6.2 RESULTS

The effect of each component of iturin on fungi or bacteria has not hitherto been elucidated. The MIC values of each component isolated by HPLC are shown in Table 2.8. The longer the side aliphatic chain R is, the lower is the MIC to the fungi. The fungi of rice diseases, *R. solani* and *P. oryzae*, were suppressed, especially in

TABLE 2.8
MIC Values of Each Component of Iturin

	Concentration of MIC (μg/ml)				
Microorganism	Peak 1	Peak 2	Peak 3	Peak 4	Peak 5
Cercospora kikuchii	25	25	12.5	6.25	6.25
Verticillium dahliae	50	25	12.5	6.25	6.25
Fusarium oxysporum f. sp. *lycopersici*	50	25	25	25	12.5
Alternaria mali	50	25	25	12.5	12.5
Rhizoctonia solani	12.5	12.5	6.25	6.25	6.25
Pyricularia oryzae	12.5	6.25	6.25	3.13	3.13
Cochliobolus miyabeanus	50	12.5	12.5	6.25	3.13
Saccharomyces cerevisiae	50	25	25	12.5	6.25
Candida tropicalis	50	25	25	12.5	12.5
Xanthomonas oryzae	12.5	6.25	6.25	3.13	3.13
Pseudomonas syringae pv. *lachrymans*	12.5	6.25	6.25	3.13	3.13
Escherichia coli	>100	>100	>100	>100	>100
Staphylococcus aureus	>100	>100	>100	>100	>100

the low concentration range. The data of peak 6 in Table 2.8 are lacking mainly because the yield of purified substances was not enough to perform the test.

2.6.3 DISCUSSION

About 40% of the compost samples collected showed a suppressive effect on the four phytopathogenic fungi. The bacteria isolated from different sources that showed a broader suppressive spectrum were *B. subtilis*. No suppressive effect on the fungi was observed in the case of the immature composts collected at the initial and intermediate stages from the composting plants. However, the concentration of the suppressive bacteria in these samples was of almost the same order as in the final product. This means that some conditions of the materials or the microbial flora alive in these stages may hinder the expression of the suppressive effect by these bacteria. These adverse factors may be eliminated through ensuring a high temperature (60°C–70°C) during composting. So far, the maturity of composts has been considered to be important, especially for avoiding damage to the growth of plants caused by rapid degradation of organic substances, such as in ammonification. This experiment suggests that thorough maturation of composts is also useful for realizing their potential to suppress phytopathogenic microorganisms.

Although four bacteria that were isolated from different composts were identified as *B. subtilis*, the degree of suppressiveness of their culture broths was slightly different. For example, the broths of NB22 and UB24 showed higher suppressiveness at a culture time of 5 days when the growth leveled off and the pH increased to 8 or 9, while YB8 and SB4 showed a lower suppressive effect at the same cultivation time mainly due to different iturin productivity (data not shown).

There have been some attempts to control plant diseases by using *B. subtilis* isolated from soil. Chang et al. (8) reported that coating kernels of corn with *B. subtilis* was as good as chemical treatment to control the seeding blight of corn caused by *F. oxysporum*. Aldrich et al. (9) isolated *B. subtilis* from rhizosphere of carnation, and dipping cuttings of carnation in a bacterial suspension was effective in decreasing *Fusarium* disease of carnation. However, neither the suppressive substances nor the suppressive spectra were obtained in these studies.

It is often said that no correlation is observed between the width of the inhibition zone *in vitro* and protection in greenhouse trials (10). However, it seems that the larger inhibition zones appearing *in vitro* by *B. subtilis* NB22, as shown in Table 2.4, reflected a stronger suppressive effect *in vivo*, as shown in Table 2.7.

The suppressive substance, iturin, purified from the culture broth of *B. subtilis* NB22, is heat resistant and pH stable in a wide range. It was newly found that this substance is effective not only to phytopathogenic fungi but also to bacteria.

Though the suppressive effects among the four *B. subtilis* isolates were slightly different, the substances produced may be presumed to be identical, while the fractions of the components produced or the protease activities excreted may not necessarily be the same. *B. subtilis* NB22 has an optimal temperature of growth at 30°C, but survives even at 40°C–45°C with least growth. In a paper (11), how mesophilic bacteria can survive, even in the thermophilic stage of composting, was demonstrated. Therefore, *B. subtilis* can be a candidate as a microbial antagonist to phytopathogenic microorganisms as an alternative to chemical fungicides.

2.7 INOCULATION OF *B. SUBTILIS* NB22 INTO COMPOST PRODUCTS OR DURING THE COMPOSTING PROCESS (12)

The application of *B. subtilis* NB22 as an agent for the biological control of plant diseases will be either by direct inoculation to the raw material used for composting or mixing with compost products. If the bacterium was able to survive in the compost and show a suppressive effect, the compost would have the new effect of suppressiveness to plant pathogens in addition to its efficacy as an organic fertilizer and soil conditioner. In this section, survivability of this bacterium in the composting reaction system or the compost products is examined in relation to expression of the suppressiveness *in vitro*.

2.7.1 MATERIALS AND METHODS

B. subtilis NB22-1 is a spontaneous streptomycin-resistant mutant from *B. subtilis* NB22. The cell mass used for inoculation to sewage sludge at the start of composting was prepared by cultivation in a 50 L jar fermentor in no. 3 medium for 24 h. The culture broth was centrifuged at $10,000 \times g$ for 30 min, washed twice with distilled water, and suspended into an appropriate volume of sterile distilled water. The cells for inoculation to the compost products was grown in a 500 ml shaking flask and prepared in the same way mentioned earlier.

2.7.1.1 Composting Material Used and Composting Operation

Dewatered sewage sludge cake from the Minamitama Wastewater Treatment Plant (Inagi City, Tokyo) was used as the composting material. The sludge was a mixture of raw settled sludge and excess sludge from the activated sludge process, containing approximately 60% water, 7.5% slaked lime, and 2.5% ferric chloride as dewatering agents. The average composition of the dried sludge was 50% volatile matter (VM), in which C, H, and N were 25%, 4%, and 3%, respectively, and 50% ash. The sludge cake was first dried to 40% of moisture content, then ground and sieved to less than 5 mesh. A part of the sludge cake was sterilized by gamma irradiation with ^{60}Co for 3 h at a dose of 1 Mrad/h. No bulking agent was added for composting.

In composting operation, the reactor was cylindrical (300 mm in diameter, 400 mm in depth) and made of thermoresistant polyvinyl chloride resin, with a perforated plate at the bottom to distribute the air supplied from a compressor. The reactor was surrounded with a cubic Styrofoam insulator to maintain minimum heat loss from the wall of the reactor. Air from the compressor was split into two streams, one at a lower flow rate of 400 ml/min and the other at a higher flow rate of 4000 ml/min. The temperature of composting was controlled automatically at any set temperature using the two flow rates; details of the control system have been previously described (3). The moisture content of the sludge cake or sterilized sludge cake was adjusted to around 60% at the start of the experiments by adding the cell suspension of *B. subtilis* NB22-1 prepared beforehand and was not controlled during the reaction.

The initial cell concentration of *B. subtilis* NB22-l, either in sewage sludge or in compost products, was adjusted by controlling the water content of the cell suspension, and the volume of the suspension added to adjust the moisture content. As the sludge contained lime, the initial pH value of the material was around 11, and a long lag period (about 3 days) was observed before microbial degradation of the raw material started. To shorten the lag time, CO_2 gas from a cylinder was supplied to the reactor to neutralize the lime. The CO_2 evolution rate was measured by a CO_2 analyzer (ZFP4, Fuji-Denki Co., Ltd., Tokyo), and conversion of VM was calculated in the same manner as described in a previous paper (11).

2.7.1.2 Compost Used

Three well-matured composts, a sewage sludge compost, a bark compost, and a cattle manure compost, all produced commercially, were collected. The moisture content of each compost was less than 20%.

2.7.1.3 Inoculation of *B. subtilis* NB22-1 to Composts

A compost product was put into a plastic bag such that a thin layer of compost (about 2 cm in thickness) was formed inside the bag. The cell suspension of *B. subtilis* NB22-1 prepared beforehand was sprayed aseptically and the bag was shaken vigorously by hand to homogenize the compost inside the bag. Spraying was repeated until the moisture content of the compost reached approximately 60%, which was estimated by relationship between the moisture content of the compost and the amount of sprayed water prepared beforehand in a preliminary experiment. The actual moisture content was determined by drying the compost samples at 80°C for 24 h. The samples were then put into plastic bottles (500 ml), which had holes at the bottom for ventilation, and placed in a 30°C incubator.

2.7.1.4 Counting the Number of *B. subtilis* NB22-1 Cells in Compost Samples

Ten grams (wet weight) of a compost sample collected periodically was suspended into 90 ml of sterile water in a sterile homogenizer cup and dispersed at $10,000 \times g$ for 10 min with a homogenizer (type EX-3, Nihon Seiki Ltd., Tokyo). After serial dilution in sterile water, the suspension was spread onto no. 3 agar-streptomycin medium plates. The plates were incubated at 30°C for 24 h and the colony numbers were counted as *B. subtilis* NB22-1. The total microbial number was counted on no. 3 agar medium plates without streptomycin.

2.7.1.5 Preparation of Suspensions of Phytopathogenic Microorganisms

As phytopathogenic fungi, *F. oxysporum* f. sp. *cucumerinum* NIAES 5117, *P. ultimum* Trow H-1, *V. dahliae* Klebahn V-3, *P. oryzae* IFO 5279, and *R. solani* NIAES5219 were used in *in vitro* suppressive test by composts inoculated with *B. subtilis* NB22-1. The fungi were grown on a potato dextrose agar medium (potato dextrose agar, 39 g; distilled water, 1 liter, pH 5.6) at 28°C for 7 days and suspended in sterile distilled water. As a phytopathogenic bacterium, *X. oryzae* IFO 3998 was used. This bacterium was grown on the PYM medium (Polypepton 10 g, yeast extract 2 g, $MgSO_4 \cdot 7H_2O$ 1 g, pH 7) at 30°C for 24 h.

2.7.1.6 Suppressiveness Testing of Composts

First, 0.1 ml suspensions of fungal or bacterial pathogens prepared as mentioned earlier were spread onto no. 3 agar medium. Then, about 0.3 g of the compost samples were placed onto the center of the plates and they were incubated at 28°C for fungi and at 30°C for the bacterium. The surface was observed for changes, and inhibitory zones formed were measured in 7 days.

2.7.2 RESULTS

2.7.2.1 Survival of *B. subtilis* NB22-1 in Compost Products

The simplest method to evaluate the suppressive effect of *B. subtilis* was to mix the bacterium with compost products already prepared. About 10^7–10^8 cells/g wet compost were inoculated into the three compost products collected, and the change in the number of the bacterium was measured periodically. Figure 2.8 shows the result for inoculation of 10^7 cells/g wet compost. The cell number of *B. subtilis* NB22-l decreased gradually in 5 days after inoculation into bark compost, but the other composts showed a rapid decline in cell number. The cell number seemed to be stabilized at around 10^5 cells/g wet compost. On the other hand, the number of indigenous microorganisms in each compost maintained the initial cell concentration or increased 10-fold, indicating that some microorganisms in the compost had an adverse effect to the survival of *B. subtilis* NB22-l. It is interesting that the cell concentration of *B. subtilis* NB22-l leveled off at around 10^5. The suppressive effect of the composts sampled at each incubation time on six phytopathogenic microorganisms is shown in Table 2.9. The original composts free from *B. subtilis* showed

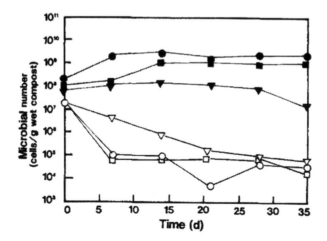

FIGURE 2.8 Changes in the number of *B. subtilis* NB22-l cells when 10^7 cells/g wet compost were inoculated to three kinds of composts. Symbols: o, *B. subtilis* NB22-l in sewage sludge compost; •, total microbial number in sewage sludge compost; □, *B. subtilis* NB22-l in cattle manure compost; ■, total microbial number in cattle manure compost; ▽, *B. subtilis* NB22-l in bark compost; ▼, total microbial number in bark compost.

TABLE 2.9
Suppressive Effect of Compost Products *In Vitro* Inoculated with *B. subtilis* NB22-1 on Six Plant Pathogens

Compost	Initial Number of *B. subtilis* NB22-1 (cells/g wet compost)	Plant Pathogens	Inhibitory Zone (mm)					
			0 Days[a]	7 Days	14 Days	21 Days	28 Days	35 Days
Sludge	10^7	*F. oxysporum*	6	1	2.5	4	7	3
		R. solani	25	10	9	10	15	18
		P. oryzae	25	20	22	26	30	23
		P. ultimum	12	0	0	0	0	0
		V. dahliae	20	9	11	10	15	10
		X. oryzae	20	9	11	10	15	10
	10^8	*F. oxysporum*	0	10	6	7	8	10
		R. solani	29	26	25	19	23	22
		P. oryzae	20	20	23	25	30	25
		P. ultimum	15	14	12	15	10	12
		V. dahliae	23	18	17	12	15	24
		X. oryzae	22	25	23	25	25	15
Bark	10^7	*F. oxysporum*	5	6	10	17	10	10
		R. solani	30	5	10	10	20	15
		P. oryzae	20	12	15	23	28	20
		P. ultimum	0	8	7	8	9	17
		V. dahliae	20	8	10	10	20	19
		X. oryzae	20	20	21	22	20	20
	10^8	*F. oxysporum*	5	10	12	18	25	22
		R. solani	40	20	21	23	23	28
		P. oryzae	36	30	32	30	30	25
		P. ultimum	20	10	9	8	14	13
		V. dahliae	20	25	20	23	23	25
		X. oryzae	50	32	28	30	20	30

[a] Incubation time of compost at 30°C (days).

no suppressive effect on the plant pathogens tested here. Bark compost inoculated with *B. subtilis* NB22-1 showed a constant suppressiveness to all plant pathogens during the period of the test. However, sludge compost showed a weak suppressive effect to *F. oxysporum* and *P. ultimum*. The growth of hypha of P. *ultimum* was sometimes so fast that the surface of the plates was covered with the hypha and no apparent inhibitory zones could be measured. In cattle manure compost, mites or nematoda appeared on the plates during incubation and preyed on the plant pathogens; thus, the suppressive effect was not measurable. When 10^8 cells/g wet compost were inoculated, a similar change in cell number was seen, as shown in Figure 2.9. The cell number of *B. subtilis* NB22-1 was again stabilized at 10^5–10^6 cells/g wet compost. Table 2.9, which presents the suppressiveness of each compost,

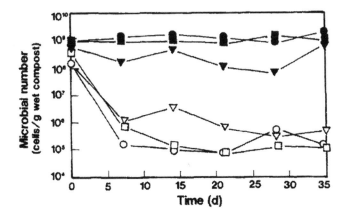

FIGURE 2.9 Changes in the number of *B. subtilis* NB22-l cells when 10^8 cells/g wet compost were inoculated to three kinds of composts. Symbols: See Figure 2.8.

shows that the degree of suppressiveness of the composts inoculated by 10^8 cells of this bacterium/g wet compost was slightly larger than that of 10^7 cells/g wet compost. As *B. subtilis* NB22-1 in bark compost seemed to maintain its concentration better than in the other composts (Figures 2.8 and 2.9), a much larger inoculation (10^9 cells/g wet compost) of *B. subtilis* NB22-1 was also carried out for bark compost, and changes in the cell number and the suppressive effect were investigated. As shown in Figure 2.10, the cell number of *B. subtilis* NB22-1 was constant, indicating that the inoculated bacterium was dominant in bark compost compared with the cell number in the compost without inoculation of *B. subtilis* NB22-1, 8.7×10^7. The suppressive effect of this compost is shown in Table 2.10. The result indicates that the degree of suppressiveness of bark compost inoculated by 10^9 cell number/g wet compost is apparently greater than that of 10^7 cells/g wet compost during incubation time. The

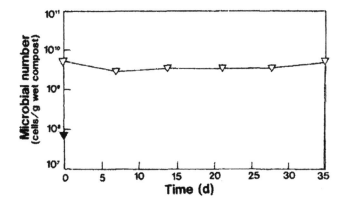

FIGURE 2.10 Changes in the number of *B. subtilis* NB22-1 cells when 10^9 cells/g wet compost were inoculated to bark compost. Symbols: $\triangledown$, *B. subtilis* NB22-1; $\blacktriangledown$, initial microbial number in bark compost.

TABLE 2.10

Suppressive Effect of Bark Compost *In Vitro* Inoculated with *B. subtilis* NB22-1 (10^9 cells/g wet compost) on Six Plant Pathogens

Plant Pathogens	Inhibitory Zone (mm)				
	7 Days[a]	14 Days	21 Days	28 Days	35 days
F. oxysporum	13	20	20	24	19
R. solani	22	32	28	41	36
P. oryzae	35	30	31	41	37
P. ultimum	8	10	14	14	7
V. dahliae	17	25	23	29	31
X. oryzae	32	37	30	35	36

[a] Incubation time of compost at 30°C (days).

diameter of the zone inhibitory to *P. ultimum* in the 10^9 cells inoculation was smaller than that in the 10^8 inoculation, but the growth of other plant pathogens was suppressed almost equally or much more severely in the 10^9 inoculation.

The interactions among microorganisms in the compost is not clear at present, but differences in composting material will reflect different chemical components, and hence different microflora, which will have an effect on the survival of inoculated *B. subtilis*, positively or negatively, depending on the environmental condition. Thus, the initial dominancy in cell number is a primarily important factor for the expression of suppressiveness.

2.7.2.2   Sterilized Composts Inoculated with *B. subtilis* NB22-1

To maintain a higher concentration of *B. subtilis* in the compost, the compost was sterilized with gamma irradiation and then the bacterium was mixed in at a concentration of 10^8 cells/g wet compost. The changes in the cell number are shown in Figure 2.11. Although the initial concentration of viable bacteria after sterilization was less than 10^2 cells/g wet compost, a rapid increase in the number of airborne microorganisms overwhelmed the *B. subtilis* NB22-1 during incubation. Nevertheless, the number of *B. subtilis* NB22-1 cells showed no significant decrease. This also indicates that the initial dominancy of *B. subtilis* NB22-1 is important for survival in the compost. This is well reflected in the suppressive effect, as shown in Table 2.11. It is clear that the degree of suppression was larger than that shown in Table 2.9.

Concerning the sterilization of compost or composting materials with gamma irradiation, intensive research has been carried out with a view to shortening the period of primary fermentation in the composting operation (13). Thus, inoculation of suppressive bacteria into sterilized compost is one means of utilizing the specific characteristics of this bacterium.

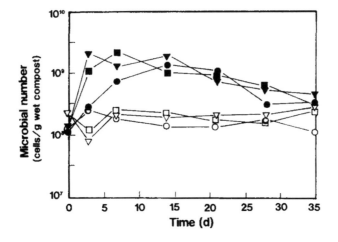

FIGURE 2.11 Changes in the number of *B. subtilis* NB22-1 cells when 10^8 cells/g wet compost were inoculated to three kinds of composts that were sterilized with gamma irradiation. Symbols: See Figure 2.8.

TABLE 2.11

Suppressive Effect of Sterilized Compost *In Vitro* Inoculated with *B. subtilis* NB22-1 (10^8 cells/g wet compost) on Six Plant Pathogens

		Inhibitory Zone (mm)				
Compost	Plant Pathogens	7 Days[a]	14 Days	21 Days	28 Days	35 Days
	F. oxysporum	11	7	10	7	10
	R. solani	20	30	35	20	28
Sludge	*P. oryzae*	30	40	35	20	28
	P. ultimum	11	2	5	5	12
	V. dahliae	20	28	26	30	32
	X. oryzae	35	35	41	37	36
	F. oxysporum	15	10	10	8	14
	R. solani	40	30	28	38	32
Bark	*P. oryzae*	40	38	36	30	30
	P. ultimum	13	10	8	6	8
	V. dahliae	18	28	25	30	30
	X. oryzae	48	40	30	30	26
	F. oxysporum	8	12	15	20	11
	R. solani	40	38	30	24	30
Cattle manure	*P. oryzae*	43	41	40	30	35
	P. ultimum	11	15	4	10	13
	V. dahliae	22	25	30	33	30
	X. oryzae	40	38	35	33	31

[a] Incubation time of compost at 30°C (days).

2.7.2.3   Survivability of *B. subtilis* NB22-I in the Composting Process

Application of suppressive *B. subtilis* NB22-1 during the composting process was investigated by inoculating the bacterium into the raw material of sewage sludge compost at the start of composting. The reasons why sewage sludge was selected as a material were the ease of collecting raw material, the relatively stable property of sewage sludge, and long experience in the operation of sewage sludge composting in my laboratory. At an initial concentration of the bacterium of 10^6 cells/g dry compost, the composting progress for nonsterilized and sterilized sewage sludge is shown in Figures 2.12 and 2.13, respectively. During the composting reaction, there were no significant differences in the change in the number of inoculated bacteria, the CO_2 evolution rate (r_{CO2}), or the conversion of volatile matter (X_{VM}) between the two composting materials. From the change of X_{VM}, it was judged that degradable organic matter was degraded and stabilized during this period (11). The total cell number increased rapidly in the sterilized composting due to rapid contamination by aerial bacteria through the air supply system. Although the percentage of *B. subtilis* NB22-1 in the total microbial number became significantly small, *B. subtilis* NB22-1 did survive, and it maintained a constant concentration even at a thermophilic temperature of 50°C. This suggests that *B. subtilis* NB22-1 contributed to the degradation of sewage sludge, as well as other major bacteria, shown as the

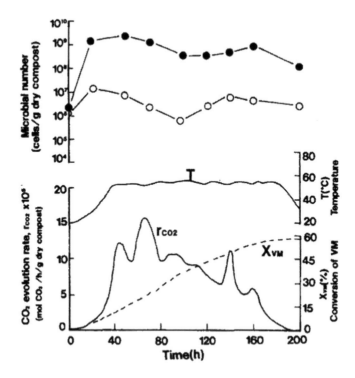

FIGURE 2.12   Time course of temperature (T), CO_2 evolution rate (r_{CO2}), conversion of volatile matter (X_{VM}), and the cell number of isolated microorganisms during composting of sewage sludge. Symbols: o, *B. subtilis* NB22- 1; •, total microbial number.

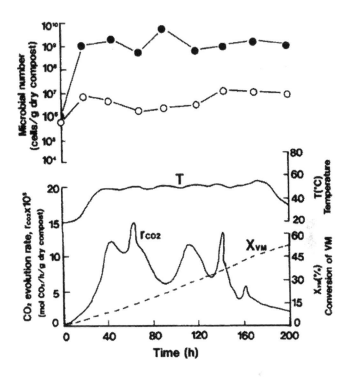

FIGURE 2.13 Time course of composting of sterilized sewage sludge. For symbols and details, see legend to Figure 2.12.

total bacterial number in Figures 2.12 and 2.13. When the initial concentration of *B. subtilis* NB22-1 was increased to 8×10^8 cells/g dry compost in non-sterilized sewage sludge, changes in the number of the bacterium, the water content (%), and temperature occurred, as shown in Figure 2.14. The decrease in water content indicates that composting progressed smoothly for 200 h. As other bacteria were not countable on no. 3 agar medium, the dominant bacterium in the compost was *B. subtilis*. Although the temperature was initially set at 50°C, the rise of temperature was so rapid that its control was impossible. When the initial cell concentration was increased to 6×10^9 cells/g dry compost and the temperature was set at 60°C, similar changes in the number of the bacterium, moisture content (%), and temperature were observed (data not shown). The cell number of *B. subtilis* decreased slightly at 60°C; however, an increase in number was observed when the temperature fell to that of the atmospheric level at the end of the composting reaction. As this bacterium can grow even at 50°C on no. 3 agar medium, *B. subtilis* NB22-1 played a major role in the degradation of organic matter in sewage sludge in the thermophilic stage. This means that a new compost inhabiting bacterium suppressive to plant pathogens is available if the initial concentration is maintained at a certain level. The suppressive effect on plant pathogens by the compost sampled at 180 h in Figure 2.14 can be expressed in terms of the diameter of inhibitory zones. These are as follows: 10 mm for *F. oxysporum*, 18 mm for *R. solani*, 25 mm for *P. oryzae*, 30 mm for *X. oryzae*,

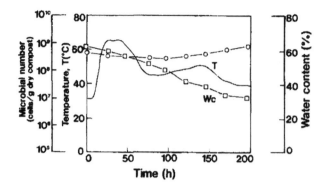

FIGURE 2.14 Time course of the cell number of *B. subtilis* NB22-1, water content (Wc), and temperature (T) during composting of sewage sludge when the initial concentration was approximately 10^9 cells/g dry compost. Symbol: o, *B. subtilis* NB22-1.

10 mm for *P. ultimum,* and 20 mm for *V. dahliae.* Although the suppressive effect of the compost with higher *B. subtilis* concentrations was not examined, it is obvious that a greater suppressive effect will be expressed.

In practical composting, maintenance of a certain level of concentration of *B. subtilis* NB22 will be secured only if a certain amount of *B. subtilis* NB22 is inoculated at the start of composting, mainly because 20% to 49% of compost products are recycled and mixed with fresh raw material as seed or as a bulking agent. The recycling ratio depends on the moisture content of the fresh raw material. Thus, the suppressive bacterium newly isolated in this study showed a clear suppressive effect on various phytopathogenic microorganisms, not only in the composting process but also when mixed with compost products. The key point for the efficient expression of the suppressiveness is the initial concentration of the bacterium.

2.8 CHARACTERIZATION OF *B. SUBTILIS* RB14

2.8.1 Suppressive Effect by RB14 *In Vitro* (14)

B. subtilis NB22 was mainly used as a suppressive bacterium (15). In this section, the basic feature of a newly isolated *B. subtilis* RB14 was investigated.

2.8.1.1 Materials and Methods

Ten phytopathogenic fungi as shown in Table 2.12 were grown on PDA medium (potato dextrose agar, 39 g; distilled water, 1 liter, pH 5.6) at 30°C for 5 days and then suspended in sterile distilled water. A portion of this suspension was mixed into L agar or PDA plate. After spotting an isolated bacterium at the center of this plate, the suppression by the bacterium was investigated by observing the sizes of inhibitory zones made on the lawn of the fungus. A phytopathogenic bacterium was grown by shaking in L broth at 30°C for 1 day. The cell suspension was diluted and mixed into L agar medium plates.

TABLE 2.12

Suppressive Effect of *B. subtilis* RB14 against Plant Pathogens in Comparison with *B. subtilis* NB22 *In Vitro*

Phytopathogens	Inhibition Zone of RB14		Inhibition Zone of NB22	
	PDA Medium (mm)	L Medium (mm)	PDA Medium (mm)	L Medium (mm)
Fungi				
Fusarium oxysporum f. sp. *lycopersici* race J1 SUF 119	13	13	9	13
Fusarium oxysporum f. sp. *lycopersici* race J2	13	16	8	12
Fusarium oxysporum f. sp. *radicis-lycopersici* KEF2R-1	11	14	8	12
Rhizoctonia salani AG-4	38	17	33	15
Cochliobolus miyabeanus NIAES 5425	32	30	31	20
Alternaria mali IFO 8984	27	23	29	21
Cercospora kikuchii NIAES 8984	27	30	24	25
Pyricularia oryzae IFO 5279	13	19	10	15
Pyricularia oryzae NIAES 5001	40	27	30	26
Verticillium dahliae Klebahn V-3	22	23	14	20
Bacteria				
Pseudomonas syringae pv. *lachrymans* P1-7415	a	12	a	9

a Not tested.

2.8.1.2   Results

All the tested microorganisms were clearly suppressed by the newly isolated *B. subtilis* RB14 as shown in Table 2.12. RB14 inhibited the growth of other phytopathogens more significantly than NB22. A phytopathogenic fungus, *F. oxysporum* f. sp. *lycopersici* race J1 SUFl 19, which causes the crown and root rot of tomato, was inhibited more significantly by this RB14 than by *B. subtilis* NB22, as shown in Figure 2.15.

2.8.2   Isolation of Iturin A and Surfactin from the Culture Broth of RB14

2.8.2.1   Materials and Methods

The suppressive substances were isolated from the culture supernatant of *B. subtilis* RB14 grown in no. 3 medium (10 g of Polypepton, 10g of glucose, 1 g of KH_2PO_4,

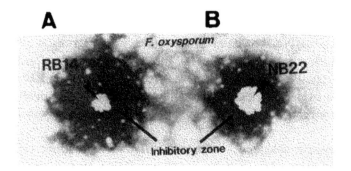

FIGURE 2.15 Suppressive effect *in vitro* of (A) a newly isolated bacterium *B. subtilis* RB14 on *F. oxysporum* in comparison with (B) the previously isolated *B. subtilis* NB22.

0.5 g of $MgSO_4 \cdot 7H_2O$ in 1 liter of distilled water [pH 7.0]) for 5 days at 30°C. The supernatant was precipitated at pH 2. The precipitate was collected by centrifugation and was extracted with methanol. The extract was analyzed by thin-layer chromatography (TLC) on silica gel plates 60F254 (Merck, Darmstadt, Germany) with chloroform–methanol–water (65:25:5, vol/vol). Components were visualized by spraying the plates with 10% H_2SO_4 and heating at 110°C for 45 min. Two main spots were detected. The R*f* values of these spots were 0.63 and 0.38. The R*f* value 0.38 corresponded to that of iturin A purified in my laboratory (1) and was confirmed to be iturin A by HPLC analysis, which was operated by a reverse-phase column Inertsil ODS-2 (GL Sciences Inc., Tokyo, 4.6 φ × 250 mm) with acetonitrile–ammonium acetate (10 mM) (2:3, vol/vol).

The R*f* value of the purified surfactin (Wako Pure Chemical Industries, Ltd., Osaka) was shown to be 0.63, which was the same R*f* value of the other spot. This substance was identified as surfactin by the following analyses. The precipitate at pH 2.0 described earlier was dissolved in distilled water (adjusted to pH 7.0 with NaOH). The suspension was kept overnight at 4°C and then extracted with the same volume of dichloromethane for 30 min at room temperature. Then, the dichloromethane phase was removed. After the extraction was repeated four times, all the dichloromethane phase was evaporated. The white precipitate was used for the further analyses. The infrared (IR) spectrum of the substance molded in KBr showed the characteristic IR absorption of surfactin previously reported (5). Strong bands characteristic of peptides at 3300 cm⁻¹, at 1650 cm⁻¹, and at 1520 cm⁻¹ were clearly observed. The bands at 2960 cm⁻¹, 2930 cm⁻¹, and 2850 cm⁻¹ reflected aliphatic chains (-CH_3, -CH_2-) of the fraction. The band at 1730 cm⁻¹ was due to lactone carbonyl absorption. These results indicated that the product had an analogous structure to surfactin.

2.8.2.2 Results

The aforementioned dichloromethane extract was analyzed by the same HPLC column used for iturin detection by monitoring at 205 nm with the solvent acetonitrile–acetic acid (1%) (68:32, vol/vol). Two main peaks were detected at retention times 53 min and 70 min. The authentic sample of surfactin was detected at retention time 70 min. Two main peaks with retention times 53 min and 70 min were collected, and the hydrolysate of each component was analyzed by an amino acid analyzer, respectively. The amino acid composition of each component was determined to be Asp, Glu, Val, and Leu in the molar ratio of 1:1:1:4, which was the same value expected from surfactin (14, 16, 17). The molecular ion peaks of the compounds with HPLC retention times 53 min and 70 min were detected by FAB-MS at 1044 and 1058, respectively. As the presence of sodium ions was detected by the yellow color in flame reaction, these molecular ion peaks were considered to be $(M+Na)^+$ formed by the attachment with Na^+, as reported by Jenny et al. (18). These results led to the molecular weight 1022 for the compound with retention time 53 min, and 1036, which is the same molecular weight as surfactin, for the compound with retention time 70 min. The mass difference of 14 units between the two was thought to be attributed to the one unit of methylene residue. All these results described indicated that the strain *B. subtilis* RB14 produced the biosurfactant surfactin as well as iturin A.

As *B. subtilis* RB14 coproduces iturin A at about l00 mg/l and surfactin at about 200 mg/l in no. 3 medium, the significant inhibition of plant pathogens by RB14 was considered to be attributed to the synergistic effect of the two compounds, iturin A and surfactin. Although surfactin itself did not inhibit the growth of a phytopathogenic fungus *F. oxysporum* at least at the concentration of 100 μg/ml, it greatly enhanced the antibiotic activity of iturin as shown in Table 2.13. The synergistic effect of iturin and surfactin is shown as the growth of a phytopathogen in Figure 2.16. The concentration of surfactin was fixed at 50 mg/l and different concentrations of iturin were mixed with surfactin and the growth of the phytopathogen *F. oxysporum* was observed on PDA plates. When the mixture of the two substances was applied, it is clear that the inhibitory effect of iturin against *F. oxysporum* was seen at significantly lower concentrations compared with that of a single iturin application.

The synergistic effects of these two lipopeptides first found in this work may explain why this bacterium showed a wide suppressive spectrum against phytopathogens, as shown in Table 2.12.

TABLE 2.13
Effects of Iturin A and Surfactin on the Growth of Phytopathogenic Fungus *F. oxysporum*

	Growth of *F. oxysporum*[a]					
	Concentration of Lipopeptide (μg/ml)					
Lipopeptide(s) Added	0	6.25	12.5	25	50	100
Iturin A	+ + +	n.t.	+	+	+	–
Surfactin	+ + +	+ + +	+ + +	+ + +	+ + +	+ + +
Iturin A and 100 μg/ml of surfactin[b]	+ + +	n.t.	+	–	–	n.t.

[a] *F. oxysporum* was spotted at the center of the plate containing lipopeptide(s), and the degrees of growth inhibition on PDA were determined after 7 days incubation at 28°C according to the following levels: + + +, no inhibition; +, significant inhibition; –, no growth; n.t., not tested.

[b] Only iturin A concentration was varied, and the surfactin was fixed at the concentration of 100 μg/ml.

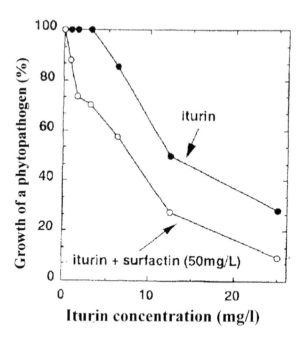

FIGURE 2.16 The synergistic effect of the mixture of iturin and surfactin. The growth of *F. oxysporum* was observed on PDA plates when iturin and surfactin (50 mg/l) were mixed. When iturin concentration was 0, the growth area of *F. oxysporum* after 3 days was set at 100%. The inhibitory zones were measured after iturin or the mixture of iturin and surfactin were applied to the plates and the growth areas of *F. oxysporum* were compared.

Coproduction of surfactin and iturin A by *B. subtilis* has been reported by others (19). However, neither the suppressive effect against phytopathogenic microorganisms nor the idea of biological control of them has been demonstrated. Surfactin is already known as a potent surface-active agent (10, 16) and as an antibiotic against some bacteria (10), but is not effective against phytopathogenic fungus *F. oxysporum* (Table 2.13). The reason why the antifungal antibiotic activity of iturin was enhanced in the presence of surfactin is being investigated by physicochemical method, but the structural similarities between the two compounds (6, 12, 20, 21) may enhance the destructive activity of iturin to the cell membrane of plant pathogens. This will help the suppressive effect against plant pathogens even in soil where this bacterium was treated as a biological control agent to replace chemical fungicides.

2.9 CHARACTERIZATION OF *B. SUBTILIS* YB8

2.9.1 PRODUCTION OF SURFACTIN (22)

2.9.1.1 Methods and Results

When YB8 formed a colony on an L agar plate on which 20 μl of tributyrin was spread, a clear halo appeared around the colony. This surface-active substance was purified and determined to be surfactin as follows: The acid precipitate of the culture supernatant was collected and extracted with methanol. The extract was separated in the HPLC system (23) previously described, and a clear peak of surfactin was detected in comparison with the authentic sample purchased from Wako Pure Chemical Industries, Ltd., Osaka, Japan. The IR spectrum, amino acid analysis, and molecular weight determination by FAB-MS of the peak sample also revealed that the substance was surfactin (data not shown). The structure of surfactin as a potent surface-active reagent (9) and antibiotic (24) is shown as follows:

$$\left(CH_3\right)_2 CH\left(CH_2\right)_9 - CHCH_2CO \rightarrow \text{L-Glu} \rightarrow \text{L-Leu} \rightarrow \text{D-Leu}$$

$$O \leftarrow \text{L-Leu} \leftarrow \text{D-Leu} \leftarrow \text{L-Asp} \leftarrow \text{L-Val}$$

2.9.2 PRODUCTION OF PLIPASTATIN

2.9.2.1 Material and Methods

The antifungal substance was isolated from the culture supernatant of *B. subtilis* YB8 cultivated in ACS medium (25) for 5 days at 30°C. This medium contained sucrose 100 g, citric acid 11.7 g, Na_2SO_4 4 g, yeast extract 5 g, $(NH_4)_2HPO_4$ 4.2 g, KCl 0.76 g, $MgCl \cdot 6H_2O$ 0.42 g, $ZnCl_2$ 0.0104 g, $FeCl_2 \cdot 6H_2O$ 0.0245 g, $MnCl_2 \cdot 4H_2O$ 0.0181 g in 1 liter, and was adjusted to pH 6.9 with NH_4OH. Purification of the antifungal substance was carried out as follows by modifying the methods developed for purification of fengycin (25) or plipastatin (26). The supernatant (4 liters) was precipitated at pH 2. The precipitate was collected by centrifugation and was extracted with 95% ethanol. The extract was mixed with 10% charcoal (w/v), stirred for 60 min and filtered through a suction funnel. The charcoal was stirred with 1 liter of chloroform–methanol–H_2O (65:25:4) for 60 min followed by the suction treatment, and the eluate was concentrated and dissolved in 4 volumes of propanol (v/w). This solution was charged on a propanol-filled column of silica gel (Silica gel 60; Merck, Darmstadt, Germany), and was followed by successive elution with one equal column-volume of propanol, 2 equal column-volumes of 90% propanol, and 2.5 equal column-volumes of 80% propanol. These fractions were bioassayed on an agar plate containing the fungus. The active eluate from the 80% propanol elution was further purified by a reversed-phase preparative HPLC using a PREP-ODS (GL Sciences, Tokyo, Japan; 2 cm × 25 cm, flow rate 10 ml/min, detection at 205 nm) with a 3:4 mixture of acetonitrile and 5 mM ammonium acetate. The main peak with the highest antifungal activity was passed through a Sephadex LH-20 (Pharmacia, Uppsala, Sweden) column with 80% methanol, and 3 mg of the purified antifungal substance was obtained.

2.9.2.2 Results

The IR spectrum of the purified antifungal substance molded in KBr showed the characteristic absorption of peptide bonds (1654 cm^{-1} and 1543 cm^{-1}) and lactone carbonyl absorption (1753 cm^{-1}). The hydrolysate of the substance was analyzed using an amino acid analyzer. The amino acid composition was determined to be comprised of Val (1), Thr (1), Glu (3), Pro (1), Ile (1), Tyr (2), and Orn (1) residues. The protonated molecular ion peak of this compound was displayed at *m/z* 1491.9 by

FAB-MS analysis. From these results, the antifungal substance was considered to coincide with plipastatin Bl, which had been reported as a new inhibitor of phospholipase A2 (26), and its structure is as follows:

$$CH_3(CH_2)_{12}\text{-}CH(OH)CH_2CO$$
$$\rightarrow \text{L-Glu} \rightarrow \text{D-Orn} \rightarrow \text{L-Tyr} \rightarrow \text{D-allo-Thr} \rightarrow \text{L-Gln} \rightarrow \text{D-Val}$$
$$| \qquad\qquad\qquad\qquad\qquad\qquad\qquad \downarrow$$
$$\text{L-Ile} \leftarrow \text{D-Tyr} \leftarrow \text{L-Glu} \leftarrow \text{L-Pro}$$

B. subtilis YB8 was the first bacterium that was the coproducer of two lipopeptide antibiotics, surfactin and plipastatin B1. The antifungal effect of plipastatin B1 was already reported as follows: This substance inhibited the growth of *Alternaria mali* (apple leaf spot pathogen), *Botrytis cinerea* (gray mold pathogen), and *Pyricularia oryzae* (rice blast pathogen), and considerable high protective value against *Helminthosporium* leaf spot was also observed in the pot test (27).

The time course of the production of surfactin and plipastatin B1 is shown in Figure 2.17. Fresh overnight culture in ACS medium was inoculated into eight of the 40 ml each of ACS medium per Erlenmeyer flask (200 ml). The cultivation was operated at 30°C, and each flask was taken out periodically and the analyses of surfactin and plipastatin B1 were conducted as follows:

The precipitate of the acidified supernatant was collected by centrifugation and was extracted with methanol. The filtered extracted solution was analyzed by reversed-phase HPLC. The HPLC condition for surfactin quantification was followed as previously (23). Plipastatin B1 was detected and quantified by the reversed-phase HPLC system newly developed in this work: The filtrate described earlier was injected into

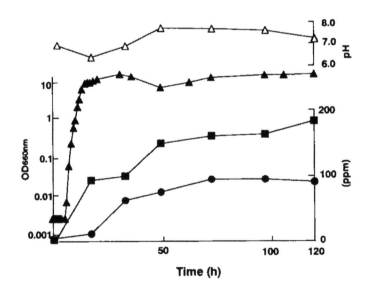

FIGURE 2.17   Time course of production of surfactin and plipastatin B1 by *B. subtilis* YB8 in ACS medium. Symbols: growth (OD_{660}) (▲), surfactin (■), plipastatin B1 (●), pH (△).

an HPLC column (ODS-2, 4.6 mm × 250 mm, GL Sciences). The system was operated at a flow rate of 1.0 ml/min with acetonitrile-ammonium acetate (10 mM) 1:1 (v/v). The elution pattern was monitored at 205 nm. The concentration of plipastatin B1 was determined by a calibration curve made by the purified sample.

Surfactin was already accumulated in the culture medium at the beginning of the stationary growth phase (16 h) while little plipastatin B1 was produced at this time. As these two substances were simultaneously extracted from the same precipitate and the same methanol extract was used for the HPLC analysis, the experimental deviation or error for this quantitative comparison was significantly small. Both substances reached a maximum at 2 to 3 days of incubation and the same level was kept at least 5 days. In the time course of the plipastatin production by *Bacillus cereus* BMG302-fF67, the maximum peak of plipastatin production was observed at one day and the gradual decrement with a pH change to alkaline was reported (26).

It has been reported that several *B. subtilis* are potential biological control agents (1). Among them, *B. subtilis* RB14 produces iturin A (12) and surfactin, and a synergistic suppressive effect of the two lipopeptides against the growth of plant pathogens was described in Section 2.8 (14). Here, *B. subtilis* YB8 will be another candidate of microbial pesticides by producing surfactin and plipastatin, which have a different suppressive spectrum to plant pathogens, and there might be a synergistic suppressive effect of the two lipopeptides against the growth of a plant pathogen.

As no biochemical or genetic analysis of a plipastatin producer is reported, molecular cloning for the production of this compound from *B. subtilis* YB8 is important for the understanding of the lipopeptide antibiotics. Usually antibiotic biosynthesis is carried out through physiologically, biochemically, and genetically complex mechanisms, thus the establishment of a host–vector system in an original antibiotic producer is important for cloning antibiotic production gene(s).

Although transformation of *B. subtilis* YB8 with plasmid DNA by a competent cell method was unsuccessful, an electroporation method (28) developed for an iturin A producer was employed for this coproducer. About 10^5 transformants were obtained from 1 μg of plasmid pC194 (29) or pUB110 (30). This transformation efficiency is high enough for shotgun cloning of the gene(s), and genetic analysis is explained in Chapter 6.

REFERENCES

1. Phae, C. G., Shoda, M., and Kubota, H. Suppressive effect of *Bacillus subtilis* and its products on phytopathogenic microorganisms. *J. Ferment. Bioeng.*, 69, 1–7 (1990).
2. Phae, C. G., Sasaki, M., Shoda, M., and Kubota, H. Characteristics of *Bacillus subtilis* isolated from composts suppressing phytopathogenic microorganisms. *Soil Sci. Plant Nutr.*, 36, 575–586 (1990).
3. Nakasaki, K., Sasaki, M., Shoda, M., and Kubota, H. Change in microbial numbers during thermophilic composting of sewage sludge with reference to CO_2 evolution rate. *Appl. Environ. Microbiol.*, 49, 37–41 (1985).
4. Winkelmann, G., Allgaier, H., Lupp, R., and Jung, G. Iturin A_L—a new long chain iturin A possessing an unusual high content of C_{16}-β-amino acids. *J. Antibiotics*, 36, 1451–1457 (1983).

5. Peypoux, F., Guinand, M., Michel, G., Delcambe, L., Das, B. C., and Lederer, E. Structure of iturin A, a peptidolipid antibiotic from *Bacillus subtilis. Biochemistry*, 17, 3992–3996 (1978).
6. Isogai, A., Takayama, S., Murakoshi, S., and Suzuki, A. Structure of β-amino acids in antibiotics iturin A. *Tetrahedron Lett.*, 23, 3065–3068 (1982).
7. Uesugi, Y. *Methods in Pesticide Science*, vol. 2. Soft Science, Tokyo (1981).
8. Chang, I., and Kommedahl, T. Biological control of seedling blight of corn by coating kernels with antagonistic microorganisms. *Phytopathology*, 58, 1395–1401 (1968).
9. Aldrich, J., and Baker, R. Biological control of *Fusarium roseum* f. sp. *dianthi* by *Bacillus subtilis. Plant Dis. Reporter*, 54, 446–448 (1970).
10. Utkhede, R. S. Antagonism of isolates of *Bacillus subtilis* to *Phytophthora cacroum. Can. J. Bot.*, 62, 1032–1035 (1983).
11. Nakasaki, K., Sasaki, M., Shoda, M., and Kubota, H. Characteristics of mesophilic bacteria isolated during thermophilic composting of sewage sludge. *Appl. Environ. Microbiol.*, 49, 42–45 (1985).
12. Phae, C. G., and Shoda, M. Expression of the suppressive effect of *Bacillus subtilis* on phytopathogens in inoculated composts. *J. Ferment. Bioeng.*, 70, 409–414 (1990).
13. Kawakami, W., Hashimoto, S., Nishimura, K., Watanabe, H., and Watanabe, H. Development of a process for radiation disinfection and composting of sewage sludge. Japan Atomic Energy Research Institute-M report 85-051 (1985).
14. Hiraoka, H., Asaaka, O., Ano, T., and Shoda, M. Characterization of *Bacillus subtilis* RB14, coproduction of peptide antibiotics iturin A and surfactin. *J. Gen. Appl. Microbiol.*, 38, 635–640 (1992).
15. Phae, C. G., Shoda, M., Kita, N., Nakano, M., and Ushiyama, K. Biological control of crown and root rot and bacterial wilt of tomato by *Bacillus subtilis* NB22. *Ann. Phytopathol. Soc. Japan*, 58, 329–339 (1992).
16. Arima, K., Kakinuma, A., and Tamura, G. Surfactin, a crystalline peptidelipid surfactant produced by *Bacillus subtilis*: isolation, characterization and its inhibition of fibrin clot formation. *Biochem. Biophys. Res. Commun.*, 31, 488–494 (1968).
17. Cooper, D. G., MacDonald, C. R., Duff, S. J. B., and Kosaric, N. Enhanced production of surfactin from *Bacillus subtilis* by continuous product removal and metal cation additions. *Appl. Environ. Microbiol.*, 42, 408–412 (1981).
18. Jenny, K., Kappeli, O., and Fiechter, A. Biosurfactants from *Bacillus licheniformis*: structural analysis and characterization. *Appl. Microbiol. Biotechnol.*, 36, 5–13 (1991).
19. Sandrin, C., Peypoux, F., and Michel, G. Coproduction of surfactin and iturin A, lipopeptides with surfactant and antifungal properties, by *Bacillus subtilis. Biotechnol. Appl. Biochem.*, 12, 370–375 (1990).
20. Kakinuma, A., Oushida, A., Shima, T., Sugino, H., Isono, M., Tamura, G., and Arima, K. Confirmation of the structure of surfactin by mass spectrometry. *Agric. Biol. Chem.*, 33, 1669–1671 (1969).
21. Isogai, I., Takayama, S., Murakoshi, S., and Suzuki, A. Structures of β-amino acids in antibiotics iturin A. *Tetrahedron Lett.*, 23, 3065–3068 (1982).
22. Tsuge, K., Ano, T., and Shoda, M. Characterization of *Bacillus subtilis* YB8, coproducer of lipopeptides and surfactin and plipastatin B1. *J. Gen. Appl. Microbiol.*, 41, 541–545 (1995).
23. Huang, C. C., Ano, T., and Shoda, M. Nucleotide sequence and characteristics of the gene, *lpa-14*, responsible for biosynthesis of the lipopeptide antibiotics iturin A and surfactin from *Bacillus subtilis* RB14. *J. Ferment. Bioeng.*, 76, 445–450 (1993).
24. Bernheimer, A. W., and Avigad, L. S. Nature and properties of a cytolytic agent produced by *Bacillus subtilis. J. Gen. Microbiol.*, 61, 361–369 (1970).

25. Vanittanakom, N., Loeffler, W., Koch, U., and Jung, G. Fengycin—a novel antifungal lipopeptide antibiotic produced by *Bacillus subtilis* F-29-3. *J. Antibiotics*, 39, 888–901 (1986).

26. Umezawa, H., Aoyagi, T., Nishikiori, T., Okuyama, A. Yamagishi, Y., Hamada, M., and Takeuchi, T. Plipastatins: new inhibitors of phospholipase A2, produced by *Bacillus cereus* BMG 302-fF67. *J. Antibiotics*, 39, 737–744 (1986).

27. Yamada, S., Takayama, Y., Yamanaka, M., Ko, K., and Yamaguchi, I. Biological activity of antifungal substances produced by *Bacillus subtilis*. *J. Pest. Sci.*, 15, 95–96 (1990).

28. Matsuno, Y., Ano, T., and Shoda, M. High-efficiency transformation of *Bacillus subtilis* NB22, an antifungal antibiotic iturin producer, by electroporation. *J. Ferment. Bioeng.*, 73, 261–264 (1992).

29. Ehrlich, S. D. Replication and expression of plasmids from *Staphylococcus aureus* in *Bacillus subtilis*. *Proc. Natl. Acad. Sci. U.S.A.*, 74, 1680–1682 (1977).

30. Keggins, K. M., Lovett, P. S., and Duvall, E. J. Molecular cloning of genetically active fragments of *Bacillus* DNA in *Bacillus subtilis* and properties of the vector plasmid pUB110. *Proc. Natl. Acad. Sci. U.S.A.*, 75, 1423–1427 (1978).

3 *In Vivo* Plant Tests

3.1 BIOLOGICAL CONTROL OF CROWN AND ROOT ROT AND BACTERIAL WILT (1)

Crown and root rot of greenhouse tomato caused by *F. oxysporum* f. sp. *radicis-lycopersici* (abbreviated as FoR) is prevalent and inflicts a great loss on tomato yield in the winter season, when the pathogenicity of the causal pathogen is expressed mostly in low temperature and the physiological activity of the host is repressed (2, 3). So far, no proper methods have been developed to prevent or cure this disease in spite of many trials (4–9).

Bacterial wilt of tomato caused by *P. solanacearum* (abbreviated as Ps) (10) is still one of the serious soilborne diseases in Japan, although various attempts to control the disease have been tentatively carried out (3). So far, no report has been published on the interaction between *B. subtilis* and Ps in the field.

In spite of many attempts to control various plant diseases biologically, utilization of *B. subtilis* as an antagonist is relatively small in number. A decline of disease incidence with *B. subtilis* by dipping plant roots in its solution (11), seed coating (7, 12–14), or mixing with soil (15–20) had been reported. The features of these papers are that *B. subtilis* was used as an antagonist against only fungal pathogens including *Fusaria*.

In the present section, the suppressive effect of NB22 on the occurrence of crown and root rot and bacterial wilt of tomato caused by FoR and Ps, respectively, was investigated in the pot test. Since NB22 is thermotolerant (21), introduction of the bacterium into infested soils combining with heat treatment by steam sterilization was carried out to enhance suppressibility of the crown and root rot of tomato caused by FoR. The experiment was also planned to assess the effect of the amendment of organic matter to soil.

3.1.1 MATERIALS AND METHODS (1)

3.1.1.1 Bacterial Culture

B. subtilis NB22 (NB22) isolated from a compost (21) was used. This bacterium was tolerant to 80°C treatment (22) and produced an antifungal peptide iturin as shown in Chapter 2. Liquid culture of NB22 was carried out using no. 3 medium (glucose 10 g, Polypepton 10 g, KH_2PO_4 1 g, $MgSO_4 \cdot 7H_2O$ 0.5 g, distilled water 1 liter [pH 7.0]) by a reciprocal shaker (100 strokes/min) at 30°C for 5 days. Cell-free culture filtrate was obtained by centrifugation of cultural suspension at 7800 × g for 30 min and then by filtration through a membrane filter (0.45 μm, Dismic-25, Advantec, Tokyo). The cell suspension of NB22 was prepared by suspending the precipitate after centrifugation in sterilized water to the final concentration of 10^8 colony-forming unit (cfu)/ml for field study.

3.1.1.2 Antagonistic Activity of NB22 *In Vitro*

Antagonistic activity of NB22 was examined against 14 species and 5 formae speciales of plant pathogenic fungi and 8 species of plant pathogenic bacteria as listed in Table 3.1. Fungi used in the experiments were subcultured on the potato dextrose agar medium (PDA; Nissui Pharmaceutical Co., Japan) (pH 5.6) at 28°C. After 5 to 7 days incubation, 2 or 3 small agar blocks with conidia were taken and were suspended into sterilized water to obtain 10^6 cfu/ml level of conidia for each fungus.

Plant pathogenic bacteria used in the experiments were also subcultured on nutrient agar (NA; Nissui Pharmaceutical Co.) for 24 h at 28°C, and one loop of the

TABLE 3.1
Suppressiveness of *B. subtilis* NB22 and Its Culture Filtrate *In Vitro*

Plant Pathogens	Inhibitory Zone by Cells (mm)	Inhibitory Zone by Filtrate (mm)
Fungi		
Alternaria mali IFO 8984	21	33
Cercospora kikuchii NIAES 5039	28	40
Phytophthora infestans IFO 5547	25	30
Phytophthora capsici 09001	23	n.d.
Botrytis cinerea Bot 1	40	34
Rhizoctonia solani NIAES 5219	40	38
Rhizoctonia solani K-1	23	n.d.
Cochliobolus miyabeanus NIAES 5425	32	33
Fusarium oxysporum f. sp. *cucumerinum* NIAES 5117	10	n.d.
Fusarium oxysporum f. sp. *lycopersici* race J1 SUF 119	21	18
Fusarium oxysporum f. sp. *radicis-lycopersici* KEF2R-1	21	20
Fusarium moniliforme H-110	21	n.d.
Fusarium roseum f. sp. *cerealis* 030201	16	n.d.
Fusarium oxysporum f. sp. *fragarariae* 02010402	16	n.d.
Fusarium oxysporum f. sp. *spinaciae* 0201501	15	n.d.
Pythium ultimum Trow H-1	18	n.d.
Verticillium dahliae Klebahn V-3	25	25
Gibberella zeae 030101	29	n.d.
Pyricularia oryzae NIAES 5001	40	35
Bacteria		
Xanthomonas oryzae IFO 3998	35	25
Xanthomonas campestris pv. *citri* QN8206	29	n.d.
Pseudomonas caryophylli A	14	n.d.
Pseudomonas solanacearum TOM-w	23	18
Pseudomonas syringae pv. *lachrymans* P1–7415	25	25
Pseudomonas glumae Ku8106	13	n.d.
Agrobacterium tumefaciens Ku7501	14	n.d.
Erwinia carotovora subsp. *carotovora* B9	14	n.d.

Note: n.d., not determined.

bacterial colonies was suspended into sterilized water to obtain about 10^8 cfu/ml of bacterial suspension.

Antagonistic effects of NB22 on both plant pathogenic fungi and bacteria were examined as follows: 0.1 ml of the conidial or the bacterial suspension was spread on PPDA medium (added 10 g of Polypepton to PDA) and/or no. 3 medium plates (added 15 g of agar to no. 3 medium), respectively, and NB22 suspension was spotted on the center of those plates and incubated at 28°C in the dark. The diameter of the inhibitory zone in each plate was measured after 3 or 5 days incubation. Three replications were conducted in each experiment.

For some tested fungi or bacteria whose growth was significantly inhibited by NB22, the suppressive effect of the cell-free culture filtrate was also examined as follows: a sterilized stainless cup (8 mm in diameter × 10 mm in height) was placed on the center of each plate where conidia or bacteria were spread beforehand. Then 0.2 ml of cell-free culture filtrate of NB22 was poured into the cup and each plate was incubated at 28°C in the dark. After 7 days of incubation, the diameter of the inhibitory zone appearing around the cup was measured.

3.1.1.3 Morphological Changes in Suppressed Hyphae of *Fusarium*

Suspensions of microconidia of *F. oxysporum* f. sp. *radicis-lycopersici* KEF2R-1 (FoR) and NB22 were streaked at the opposite sides on PDA plates and incubated at 28°C in the dark. After 5 days, the tips of the growing hyphae of FoR were stained with 0.5% lactophenol cotton blue for 3 min. After being rinsed with distilled water three times, they were observed under a microscope.

3.1.1.4 Effect of *B. subtilis* NB22 on the Suppression of Crown and Root Rot of Tomato

Figure 3.1 shows the experimental flow diagram. Naturally infested soils with FoR were taken from the field in a greenhouse at Kanagawa Horticultural Experiment Station, Japan. The average concentration of FoR in the soil was 4.6×10^2 cfu/g of dry soil. Ten liters of the soil were mixed with 50 g of ground dolomitic limestone and then with 20 g of cyclo diurea (CDUs$_{555}$) before treatment. Then, 1 liter of cultural broth of NB22 (3.1×10^9 cfu/ml) was mixed into soil and put into a 1/2000 a Wagner pot.

Chopped rice straw and bark compost (Tokachi Bark, Mori Co., Japan) were used as soil amendments. The 200 g of chopped rice straws that were cut into 0.5–10 cm length and immersed in 1 liter of cultural suspension of NB22 for 24 h at 30°C were mixed with the infested soil in each pot. The 200 g of bark compost were also immersed in a 1 liter cultural suspension of NB22 and treated in a similar method to the rice straws. In each case, rice straw or bark compost without treatment was prepared for the control. Ten pots were prepared in each treatment and then half of them were steam-sterilized for 1 h at 80°C using a soil sterilizer (Tomoe Boiler Co., Japan) after treatment with NB22.

After these treatments, one tomato seedling ("Zuishu," Sakata Seed Co., Japan), sown in sterilized soil and grown in a greenhouse for 60 days, was planted in each pot. After planting, the temperature was maintained at 25°C during the day and 10°C at night. After 3 months of planting, plants were taken out from the pots, and the

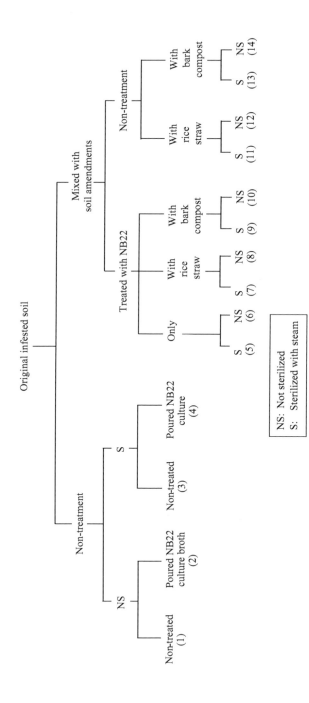

FIGURE 3.1 Experimental procedure for examination of the effect of *B. subtilis* NB22 on the occurrence of crown and root rot of tomato. Numbers in parentheses correspond to those in Table 3.2.

roots were washed in water to evaluate the root rot. Then, they were cut at the bottom of the stem to determine the index of incidence of xylem browning. The total yield per plant was also measured.

At 0, 30, and 60 days after planting, soils were taken from every pot and numbers of thermotolerant bacteria (TTB) and fungi were determined by decimal dilution of soil suspension. For fungi, PDA medium with 0.033 g/liter of Rose Bengal was used. The colonies that appeared on the PDA medium were randomly sampled, and the pathogenicity of each colony was tested by using the tomato seedlings to identify FoR. For TTB, soil suspension was heated at 80°C for 10 min in a water bath and spread onto the no. 3 medium plates. To identify *B. subtilis* among TTB, colonies on the plate were randomly picked up and subjected to morphological and physiological tests by following the procedure of Claus et al. (23). Five plates were used for each treatment. Some of *B. subtilis* identified were further cultivated in no. 3 medium, and cell-free cultural liquid was subjected to high-performance liquid chromatography (HPLC) analysis to identify iturin production (21).

3.1.1.5 Effect of NB22 on the Suppression of Bacterial Wilt of Tomato

Cut stems of tomatoes (0.5 to 1 cm) that were naturally infected by Ps were immersed in sterilized water for 24 h and the suspension of Ps was prepared. The concentration of the Ps suspension was counted on the selective medium reported by Tanaka and Fukuda (24). To prepare severely infested soil, 1 liter of the bacterial suspension was poured into 10 liters of naturally infested soil taken from a field in Kanagawa Horticultural Experiment Station. Tomato seedlings "Okitu No. 3" (National Research Institute of Vegetables, Ornamental Plants and Tea, Japan), sown in sterilized soil, were taken out at the stage of five to six true leaves. After being washed thoroughly with distilled water, the roots were dipped into either the culture broth containing 10^8 cells/ml of NB22 or the suspension of NB22 (10^8 cfu/ml) for 1 h and planted in the infested soil. The plants were grown in a greenhouse where the soil temperature was maintained at 30°C by using an electric heater. The number of the dead plants was counted periodically until 29 days after planting.

3.1.2 Results

3.1.2.1 Antagonistic Activity of *B. subtilis* NB22 *In Vitro*

NB22 suppressed the growth of all the plant pathogens used in the experiment (Table 3.1). Conspicuous suppressiveness was observed in the pathogens such as *F. oxysporum*, *R. solani*, *P. caryophylli*, and *P. solanacearum*. Degree of suppressiveness did not differ among the five formae speciales of *F. oxysporum*. When FoR was cultured at the opposite side of NB22, the growth of FoR was suppressed and the tips of hyphae of the FoR became malformed and/or burst partly as shown in Figure 3.2, when observed under a microscope. NB22 also showed a clear inhibitory zone to *P. solanacearum* TOM-w (Figure 3.3), which was purified on the selective medium (24) from the extracted suspension of the tomato stem suffering from the bacterial wilt.

Cell-free culture filtrate also suppressed the growth of nine species of pathogenic fungi and two formae speciales of *F. oxysporum* and two species of pathogenic

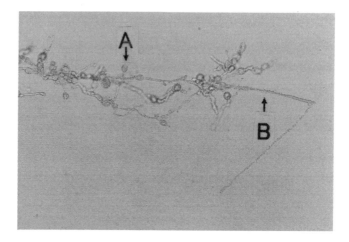

FIGURE 3.2 Morphological changes of hyphae of *F. oxysporum* f. sp. *radicis-lycopersici* caused during co-cultivation with *B. subtilis* NB22 on glucose-supplemented PDA medium. (A) Swelling hyphae. (B) Burst hyphae.

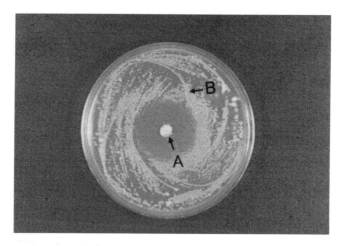

FIGURE 3.3 Suppressive effect of (A) *B. subtilis* NB22 on (B) *P. solanacearum* on no. 3 medium.

bacteria. Suppressiveness of the culture filtrate to some fungi was higher than that of the cells of NB22.

3.1.2.2 Effect of NB22 and Steam Sterilization on Crown and Root Rot of Greenhouse Tomato

The effect of NB22 was enhanced especially when combined with the steam sterilization (Table 3.2). This was well reflected in total yield. Although the treatment of rice straw per se was effective to suppress the incidence of xylem browning and root rot, reduction of the disease incidence and increase of total yield were the most significant when soil was treated with rice straws immersed in NB22 culture broth.

TABLE 3.2

Effect of *B. subtilis* NB22 on the Incidence of Xylem Browning and Root Rot and the Total Yield of Tomato with or without Steam Sterilization of Soil

Treatment	Index of Incidence of Xylem Browning (%)[a]		Index of Incidence of Root Rot (%)[b]		Total Yield (g/plant)	
	Unsterilized	Sterilized	Unsterilized	Sterilized	Unsterilized	Sterilized
Control[c]	95(1)[d]	55(3)	95(1)	53(3)	372(1)	1205(3)
B. subtilis poured	75(2)	45(4)	85(2)	45(4)	681(2)	1280(4)
B. subtilis mixed	45(6)	25(5)	75(6)	30(5)	651(6)	1479(5)
Rice straw mixed	28(12)	10(11)	98(12)	38(11)	710(12)	1136(11)
B. subtilis plus rice straw mixed	43(8)	10(7)	78(8)	25(7)	1007(8)	1504(7)
Bark compost mixed	20(14)	20(13)	88(14)	40(13)	405(14)	1121(13)
B. subtilis plus bark compost mixed	23(10)	18(9)	75(10)	45(9)	708(10)	1298(9)

[a] Index of ratio of the area of xylem browning at the bottom of the stem of tomato plant to the whole area of the stem was calculated by Equation 1 by scoring: 0, no symptom; 1, 1–25%; 2, 26–50%; 3, 51–75%; 4, 76–100%.

[b] Index of degree of root rot was calculated by Equation 1 by scoring: 0, no symptom; 1–4, slightly to severely infected.

[c] Neither treatment of *B. subtilis* nor amendment of organic matter.

[d] Numbers in parentheses correspond to those in Figure 3.1.

$$\text{Equation (1)} = \frac{\Sigma(\text{score of disease} \times \text{number of plants}) \times 100}{4 \times \text{number of plants}}$$

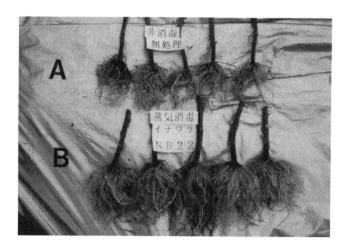

FIGURE 3.4   Effect of *B. subtilis* NB22 on the growth of roots of tomato. (A) Roots of tomato grown in unsterilized soil without *B. subtilis* NB22. (B) Roots of tomato grown in steam-sterilized soil mixed with *B. subtilis* NB22 and chopped rice straw.

The condition of roots 60 days after treatment is shown in Figure 3.4. A similar effect on the incidence of xylem browning was observed in bark compost treatment. However, the bark compost treatment did not effectively suppress the occurrence of the root rot.

Numbers of and fungi in each soil are shown in Table 3.3. The number of TTB was larger in general when NB22 was inoculated and incubation time became longer. Although the same level of TTB was observed, the disease was not suppressed without treatment of NB22. This verifies that introduction of NB22 was effective in spite of the variation of number of TTB in soil.

By random sampling of soils treated with NB22, almost all the TTB were identified as *B. subtilis* by the method described in Section 3.1.1. Most of those *B. subtilis* grown in no. 3 medium produced iturin, which was verified by HPLC analysis (21).

3.1.2.3   Effect of *B. subtilis* NB22 on Bacterial Wilt of Tomato

Approximately 90% of tomato seedlings were dead within 24 days after planting in the soil infested with Ps where no NB22 was applied (Figure 3.5). On the contrary, in the soil where culture broth of NB22 was applied, only 10% of tomato seedlings were dead even in 24 days after planting. In soil inoculated with only the cells of NB22 per se, the percentage of dead seedlings was 50% until 24 days after planting (Figure 3.6).

3.1.3   Discussion

This is the first report in which *B. subtilis* NB22 showed a suppressive effect on both fungal and bacterial pathogens, not only *in vitro* but also *in vivo*. The antifungal and antibacterial activity of NB22 was confirmed clearly by the *in vitro* tests, as shown in Table 3.1, indicating that NB22 had a broad suppressive spectrum against both pathogenic fungi and bacteria. The suppressiveness expressed only by the culture filtrate *per se*

TABLE 3.3

Total Number of ThermoTolerant Bacteria (TTB),[a] Fungi in the Naturally Infested Soil[b] at 0, 30, and 60 Days after Planting of Tomato

Treatment		Number of TTB-Bacteria			Number of Fungi		
		0[c]	30	60	0	30	60
Control	Steam	$\times 10^5$	$\times 10^5$	$\times 10^5$	$\times 10^2$	$\times 10^2$	$\times 10^2$
	Sterilized	3	39	203	4	23	25
	Unsterilized	42	53	141	2408	2478	1836
Mixed with *B. subtilis* NB22	Sterilized	—[d]	—	—	—	—	—
	Unsterilized	—	—	—	—	—	—
Poured with *B. subtilis* NB22	Sterilized	—	—	—	—	—	—
	Unsterilized	—	—	—	—	—	—
Mixed with rice straw	Sterilized	38	137	173	173	95	734
	Unsterilized	26	42	140	1823	1028	2105
Rice straw with *B. subtilis* NB22	Sterilized	126	241	434	229	193	1166
	Unsterilized	0	21	123	1790	1695	2523
Bark compost	Sterilized	57	145	184	13	44	171
	Unsterilized	35	37	170	1306	824	2547
Bark compost with *B. subtilis* NB22	Sterilized	1	44	132	831	24	291
	Unsterilized	13	9	86	1662	1047	1048

[a] Refer to the text.

[b] By *F. oxysporum* f. sp. *radicis-lycopersici*.

[c] Days after planting.

[d] Not determined.

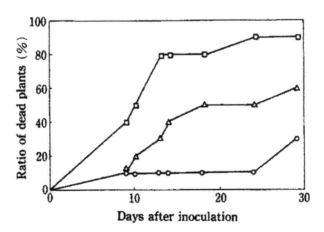

FIGURE 3.5 Suppressive effect of *B. subtilis* NB22 and its culture broth on bacterial wilt caused by *P. solanacearum*. □, control; △, supplied with cells of NB22; ○, supplied with culture broth of NB22.

FIGURE 3.6   Suppressive effect of *B. subtilis* NB22 and its culture broth on bacterial wilt. (A) Treated with culture broth of *B. subtilis* NB22. (B) Treated with cells of *B. subtilis* NB22. (C) Control.

indicated that the extracellular production of iturin (7) was associated with the effect. The lytic effect on the cell wall and or cell membrane of the fungi by iturin was considered to cause the ill-formation of the hyphae as shown in Figure 3.2. Tsuyumu et al. (25) also reported similar morphological changes in *Fusarium* hyphae *in vitro* in co-cultivation with an avirulent strain of *P. solanacearum* having antifungal activity.

Various microorganisms besides *B. subtilis* (2, 11, 12) have been used to control *Fusarium* diseases. Kijima et al. (26) showed that the crown and root rot of tomato was suppressed by mixed cropping with Chinese chives (*A. tuberosum*) treated with *Pseudomonas gladioli*, which showed a strong antifungal activity to FoR. Ogawa and Komada (27) demonstrated a remarkable suppressiveness of *Fusarium* wilt of sweet potato by induced resistance using nonpathogenic *Fusarium*. Although these trials were carried out using fields for several months; their evaluation was primarily on degree of disease occurrence. In the present study, it was evaluated that the suppressiveness was not only by disease occurrence but also by the total yield in addition to analysis of the microbial changes. Commercially acceptable yield by amendment of rice straw mixed with NB22 followed by steam sterilization was obtained.

The combination of NB22 treatment with steam sterilization was an effective method to suppress the disease of crown and root rot of tomato. Baker et al. (15, 19, 28) showed that steam sterilization of *Phytophtora-* or *Rhizoctonia*-infested soil at 60°C for 1 h was most effective suppressing the disease incidence. They postulated that the dominance of heat-tolerant microorganisms, such as *Bacillus* sp., resulting from heat treatment might be responsible for the suppressiveness. In this experiment, a similar phenomenon will have happened in steam-sterilized soils in which rice straw with NB22 were amended. Bark compost, however, showed less effectiveness than rice straw in suppression (Table 3.2). The difference in suppressive effect between rice straw and bark compost indicated that the characteristic or quality of organic amendments was also an important factor to effectively cooperate with microbial antagonists.

Besides heat treatment, co-treatment of chemical soil sterilization and organic matter was reported to be effective to reduce *Fusarium* disease (29). If NB22 is resistant to chemical sterilizers, co-treatment of NB22 with these chemicals may give another means to biological control.

So far, *B. subtilis* was considered to be not effective to bacterial pathogens, as Podile reported that *B. subtilis* showed no inhibitory effect on Ps *in vitro* (30). However, this study clearly demonstrated that NB22 suppressed not only the growth of Ps *in vitro* (Figure 3.3) but also the occurrence of the disease effectively by the application of the culture broth of NB22 to soil (Figures 3.5 and 3.6). These results suggested that an antimicrobial substance, iturin, which was presumably secreted in the infested soil, was responsible for the suppression of bacterial wilt of tomato. Further experiments must be conducted to confirm the secretion of iturin in the soil.

Concerning the introduction of *B. subtilis* into soil, the possibility of causing any adverse effect in soil seems to be low, mainly because no pathogenic *B. subtilis* was found and the safety of the bacterium is guaranteed. Dominancy of this bacterium in soil was found to be relatively lower compared with other soil microorganisms (31). The acute toxicity and chronic toxicity tests for *B. subtilis* NB22 and iturin have been under progress, and so far no acute toxicity for both has been found (data not shown). It has already been reported that minimum inhibitory concentration (MIC) of iturin to nonpathogenic microorganisms was more than 100 times larger than that to pathogenic microorganisms, indicating that iturin is specifically active to plant pathogenic microorganisms (see Chapter 2, Table 2.8) (21).

Several methods have been demonstrated to utilize NB22 to draw its potentiality as an agent of biological control. NB22 was easily distinguishable from other bacteria both by its morphological feature on no. 3 medium as described in detail in Chapter 2 and by a simplified identification method (23). As a spontaneous mutant with streptomycin resistant to NB22, (31) has been available to increase the sensitivity and reliability to detect its concentration in soil, further investigation on interactions and population dynamics between NB22 and various plant pathogens in soil will be possible to make the role of NB22 in biological control clear.

3.2 BIOLOGICAL CONTROL OF *RHIZOCTONIA SOLANI* DAMPING-OFF

B. subtilis RB14, which showed antibiotic activities against several phytopathogens *in vitro* by producing the antibiotics iturin A and surfactin (Chapter 2), was subjected to a pot test to investigate its ability to suppress damping-off of tomato seedlings caused by *Rhizoctonia solani*.

3.2.1 Materials and Methods (32)

3.2.1.1 Microorganisms and Plasmid

The bacterial strains and plasmid used are listed in Table 3.4. *B. subtilis* RB14, which was isolated originally from compost, produces iturin A and surfactin (33). To facilitate recovery from soil, *B. subtilis* RB14-C, a spontaneous streptomycin-resistant

TABLE 3.4

Bacterial Strains and Plasmid

| | | Production ($\mu g/ml$)[b] | | |
| | Phenotype or | | | |
Strain or Plasmid	Plasmid Marker[a]	Iturin A	Surfactin	Reference
B. subtilis				
RB14		140	335	(33)
RB14-C	Sm^r	127	308	This work
RΔ1	Em^r	0	0	(34)
RΔ1(pC115)	Em^r Cm^r	152	41	(35)
Plasmid pC115	Cm^r , 4.2 kb, *lpa-14*$^+$			(35)

[a] Sm^r, streptomycin resistance; Em^r, erythromycin resistance; Cm^r, chloramphenicol resistance.

[b] Each strain was cultivated in no. 3 medium at 30°C for 5 days, and concentrations of iturin A and surfactin were determined by HPLC as described in Section 3.2.1.7.

mutant of RB14, was selected. The growth rate and antifungal activity of RB14-C were confirmed to be the same as those of the parental strain RB14.

B. subtilis RΔ1, derived from RB14 (34), is deficient in the *lpa-14* locus associated with the production of iturin A and surfactin (Chapter 5) (35) and therefore does not produce either antibiotic.

B. subtilis RΔ1(pC115), a transformant of RΔ1 with the plasmid pC115 carrying *lpa-14* (35), restored the production of iturin A and surfactin. The average concentrations of iturin A and surfactin produced in a liquid medium by the bacterial strains in several experiments are shown in Table 3.4. The level of iturin A production by RΔ1(pC115) was almost the same as that of RB14, but that of surfactin was only about one-eighth that of RB14.

The detail of *B. subtilis* RΔ1, *B. subtilis* RΔ1(pC115), and the *lpa-14 gene* is explained in Chapter 5.

The pathogenic fungus *R. solani* K-1 was isolated from cockscomb at the Kanagawa Horticultural Experiment Station, Kanagawa, Japan (1). The anastomosis grouping of K-1 is AG-4. *R. solani* K-1 falls into the praticola type as a severe damping-off pathogen of many plants (36, 37).

3.2.1.2 Media

L medium contains (per liter) 10 g of Polypepton (Nihon Pharmaceutical Co., Tokyo, Japan), 5 g of yeast extract, and 5 g of NaCl, adjusted to pH 7.0. Number 3 medium used for antibiotic production contains (per liter) 10 g of Polypepton, 10 g of glucose, 1 g of KH_2PO_4, and 0.5 g of $MgSO_4 \cdot 7H_2O$ adjusted to pH 6.8. To cultivate the strains or select the transformants from soil, chloramphenicol, erythromycin, or streptomycin was added at a concentration of 5, 5, or 100 $\mu g/ml$, respectively. Each medium was solidified with 2.0% agar, when necessary. PDA medium (Eiken Chemical Co.,

Tokyo, Japan) containing (per liter) 200 g of potato infusion, 20 g of glucose, and 15 g of agar, adjusted to pH 5.6, was used for stock culture of *R. solani* K-1. PDP medium containing (per liter) 200 g of potato infusion, 20 g of glucose, and 1 g of Polypepton (pH 5.6) was used for cultivation of *R. solani* K-1.

3.2.1.3   Soil Treatments

The soil used in this study was a low humic andosol taken from a field at the Kanagawa Horticultural Experiment Station, Japan (1). The soil was sieved through an 8-mesh (about 2 mm pore size) screen and air-dried. The soil and vermiculite were mixed in the ratio of 4:1 (wt/wt) and nutrient amended so that the final concentrations of N, P_2O_5, and K_2O were 70 mg, 240 mg, and 70 mg per 100 g of dry soil, respectively. The prepared soil was kept in plastic bags at room temperature. The soil was put into a sterilizable polypropylene bag and autoclaved for 60 min at 121°C four times at 12 h intervals. The main characteristics of the soil thus prepared were as follows: texture, low humic andosol; moisture content, 12.7%; maximum water-holding capacity, 137 g/100 g of dry soil; pH 5.9; and bulk density, 0.522 g/cm³. The measurement of these properties followed previously described methods (38). Sterilized soil (150 g) was put into a plastic pot with a diameter of about 90 mm, and the moisture was kept at 60% of the maximum water-holding capacity by the daily addition of sterilized water.

3.2.1.4   Inoculation of *R. solani* into Soil

Fifty milliliters of sterile PDP medium in 200 ml Erlenmeyer flasks was inoculated with 5 mm plugs taken from a PDA petri dish culture of *R. solani* K-1. The flasks were incubated without shaking in the dark at 30°C for 1 week. The mycelial mats, on the surface of the medium, were then homogenized (4000 rpm, 2 min) in sterile water and inoculated into the soil at the ratio of one-eighth piece of the mat to one pot 5 days before planting the germinated tomato seeds. After inoculation, the pots were incubated at 30°C and the moisture was kept at 60% of the maximum water-holding capacity.

3.2.1.5   Application of RB14-C and Its Derivative Strains to Soil

B. subtilis RBl4-C was incubated for 16 h in L medium with streptomycin at 30°C in a shaker, and then 1 ml of the culture broth was inoculated into 100 ml of no. 3 medium with streptomycin in an Erlenmeyer flask. The flask was shaken for 5 days at 30°C. Twenty milliliters of culture broth of RBI4-C were mixed with 150 g of soil in a pot. For the treatment consisting only of cell suspension, the culture broth was centrifuged (8000 × *g*, 10 min, 4°C) and the sedimented cells were washed in 0.85% NaCl (pH 7.0) and then centrifuged again under the same conditions. The washed cells were suspended in sterile distilled water, and 20 ml of the cell suspension was mixed with 150 g of soil. For the treatment consisting of centrifuged culture broth, 20 ml of supernatant, after the centrifugation described earlier, was mixed with 150 g of soil. The three inoculations into soil were carried out simultaneously 3 days before planting the germinated tomato seeds. For each treatment, two to six pots were prepared and experiments were repeated at least three times.

For comparing the suppressive effects of cell suspensions of RB14-C, RΔ1, and RΔ1(pC115), each strain was cultivated as follows: 3 ml of each culture broth, after 16 h at 30°C in L medium with appropriate antibiotics, was inoculated into 300 ml of fresh L medium with the antibiotics and cultivated for 24 h. The cells were collected by centrifugation (8000 × g, 10 min, 4°C), washed in 0.85% NaCl solution, and then centrifuged again. Then, 20 ml of each cell suspension in 0.85% NaCl solution was inoculated into the 150 g of soil. These inoculations were also carried out 3 days before planting the germinated tomato seeds.

3.2.1.6   Cell Recovery from Soil and Counting of Viable Cell Number

The soils were sampled both before planting and at 14 days after planting. In the latter sampling, the roots and surrounding soil were sampled together. Three grams of soil was suspended in 8 ml of 0.85% NaCl solution (pH 7.0) in a 50 ml Erlenmeyer flask and then shaken for 15 min at 140 strokes per min at room temperature. The suspension was serially diluted in 0.85% NaCl solution and plated onto L agar plates containing the proper antibiotics. The plates were incubated at 37°C, viable cells were counted after 12 h, and the cell number was expressed as the total cell number. To determine the number of spores, 1 ml of the suspension was heated for 15 min at 80°C serially diluted and spread onto L agar plates. All data are expressed as log cfu (colony-forming unit) per gram of dry soil.

3.2.1.7   Quantitative Analysis of Iturin A and Surfactin Recovered from Soil

Three grams of soil was suspended in 21 ml of a mixture of acetonitrile–3.8 mM trifluoroacetic acid (4:1 [vol/vol]) in a 50 ml Erlenmeyer flask and then shaken for 1 h (140 strokes per min, room temperature). The soil in the suspension was then removed with filter paper (Toyo Roshi Co., Ltd., Tokyo, Japan), and the filtrate was evaporated. The precipitate was extracted with 2 ml of pure methanol for 2 h. After the extracted solution was centrifuged at 10,000 × g for 2 min, the supernatant was filtered through a 0.2 μm pore size polytetrafluoroethylene membrane (JP020; Advantec, Ltd., Tokyo, Japan) and injected into an HPLC column (octyldecyl silanolate-2, 4.6 mm [diameter] by 250 mm; GL Sciences, Tokyo, Japan). The system was operated at a flow rate of 1.0 ml/min with acetonitrile–10 mM ammonium acetate (3:4 [vol/vol]) for measurement of iturin A. The elution was monitored at 205 nm. Surfactin was analyzed with the same HPLC column used for identification of iturin A. The system was operated at a flow rate of 1.5 ml/min and monitored at 205 nm with the solvent acetonitrile–3.8 mM trifluoroacetic acid (80:20 [vol/vol]) (35). The concentration of iturin A was determined with a calibration curve made with each purified component, and the total amount of five homologues of iturin A, as shown in Figure 3.7, was used as the concentration of iturin A. The concentration of surfactin was determined with authentic surfactin purchased from Wako Pure Chemical Industries, Ltd., Osaka, Japan. Lower limits for detection of iturin A and surfactin by HPLC were about 0.30 μg/ml. Therefore, when the concentrations of both antibiotics were lower than the limit of detection, they were judged to be not detectable. It had been confirmed beforehand that no peak for iturin A or surfactin was observed in nontreated soil, and about 70% to 90% of iturin A and surfactin could be recovered from the soil by this method.

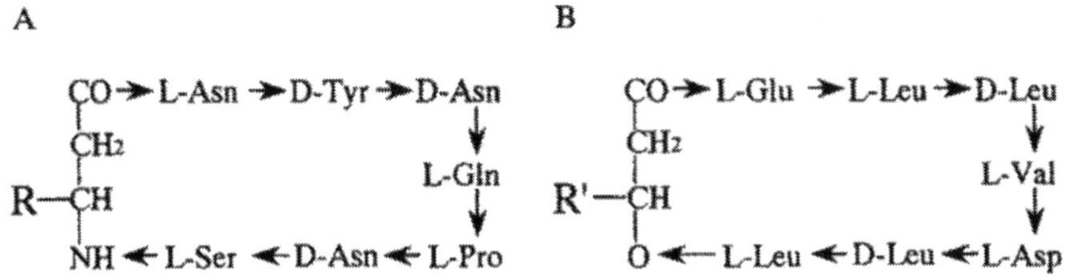

A

$$CO \rightarrow L\text{-}Asn \rightarrow D\text{-}Tyr \rightarrow D\text{-}Asn$$

$$CH_2 \qquad\qquad\qquad \downarrow$$

$$\qquad\qquad\qquad\qquad L\text{-}Gln$$

$$R\text{---}CH \qquad\qquad\qquad \downarrow$$

$$NH \leftarrow L\text{-}Ser \leftarrow D\text{-}Asn \leftarrow L\text{-}Pro$$

B

$$CO \rightarrow L\text{-}Glu \rightarrow L\text{-}Leu \rightarrow D\text{-}Leu$$

$$CH_2 \qquad\qquad\qquad \downarrow$$

$$\qquad\qquad\qquad\qquad L\text{-}Val$$

$$R'\text{---}CH \qquad\qquad\qquad \downarrow$$

$$O \leftarrow L\text{-}Leu \leftarrow D\text{-}Leu \leftarrow L\text{-}Asp$$

FIGURE 3.7 Structures of (A) iturin A and (B) surfactin. R denotes one of the following aliphatic chains: $CH_3(CH_2)_{10}$, $CH_3CH_2CH(CH_3)(CH_2)_8$, $(CH_3)_2CH(CH_2)_9$, $CH_3(CH_2)_{12}$, or $(CH_3)_2CH(CH_2)_{10}$. R' denotes $(CH_3)_2CH(CH_2)_9$.

3.2.1.8 Plant Growth

Seeds of tomato (Ponderosa) were surface disinfected for 1 min with 70% ethanol, rinsed five times with sterile distilled water, and then surface disinfected again for 5 min with 0.5% sodium hypochlorite. After at least 10 rinses with sterile distilled water, the seeds were germinated on a 2% agar plate at 30°C for 2 days. Each pot was sown with nine germinated seeds and placed in a growth chamber at 30°C with 80% relative humidity under 16 h of light (about 12,000 lux). At least two pots were prepared with each treatment. The moisture content was kept at 60% of the maximum soil water-holding capacity. After 2 weeks, the percentage of diseased seedlings per pot was determined. Furthermore, the shoots were clipped off at the soil surface level and the lengths and dry weights of the shoots were measured.

3.2.1.9 Stability of Iturin A and Surfactin in Soil

To observe the stability of iturin A and surfactin in soil without plants, 20 ml of the centrifuged culture broth of RB14-C, which was prepared as described earlier, was mixed with 150 g of soil in a pot and incubated in a growth chamber as with the plant test described earlier. Iturin A and surfactin were recovered from two samples taken periodically from one pot, and the concentration of each was determined. The deviation of the concentrations for the duplicate samples was less than 5%, and the average concentrations are shown in Figure 3.8.

3.2.1.10 Determination of Plasmid Stability

The stability of pC115 in RΔ1(pC115) in soil was determined as follows: 100 colonies that appeared on the L agar medium with erythromycin and were judged to be RΔ1(pC115) by their surface appearance were randomly selected and transferred by replica plating onto L agar medium with and without chloramphenicol. That a fraction of bacteria expressed chloramphenicol resistance (Cm^r) and therefore carried pC115 among the 100 colonies was considered the indicator of pC115 stability. The existence of plasmid pC115 in the Cm^r colonies was confirmed by extraction of the plasmid from 10 randomly chosen colonies.

3.2.1.11 Statistical Analysis

Each plant test was repeated at least three times, and each mean of data was analyzed by Fisher's analysis of variance. The means ± standard deviations of the population densities were calculated with logarithmically transformed values.

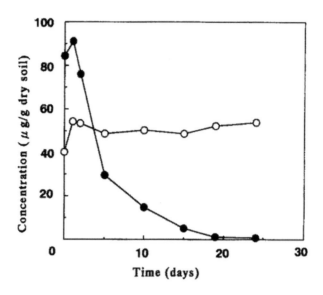

FIGURE 3.8 Stability of iturin A (•) or surfactin (o) in soil. The centrifuged culture broth of RB14-C was mixed with soil in a pot and incubated at 30°C. Sterilized water was supplied daily to maintain soil water content at 60% of the maximum water-holding capacity. On each day, 3 g of soil was sampled and suspended in 21 ml of a mixture of acetonitrile–3.8 mM trifluoroacetic acid (4:1 [vol/vol]) to extract iturin A and surfactin. Each concentration was determined by HPLC as described in Section 3.2.1.7.

3.2.2 Results

3.2.2.1 Suppressive Effect of Culture Broth of *B. subtilis* RB14-C

The suppressive effect of the cells and culture broth of *B. subtilis* RB14-C on the damping-off of tomato seedlings caused by *R. solani* is shown in Table 3.5. In the pots that were not infested with *R. solani*, all tomato seedlings grew normally and no disease appeared. However, in the pots infested only with *R. solani*, the percentage of diseased plants was 85.2% and the tomato shoot dry weights and shoot lengths were markedly decreased. With the RB14-C culture broth treatment, the percentage of diseased plants decreased to 16.7%, and both the shoot dry weights and the shoot lengths significantly increased. The differences in shoot length, shoot weight, and percent diseased plants after five treatments were statistically significant at $P = 0.05$.

The cells of RB14-C were recovered from the soil in the pots. On day 0, the total cell number was 7.76×10^8 cfu/g of dry soil and the spore number was 1.25×10^8 cfu/g of dry soil. On day 14, the final day of the test, the total cell number was 4.56×10^7 cfu/g of dry soil and the spore number was 4.78×10^7 cfu/g of dry soil. A slight decrease in cell number was observed, and most of the viable cells of RB14-C existed as spores.

The amounts of iturin A and surfactin recovered from soils on days 0 and 14 are shown in Table 3.6. Each value is an average of results from two pots in one experiment. Significant amounts of iturin A and surfactin were recovered on day 14. The decrease in the amount of iturin A was larger than that of surfactin in spite

TABLE 3.5
Effects of Different Treatments of *B. subtilis* RB14-C Cultures on the Suppression of Damping-Off of Tomato Plants Caused by *R. solani* K-1 14 Days after Planting*

Treatment		Shoot Length	Shoot Weight	Diseased Plants
R. solani	RB14-C	(mm)	(mg [dry wt]/pot)	(%)
+		13.1[a]	77[a]	85.2[a]
+	Culture broth	77.4[c]	502[b]	16.7[b]
+	Cell suspension	66.9[b]	574[b]	27.8[b]
+	Centrifuged culture broth	82.4[c]	651[b]	11.1[c]
−		83.1[c]	626[b]	0[d]

* For each treatment, each datum is an average of results from two to six pots (which contain nine seeds per pot) from experiments repeated three times. Means in any column with different letters are significantly different ($P = 0.05$) according to Fisher's protected least significant difference analysis.

of the existence of similar amounts of the two substances on day 0. As presented in Figure 3.8, the level of recovery of surfactin was stable over the duration of the test period (24 days), whereas that of iturin A gradually decreased.

3.2.2.2 Suppressive Effects of Cell Suspension and Centrifuged Cell-Free Culture Broth of *B. subtilis* RB14-C

Cell suspensions or the centrifuged cell-free culture broth of *B. subtilis* RB14-C was inoculated into soil, and the suppressive effects were compared (Table 3.5). When

TABLE 3.6
Concentrations of Iturin A and Surfactin Recovered from Treated Soils 0 and 14 Days after Treatment

	Concentration of Indicated Antibiotic			
	(μg/g of dry soil)[a]			
	0 Days		14 Days	
Treatment with RB14-C	Iturin A	Surfactin	Iturin A	Surfactin
Culture broth	19.4	18.1	5.3	18.1
Cell suspension	0.58	4.73	1.78	0.89
Centrifuged culture broth	19.5	18.0	1.24	10.7
None	ND[b]	ND	ND	ND

[a] Each value is an average of results from two pots in one experiment.
[b] ND, not detected. No peaks for iturin A or surfactin were observed by HPLC.

the centrifuged culture broth was inoculated, the suppressive effect was the same as that of the nonfractionated culture broth. The percentage of diseased plants was 11.1%, and the level of plant growth was restored to that of the control without *R. solani*. The concentration of iturin A decreased more rapidly than that of surfactin in both culture broth and centrifuged cell-free culture broth treatments (Table 3.6).

When the cell suspension of RB14-C was inoculated into soil, a suppressive effect was observed (Table 3.5). On day 0, the total cell number was 3.91×10^7cfu/g of dry soil and the spore number was 3.34×10^7cfu/g of dry soil. On day 14, the total cell number was 2.73×10^7 cfu/g of dry soil and the spore number was 3.06×10^7 cfu/g of dry soil. The slight decrease in cell number and the existence of most of the viable cells of RB14-C as spores were similar to the results of the culture broth treatment (data not shown).

3.2.2.3   Comparison of Suppressive Effects by RB14-C, RΔ1, and RΔ1(pC115)

To ascertain that the suppressive effect of RB14-C was related to iturin A and surfactin, cell suspensions of RB14-C, RΔ1, and RΔ1(pC115) were inoculated into soil and the suppressive effects were compared (Table 3.7). Washed cell suspensions were used in this experiment to minimize the effect of carryover of iturin A and surfactin from the culture broth and to clarify the effect of the cells themselves. In soil infested with *R. solani*, the percentage of diseased plants was 85.2% and the tomato shoot dry weights and the shoot lengths were significantly less than those of the healthy control. When RB14-C was inoculated into soil, the percentage of diseased plants was reduced to 24.1% and the shoot dry weights and shoot lengths were significantly greater than those of the nontreated pots infested with *R. solani*. In soil inoculated with RΔ1, the percentage of diseased plants was 64.8%, which was significantly

---

TABLE 3.7

Effects of Strains of RB14-C and Its Derivatives on the Suppression of Damping-Off of Tomato Plants Caused by *R. solani* K-1 14 Days after Planting*

Treatment		Shoot Length	Shoot Weight	Diseased Plants
R. solani	Bacterium	(mm)	(mg [dry wt]/pot)	(%)
+		17.2[a]	129[a]	85.2[a]
+	RB14-C	70.7[b]	560[b]	24.1[b]
+	RΔ1	37.2[a]	312[a]	64.8[a]
+	RΔ1(pC115)	69.2[b]	536[b]	29.4[b]
−		95.9[b]	704[b]	0[b]

* For each treatment, each datum is an average of results from experiments repeated three times with two pots (which contain nine seeds per pot). Means in any column with different letters are significantly different ($P = 0.05$) according to Fisher's protected least significant difference analysis.

higher than that of pots treated with RB14-C, and the shoot dry weights and the shoot lengths also were significantly reduced.

When RΔ1(pC115) was inoculated, the disease was considerably suppressed, with 29.6% diseased plants. Both shoot dry weight and shoot length were restored to values comparable to those found after the RB14-C treatment, showing that the suppressive effect of RΔ1(pC115) was equivalent to that of RB14-C. The values for shoot length and shoot weight and the percent diseased plants after three treatments also were statistically significant at $P = 0.05$.

The populations of these strains at the beginning and the end of the plant test are shown in Table 3.8. The initial cell numbers showed only a slight difference, and most of the viable cells existed as spores both on day 0 and on day 14. Spores of all three strains were stable in soil, with similar persistence in the rhizosphere.

The stability of plasmid pC115 in RΔ1(pC115) added to soil also was determined. On day 0, its stability was 82.7% in total cells and 83.3% in spores, whereas on day 14 the plasmid was present in 76.3% of total cells and in 79.7% of spores. This indicates that pC115 was stably maintained in RΔ1 in soil.

Iturin A and surfactin were recovered from the soil as shown in Table 3.9. Significant amounts were recovered from soil both on day 0 and on day 14 when the cell suspension of either RB14-C or RΔ1(pC115) was introduced. However, neither iturin A nor surfactin was detected by HPLC when the cell suspension of RΔ1 was inoculated into soil.

3.2.3 DISCUSSION

The culture broth, cell suspension, and centrifuged cell-free culture broth of *B. subtilis* RB14-C clearly showed a suppressive effect on the incidence of damping-off of tomato caused by *R. solani in vivo* (Table 3.5). When plants were treated with culture broth of RB14-C and without *R. solani*, the growth of the tomato plants was the same as that of the control, indicating that RB14-C imparts no growth-enhancing activity to the plant (data not shown).

TABLE 3.8

Populations of *B. subtilis* RB14-C, RΔ1, and RΔ1(pC115) in Soils

	Cell Number (log cfu/g of dry soil)[a]			
	0 Days		**14 Days**	
Strain	**Total**	**Spore**	**Total**	**Spore**
RB14-C	8.12 ± 0.23	8.08 ± 0.21	8.10 ± 0.09	7.94 ± 0.14
RΔ1	7.81 ± 0.07	7.84 ± 0.09	7.73 ± 0.26	7.79 ± 0.18
RΔ1(pC115)	7.83 ± 0.12	7.69 ± 0.18	7.67 ± 0.20	7.65 ± 0.27

[a] Each value is the mean (± the standard deviation) from three independent experiments with two replicates (pots) per experiment.

TABLE 3.9

Concentrations of Iturin A and Surfactin Recovered from Soil When Cell Suspensions of RB14-C, RΔ1, and RΔ1(pC115) Were Introduced into Soil

| | Concentration of Indicated Antibiotic (µg/g of dry soil)[a] | | | |
| | 0 Days | | 14 Days | |
Strain	Iturin	Surfactin	Iturin	Surfactin
RB14-C	0.73 ± 0.31	3.75 ± 0.16	0.49 ± 0.21	4.36 ± 0.15
RΔ1	ND[b]	ND	ND	ND
RΔ1(pC115)	0.61 ± 0.4	1.73 ± 0.1	0.40 ± 0.2	2.02 ± 1.0

[a] Each value is the mean (± the standard deviation) from three independent experiments with two pots for each treatment.

[b] ND, not detected. The determined concentrations were lower than the limit of detection by HPLC.

When culture broth or centrifuged culture supernatant was added to the soil, the initial concentration of iturin A and surfactin was about 20 µg/g of dry soil. Both substances were recovered from the soil after 14 days. However, the concentration of iturin A was reduced to 1 to 5 µg/g of dry soil and the concentration of surfactin was maintained at more than 10 µg/g of dry soil. This suggests a difference in persistence in soil. Surfactin was again more persistent than iturin A in soil. We do not know if the reduced amount of iturin A was caused by leaching from the soil during watering, by possible biodegradation by airborne microorganisms or residual *B. subtilis* cells introduced during the experiment, or by irreversible binding to soil materials or humic acids. Since concentrations of iturin A and surfactin in centrifuged culture supernatant were stably maintained for more than 1 month, such instability was probably related to the soil environment. In spite of the lower concentration of iturin A over time, our results showed that iturin A is effective for suppression of the disease. The persistence of surfactin in soil may contribute to suppressiveness by its synergistic effect on the antifungal activity of iturin A (33, 39).

The contributions of iturin A and surfactin to the suppressiveness of RB14-C in soil were confirmed with a mutant, RΔ1, which is deficient in the production of iturin A and surfactin. RΔ1 exhibited weak suppressiveness of damping-off as shown in Table 3.7. RΔ1(pC115), a transformant of RΔ1 containing the plasmid pC115, which carries the *lpa-14* gene, was restored for production of iturin A and partially restored for production of surfactin (Table 3.9). The complemented mutant exhibited a level of suppressive activity almost equal to that of the parental strain RB14-C. From these experiments, we infer that the major mechanism of suppression by RB14-C is production of the antibiotics iturin A and surfactin. Although the amount of surfactin detected in soil inoculated with RΔ1(pC115) was slightly lower than that of soil inoculated with RB14-C (Table 3.9), the suppressive ability of the transformant was almost the same as that of RB14-C. This also suggests that iturin A plays a role more important than that of surfactin in the suppression of *R. solani* damping-off in soil.

The percentage of diseased plants in soil treated with RΔ1 (64.8%) was significantly lower than that of the control treatment of soil with *R. solani* (85.2%), and the shoot lengths and shoot weights were significantly higher than those of the control (Table 3.7). As neither iturin A nor surfactin was detected on days 0 and 14 (Table 3.9), other determinants, such as unknown antibiotics, lytic enzymes, or siderophores (40), may be involved in the activity of RΔ1; this may be clarified by more complete characterization of the mutant. Similar phenomena were reported for *Pseudomonas fluorescens* strains which produce siderophore (41) or the antibiotic phenazine-1-carboxylate (42,43) in that they retained suppressiveness even when mutants lacking the production of these substances were prepared. The different plant appearances in different derivatives of *B. subtilis* RB14-C are shown in Figure 3.9.

The cells of RB14-C recovered from soil existed mostly as spores, and the population of the bacterium in the soil was on the order of 10^7 cfu/g of dry soil after 14 days. When vegetative cells of *B. subtilis* were inoculated into soil, sporulation occurred quickly, mainly because of the oligotrophic condition of the soil. The population was stabilized at the concentration of the spore number, and almost no change was observed for 30 days (36). In this experiment, the initial populations of RB14-C and its derivatives were also well maintained in soil, and most of the cells existed as spores (Table 3.8).

Although *B. subtilis* is not considered a representative rhizosphere species, like *Pseudomonas* spp., the rhizosphere population density, as well as the persistence of the bacterium in soil, is an important factor in the suppression of damping-off caused by *R. solani*. As the difference between the total cell number and spore number, which corresponds to the vegetative cell number, was small and the inoculation level of 10^7 to 10^8 cfu/g of soil in these experiments was rather large (Table 3.8), it is difficult to estimate a threshold population density of RB14-C to show significant

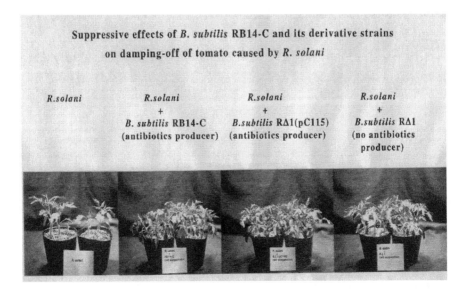

FIGURE 3.9   The tomato plants treated with different derivatives of *B. subtilis* RB14-C.

suppression of plant pathogens, which was reported as 10^5 cfu/g of root in the case of *Pseudomonas* spp. (42). However, the fact that most of the viable cells recovered from soil were in the form of spores and that spores have no ability to produce antibiotics suggest that the spores germinate in the rhizosphere and that eventually the antibiotics are produced, although the exact number of vegetative cells was not determined.

B. subtilis organisms are stable in soil as spores, and this is advantageous for the use of this bacterium as a biocontrol agent mainly because of the spores' stability and ease of handling. This study showed that treatment with the culture broth, cell suspension, or centrifuged culture broth will be effective as a biological control.

3.3 BIOLOGICAL CONTROL OF DAMPING-OFF OF TOMATO SEEDLINGS AND CUCUMBER *PHOMOPSIS* ROOT ROT (44)

Various types of composts have been developed with the specific effects of antagonism, competition, parasitism, elicitation of plant defense responses, or plant growth promotion (45–48). As for microbial antagonism, *B. subtilis*, which was isolated from compost, produces an antifungal peptide, iturin A (21, 32) and has a wide range of antifungal and antibacterial activities that can be used as a prominent biological control agent (1, 21, 32). However, application conditions, the spectrum of soilborne diseases, and persistency of the bacterium in the soil are major problems that still remain to be solved, because the suppressive effect is not stable and does not persist long enough to sustain the expected yield. In this section, the results of an evaluation of the suppressive ability of *B. subtilis* RB14-C, against damping-off of tomato seedlings and *Phomopsis* root rot of cucumber are presented.

3.3.1 BIOLOGICAL CONTROL OF DAMPING-OFF OF TOMATO SEEDLINGS

3.3.1.1 Materials and Methods

3.3.1.1.1 Plants

Tomato seeds of "Ponderosa" were surface-sterilized by 70% ethanol followed by 0.5% sodium hypochlorite treatment for 10 min, washed thoroughly by sterilized water for 5 min, placed on a plate, and kept at 28°C in the dark. After 24 h, one-third of the pregerminated seeds were transferred to a new plate for the germinated seed treatments described next. The remaining seeds were further incubated for 48 h. Thus, in both cases, 3-day-old, synchronously germinated seeds were used for the experiments.

3.3.1.1.2 Preparation of Infested Soil

Rhizoctonia solani K-1 (Chapter 2) was grown on potato dextrose agar (PDA) plates at 28°C in the dark for 5 days. The mycelia were recovered from five plates with PDA medium, homogenized with 500 ml of sterilized water and then mixed with 1 kg of soil wheat bran medium (a mixture of clay soil and wheat bran at the ratio of 4:1) that had been autoclaved at 1 20°C with 1.2 kg/cm^2 pressure for 1 h. After 20-day incubation at 28°C in the dark, 1 kg of the soil–wheat bran inoculum was mixed with

5 liters of clay soil preautoclaved twice at 120°C, 1.2 kg/cm² for 1 h every other day. The infested soil was then put in a plastic box (depth × width × length = 10 × 5 × 10 cm) and used for the experiments. As for the control, the infested soil was reautoclaved at 120°C and 1.2 kg/cm² pressure for 1 h.

3.3.1.1.3 Preparation of B. subtilis RB14-C and Flutolanil

B. subtilis RB14-C, which is a spontaneous streptomycin-resistant mutant of strain RB14 (33) and a coproducer of the antifungal substances iturin A and surfactin (32, 49), was grown in Luria-Bertani (LB) medium supplemented with 100 mg/l of streptomycin at 37°C for 4 days according to the method described in the previous chapter. To exclude the antifungal substances secreted to the culture fluid, the cells were collected by centrifuging at 6000 × g, washed and resuspended with sterilized water to give a concentration of ca. 10⁸ cfu/ml determined by measuring the optical density at 660 nm. The actual number of cells in the applied cell suspension was determined by spreading the decimally diluted cell suspension on LB-agar plates (LB plus 1.5% agarose) supplemented with 100 mg/l of streptomycin at the time of application. Moncut® granules (Nihon Nohyaku Co., Ltd., Tokyo, Japan) containing 7% of flutolanil were ground, sieved through 0.5 mm mesh and suspended with either sterilized water or the RB14-C cell suspension to give the final concentration of 500 mg/l.

3.3.1.1.4 RB14-C Cell and Flutolanil Treatments

As for the pouring treatments, 20 germinated tomato seeds were sown at 1 cm depth and then 200 ml of RB14-C cell suspension, flutolanil suspension, and the mixture of the two were gently applied by pouring onto the soil surface, respectively. In the germinated seed treatments, two sheets of filter paper were immersed in RB14-C cell suspension of 10⁸ cfu/ml, put in a plate, and then surface-sterilized tomato seeds that had been pregerminated at 28°C for 24 h in the dark were placed on the surface. After 48 h incubation at 28°C in the dark, 20 germinated seeds were then sown at 1 cm depth and then either 200 ml of sterilized water or flutolanil suspension was applied. Both pouring and germinated seed treatments were done on the same day in the same greenhouse. Sterilized water was used for the negative controls and each treatment was carried out in three replications with triplicates.

3.3.1.1.5 Detection of Iturin A from Soil

As RB14-C has been shown to produce five homologous iturin A and can be detected in the soil that suppresses damping-off of tomato seedlings (49), iturin A was extracted from the soil of each treatment according to the method described in a previous section. The crude extracts from the soil were evaporated into dryness under reduced pressure, dissolved with 2 ml of methanol through 0.2 μm mesh for HPLC analysis. HPLC was preformed using SMART system (Amersham Pharmacia Biotech Inc., Tokyo, Japan) with sephasil CI8SC2. 1/10 column (2.1 mm in diameter and 100 mm in length) and run with 10 mol/m³ ammonium acetate and acetonitrile as the developing solvent. The extracts were first absorbed to the resin after equibilizing with 10 mol/m³ ammonium acetate for 5 min and then eluted with ammonium acetate by a gradient from 35% to 55% within 22 min at the elution speed of 200 μl per min. The eluents were detected by absorbance at 205 nm. For the HPLC standard, iturin A was purified from the culture fluid of RB14-C as described in a previous section.

3.3.1.2   Results

3.3.1.2.1   Effects of RB14-C and Flutolanil on the Occurrence of Damping-Off of Tomato Seedlings

Effects of pouring and germinated seed treatments of RB14-C cell suspension, flutolanil, and the mixture of the two on the occurrence of damping-off of tomato seedlings were determined 14 days after treatment, and the results are summarized in Table 3.10. In the pouring treatment of RB14-C cell suspension, there was no significant difference in the occurrence of the damping-off with that of the positive control (Figure 3.10B and C), indicating that the single RB14-C cell suspension treatment has no suppressive effect on the disease occurrence. In contrast, flutolanil treatment effectively suppressed the damping-off (Figure 3.10D), confirming the anti-rhizoctonial activity. In the case of the combination treatment of RB14-C and flutolanil, the level of the suppressiveness was enhanced compared with the single flutolanil treatment, resulting in the lower disease occurrence (Figure 3.10E). When the germinated seeds were treated with RB14-C cell suspension before seeding, aiming to let RB14-C cells colonize the tomato rhizosphere, the occurrence of damping-off was significantly lower than that of the positive control in the pouring treatments (Figure 3.10F). No significant difference was observed in the disease occurrence between the negative controls of pouring and germinated seed treatments, indicating that RB14-C cells did not affect the vigor of the germinating tomato seeds. Interestingly, RB14-C application in both pouring and germinated seed treatments enhanced the level of the suppressiveness of the flutolanil application.

TABLE 3.10
Effect of *B. subtilis* and Flutolanil on the Occurrence of Damping-Off of Tomato Seedlings Caused by *R. solani*

Treatment	Damping-Off (%)		
Pouring treatment			
Disinfested soil + water	0.6[c]	±	1.0
Infested soil + water	23.9[a]	±	31.3
Infested soil + RB14-C	29.4[a]	±	34.3
Infested soil + flutolanil	10.5[b]	±	7.5
Infested soil + RB14-C + flutolanil	6.2[c]	±	3.1
Germinated seed treatment			
Disinfested soil + water	4.2[c]	±	3.5
Infested soil + RB14-C	15.8[b]	±	4.4
Infested soil + RB14-C + flutolanil	5.0[c]	±	2.4

Notes: Data represent the average and standard deviation. The same letters within a column are not significantly ($P < 0.05$) different from each other according to Fisher's protected least significant difference test.

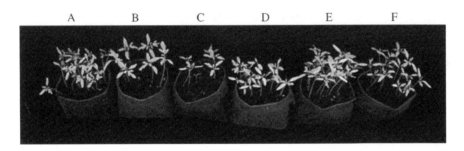

FIGURE 3.10 Effects of *B. subtilis* RB14-C and flutolanil on the occurrence of damping-off of tomato seedlings caused by *R. solani*. Tomato seeds were pregerminated at 28°C in the dark for 72 h and sown in (A) clay soil disinfested by autoclave, and (B–F) clay soil infested with *R. solani*, followed by application of (A and B) water, (C) RB14-C, (D) flutolanil, and (E) combination of RB14-C and flutolanil. (F) The result of germinated seed treated by RB14-C cell suspension (10^8 cfu/mL) at the time of pregermination for 48 h in the same condition.

3.3.1.2.2 Detection of Iturin A from Soil

HPLC analysis in the present study could detect five peaks comprising iturin A at the lowest concentration of 2 mg/ml using the standard compound (Figure 3.11A). When tomato rhizosphere soil samples were taken 8 days after pouring and germinated seed treatments, no noticeable peaks of iturin A were detected in any of the soil extracts including the soil treated with RB14-C cell suspension (Figure 3.11B). In contrast, the positive control soil, in which 3 ml of 10 mg/ml iturin A was premixed with the infested soil just before sampling, showed five typical peaks corresponding to the five homologous iturin A (Figure 3.11C). In this case, the recovery rate was approximately 80%.

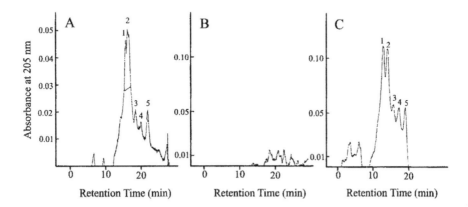

FIGURE 3.11 Detection of iturin A by HPLC. (A) Standard chromatogram of 2 ppm iturin A purified from culture fluid of *B. subtilis* RB14-C. (B) Chromatogram of extract from clay soil infested with *R. solani*, in which tomato seedlings have been grown for 30 days after pouring *B. subtilis* RB14-C cell suspension (10^8 cfu/ml). (C) Chromatogram of clay soil extract from 10 g of the same infested clay soil just after pouring 3 ml of 10 ppm iturin A standard solution. (1–5) Five homologous iturin A.

3.3.2 BIOLOGICAL CONTROL OF *PHOMOPSIS* ROOT ROT OF CUCUMBER BY RB14-C

3.3.2.1 Materials and Methods

3.3.2.1.1 Plants

Cucumber seeds of "Sagami-Hanjiro" were sown in pots and grown in a greenhouse at 25°C/15°C (day/night) for 10 days. Hyponex liquid fertilizer (6–10–5 for all purpose; HYPONEX JAPAN Corp., Ltd., Osaka, Japan) was applied at a 2000-fold dilution every other day. When the first true leaves were fully developed, 10 seedlings were taken for each treatment and the roots were soaked in tap water to gently remove the attached soil.

3.3.2.1.2 Preparation of Infested Soil

Phomopsis sp. M-1 isolated from infected summer squash (*Cucurbita maxima*) (Uekusa 1992, unpublished) was grown on PDA plates at 25°C for 14 days in the dark. The mycelia were recovered from five plates with PDA medium, homogenized with 500 ml of sterilized water and then mixed with 5 liters of sterilized clay soil and incubated at 25°C for 24 h in the dark before use. As for the control, the infested soil was autoclaved at 1.2 kg/cm², 120°C for 1 h.

3.3.2.1.3 Root Immersion Treatment

B. subtilis RB14-C cell suspension of 10^8 cfu/ml was prepared as mentioned earlier. Roots of the cucumber seedlings were immersed in the cell suspension, kept for 30 min and planted in pots with 500 ml of infested or sterilized soil. Sterilized water was used as the control. After giving 200 ml of sterilized water per each pot, the seedlings were grown in a greenhouse at 25°C/15°C (day/ night). Hyponex liquid fertilizer (6–10–5 for all purpose) was applied at a 2000-fold dilution every other day. Three replications with triplicates using 10 seedlings per each treatment were performed.

3.3.2.2 Results

3.3.2.2.1 Suppression of Phomopsis *Root Rot of Cucumber*

The effect of RB14-C on the suppressiveness of *Phomopsis* root rot of cucumber was determined. When cucumber seedlings were taken, their roots washed gently, and immersed in the RB14-C cell suspension before planting in the infested soil, the growth of the seedlings was retarded at first for several weeks probably due to the *Phomopsis* infection. Accordingly, as shown in Figure 3.12A, on 50 days after treatment, RB14-C-treated seedlings were apparently small in size compared with the negative control but all the plants recovered and started to grow vigorously thereafter. In contrast, the growth of the positive control plants was severely affected. Plant shoot weight, root weight, and leaf number 50 days after treatment reflected the difference as shown in Figure 3.13.

The shoot weight and leaf number clearly represented a difference among the treatments, indicating that the root damage induced by *Phomopsis* infection greatly affected the plant growth. Roots of the RB14-C treated seedlings, however, showed almost the same mass as that of the disinfested, negative control soil in spite of the

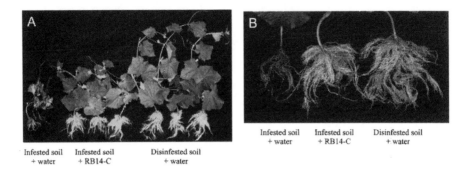

Infested soil   Infested soil   Disinfested soil
+ water   + RB14-C   + water

Infested soil   Infested soil   Disinfested soil
+ water   + RBI4-C   + water

FIGURE 3.12 Suppression of cucumber root rot caused by *Phomopsis* sp. 50 days after root immersion treatment of *B. subtilis* RB14-C cell suspension (10^8 cfu/ml). (A, from left) Growth of cucumber treated with water in the infested soil, RBI4-C in the infested soil, and water in disinfested soil. (B, from left) Roots of cucumber treated with water in the infested soil, RB14-C in the infested soil, and water in disinfested soil.

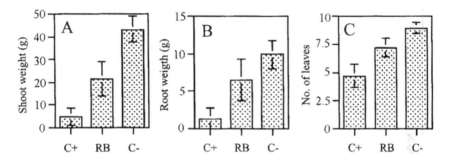

FIGURE 3.13 Suppression of cucumber root rot caused by *Phomopsis* sp. 50 days after root immersion treatment of *B. subtilis* RB14-C cell suspension (10^8 cfu/ml). (A) Shoot weight. (B) Root weight. (C) Leaf number. C+, plants grown in the infested soil after the roots were treated with water. RB, RB14-C cell suspension. C−, disinfested soil treated with water. Data represent the average and standard deviation (bars) obtained from the results of three replications with triplicates using 10 plants per each treatment.

growth retardation. Root browning was also significantly reduced whereas the roots of the positive control turned brown and severely rotted (Figure 3.12B). The difference in the average root weight apparently reflected the growth difference among the treatments, but there was statistically no significant difference between RB14-C treatment and the negative control (Figure 3.13B).

3.3.3 DISCUSSION

In the damping-off diseases caused by *R. solani*, Asaka and Shoda (32) have shown by using antagonistic *B. subtilis* that effective suppression could be achieved only when the bacterial culture filtrates containing iturin A were present and thus the disease suppression depends on the antibiotic activity of iturin A. They also found that most of the viable cells of *B. subtilis* in the soil are in the form of spores that have no ability to produce antibiotics. In the present study of the pouring treatment of the RB14-C cell suspension, in which culture fluid containing iturin A was removed, iturin A could not be detected from the treated soil and, accordingly, the occurrence of damping-off of tomato seedlings caused by *R. solani* could not be suppressed even though high densities of RB14-C cells were applied. On the contrary, seed treatment of RB14-C cells at the time of germination reduced the disease occurrence as reported by other studies (7, 50). In addition, the anti-rhizoctonial effect of flutolanil was enhanced by the simultaneous application of the RB14-C cell suspension and the use of germinated seeds pretreated with RB14-C cells. Although the detailed mechanisms of the enhanced suppressiveness are unclear, these results suggest that the germinated seed treatment of RB14-C can be used as a simple biological control practice especially when used in combination with flutolanil application (51).

Phomopsis sp. causal agents of root rot disease of cucurbitaceae grow slowly and are known to have relatively weak pathogenicity (52). Actually, it takes a longer duration until they successfully colonize cucumber roots and induce root rot symptoms than other soilborne disease pathogens such as *R. solani* used in the present study. This time lag for the expression of pathogenicity, in turn, is considered to be beneficial to antagonistic microorganisms in terms of their successful colonization in the rhizosphere and/or root surface. Moody and Gindrat (53) isolated *Gliocladium roseum* that parasitizes a cucumber pathogen, *Phomopsis sclerotioides*, and showed that *G. roseum* could effectively suppress the occurrence of the disease. In the present study, roots of cucumber seedlings were immersed in RB14-C cell suspension at the time of transplanting to let the bacterium colonize the cucumber roots. As expected, the root rot was suppressed resulting in growth recovery 50 days after treatment even though initial growth was retarded probably due to the *Phomopsis* infection. Further experiments will be carried out to establish a systematic disease control practice that can be employed in commercial production using antagonistic *B. subtilis*.

In summary, we have shown that germinated tomato seed treatment of RB14-C cells and root immersion treatment using RB14-C cell suspension in cucumber seedlings at the time of transplanting reduced the occurrence of the damping-off of tomato seedlings and *Phomopsis* root rot of cucumber, respectively. These results suggest that germinated seed treatment of RB14-C cells and root immersion treatment with RB14-C cell suspension can be applied as promising biological control practices of the damping-off of tomato seedlings and the cucumber *Phomopsis* root rot, respectively.

3.4 RESULTS OF OTHER PLANT TESTS

Preliminary experiments were conducted on some other plants to observe suppressive effects of *B. subtilis*.

Figure 3.14 shows that *B. subtilis* NB22 suppresses the *Fusarium* disease of cucumber caused by *F. oxysporum* f. sp. *cucumerinum*.

Figure 3.15 shows suppression of melon root rot caused by *Phomopsis* sp. after introduction of *B. subtilis* NB22 into infested soil.

B. *subtilis* added   No B. *subtilis* added

FIGURE 3.14 Suppression of cucumber *Fusarium* disease caused by *F. oxysporum* after *B. subtilis* NB22 cell suspension was added.

FIGURE 3.15 Suppression of melon root rot caused by *Phomopsis* sp. after culture broth of *B. subtilis* NB22 was introduced into infested soil.

REFERENCES

1. Phae, C. G., Shoda, M., Kita, N., Nakano, M., and Ushiyama, K. Biological control of crown and root rot and bacterial wilt of tomato by *Bacillus subtilis* NB22. *Ann. Phytopath. Soc. Japan*, 58, 329–339 (1992).
2. Jarvis, W. R., Thorp, H. J., and Macneil, B. H. A foot and root rot of diseases of tomato caused by *Fusarium oxysporum. Can. Plant Dis. Surv.*, 55, 25–26 (1975).
3. Marois, J. J., and Mitchell, D. J. Effect of fungal communities on the pathogenic and saprophytic activities of *Fusarium* oxysporum f. sp. *radicis-lycopersici. Phytopathology*, 71, 1251–1256 (1981).
4. Kita, N., Ushiyama, K., and Aono, N. Crown and root rot of tomatoes in Kanagawa. *Bui. Kanagawa Hortic. Exp. Stn.*, 35, 1–6 (1988).

5. Kommedahl, T., and Mew, I. C. Biocontrol of corn root infection in the field by seed treatment with antagonists. *Phytopathology*, 65, 296–300 (1975).

6. Marois, J. J., and Mitchell, D. J. Effects of fumigation and fungal antagonists on the relationships of inoculum density to infection incidence and disease severity in *Fusarium* crown rot of tomato. *Phytopathology*, 71, 167–170 (1981).

7. Merriman, P. R., Price, R. D., Kollmorgen, F., Piggot, T., and Ridge, E. H. Effect of seed inoculation with *Bacillus subtilis* and *Streptomyces griseus* on the growth of cereals and carrots. *Aust. J. Agric. Res.*, 25, 219–226 (1974).

8. Rowe, R. C., Farley, J. D., and Coplin, D. L. Airborne spore dispersal and recolonization of steamed soil by *Fusarium oxysporum* in tomato green houses. *Phytopathology*, 67, 1513–1517 (1977).

9. Rowe, R. C., and Farley, J. D. Strategies for controlling *Fusarium* crown and root rot in greenhouse tomatoes. *Plant Dis.*, 65, 107–112 (1981).

10. Nesmith, W. C., and Jenkins, Jr., S. F. Influence of antagonists and controlled matric potential on the survival of *Pseudomonas solanacearum* in four North Carolina soils. *Phytopathology*, 75, 1182–1187 (1985).

11. Aldrich, J., and Baker, R. Biological control of *Fusarium roseum* f. sp. *dianthi* by *Bacillus subtilis*. *Plant Dis. Rep.*, 54, 446–448 (1970).

12. Chang, I. P., and Kommedahl, T. Biological control of seedling blight of corn by coating kernels with antagonistic microorganisms. *Phytopathology*, 5, 1395–1401 (1968).

13. Filippi, C., Bagnoli, G., and Picci, G. Antagonistic effects of soil bacteria on *Fusarium oxysporum* f. sp. *dianthi* (Prill and Del.) Synd and Hans. IV-studies on controlled *Solanum lycopersicum* L. and *Dianthus caryophyllus* L. rhizosphere. *Agr. Med.*, 119, 327–336 (1989).

14. Gueldner, R. C., Reilly, C. C., Pusey, P. L., Costello, C. E., Arrendale, R. F., Cox, R. H., Himmelsbach, D. S., Crumley, F. G., and Cutler, H. G. Isolation and identification of iturin as antifungal peptides in biological control of peach brown rot with *Bacillus subtilis*. *J. Agric. Food Chem.*, 36, 366–370 (1988).

15. Baker, C. J., Stavely, J. R., and Mock, N. Biocontrol of bean rust by *Bacillus subtilis* under field conditions. *Plant Dis.*, 69, 770–772 (1985).

16. Bochow, H. Use of microbial antagonists to control soil borne pathogens in greenhouse crops. *Acta Hortic.*, 255, 271–280 (1989).

17. Broadbent, P., Baker, K. F., Franks. N., and Holland, J. Effect of *Bacillus* spp. on increased growth of seedlings in steamed and in nontreated soil. *Phytopathology*, 67, 1027–1034 (1977).

18. Gupta, V. K., and Utkhede, R. S. Nutritional requirement for production of antifungal substance by *Enterobacter aerogenes* and *Bacillus subtilis* antagonists of *Phytophthora cactorum*. *J. Phytopathol.*, 120, 143–153 (1987).

19. Olsen, C. M., and Baker, K. F. Selective heat treatment of soil and its effect on the inhibition of *Rhizoctonia solani* by *Bacillus subtilis*. *Phytopathology*, 58, 79–83 (1968).

20. Yuen, G. Y., Schroth, M. N., and McCain, A. H. Reduction of *Fusarium* wilt of carnation with suppressive soil and antagonistic bacteria. *Plant Dis.*, 69, 1071–1075 (1985).

21. Phae, C. G., Shoda, M., and Kubota, H. Suppressive effect of *Bacillus subtilis* and its products on phytopathogenic microorganisms. *J. Ferment. Bioeng.*, 69, 1–7 (1990).

22. Phae, C. G., and Shoda, M. Expression of the suppressive effect of *Bacillus subtilis* on phytopathogens in inoculated composts. *J. Ferment. Bioeng.*, 70, 409–414 (1990).

23. Claus, D., and Berkeley, R. C. W. Genus Bacillus Cohn 1872, 174[AL] in Section 13, Endospore-forming Gram. Positive Rods and Cocci, pp. 1105–1139. In Sneath, P. H. A., Mair, N. S., Sharpe, M.E., and Holt, J.G,. (eds), *Bergey's Manual of Systematic Bacteriology*, vol. 2. Williams & Wilkins, Baltimore, MD (1986).

24. Tanaka, H., and Fukuda, N. Detection of *Pseudomonas solanacearum* in +infested soils by a selective medium and a bioassay. *Bull. Utsunomiya Tobacco Exp. Sta.*, 19, 1–10 (1982).
25. Tsuyumu, S., Tsuchida, S., Nakano, T., and Takikawa, Y. Antifungal activity in cell-free culture fluid of *Pseudomonas solanacearum. Ann. Phytopath. Soc. Japan*, 55, 9–15 (1989).
26. Kijima, T., Arie, T., Kimura, S., Minegishi, N., Tezuka, N., Rashida, K., and Hukuda, T. Biological control of soil-borne diseases by mixed cropping with associate crops inoculated with *Pseudomonas gladioli* strain M-2196. *Bull. Tochigi Agr. Exp. Stn.*, 35, 95–128 (1988).
27. Ogawa, K., and Komada, H. Biological control of *Fusarium* wilt of sweet potato by non-pathogenic *Fusarium oxysporum. Ann. Phytopath. Soc. Japan*, 50, 1–9 (1984).
28. Baker, K. F., Flentje, N. T., Olsen, C. M., and Stretton, H. M. Effect of antagonists on growth and survival of *Rhizoctonia solani* in soil. *Phytopathology*, 57, 591–597 (1967).
29. Sekiguchi, A., Akanuma, R., Nakazawa, H., and Minami, M. Effect of control of the Chinese yam brown rot disease by bark compost with soil sterilization chemicals. *Bull. Nagano Veg. Ornam. Exp. Sta. Japan*, 1, 29–38 (1981).
30. Podile, A. R., Prasad, G. S., and Dube, H. C. *Bacillus subtilis* as antagonist to vascular wilt pathogens. *Curr. Sci.*, 54, 864–865 (1985).
31. Tokuda, Y. Behavior of *B. subtilis* NB22 inoculated into various soils, Master's thesis, Tokyo Institute of Technology (1991).
32. Asaka, O., and Shoda, M. Biocontrol of *Rhizoctonia solani* damping-off of tomato with *Bacillus subtilis* RB14. *Appl. Environ. Microbiol.*, 62, 4081–4085 (1996).
33. Hiraoka, H., Asaka, O., Ano, T., and Shoda, M. Characteristics of *Bacillus subtilis* RB14, coproducer of peptide antibiotics iturin A and surfactin. *J. Gen. Appl. Microbiol.*, 38, 635–640 (1992).
34. Hiraoka, H., Ano, T., and Shoda, M. Molecular cloning of a gene responsible for the biosynthesis of the lipopeptide antibiotics iturin and surfactin. *J. Ferment. Bioeng.*, 74, 323–326 (1992).
35. Huang, C. C., Ano, T., and Shoda, M. Nucleotide sequence and characteristics of the gene, *lpa-14*, responsible for biosynthesis of the lipopeptide antibiotics iturin A and surfactin from *Bacillus subtilis* RB14. *J. Ferment. Bioeng.*, 76, 445–450 (1993).
36. Tokuda, Y., Ano, T., and Shoda, M. Survival of *Bacillus subtilis* NB22 and its transformant in soil. *Appl. Soil Ecol.*, 2, 85–94 (1995).
37. Ushiyama, K., Nishimura, J., and Aono, N. Damping-off of feather cockscomb (*Celosiaargentea* L. var. *cristata* O. Kuntze) caused by *Rhizoctonia solani* Kuhn. *Bull. Kanagawa Hortic. Exp. Stn.*, 34, 33–37 (1987).
38. Katayama, A., Hirai, M., Shoda, M., and Kubota, H. Factors affecting the stabilization period of sewage sludge in soil with reference to the gel chromatographic pattern. *Soil Sci. Plant Nutr.*, 32, 383–395 (1986).
39. Maget-Dana, R., Thimon, L., Peypoux, F., and Ptak, M. Surfactin/iturin A interactions may explain the synergistic effect of surfactin on the biological properties of iturin A. *Biochimie*, 74, 1047–1051 (1992).
40. Ito, T., and Neilands, J. B. Products of "low iron fermentation" with *Bacillus subtilis*: isolation, characterization and synthesis of 2,3-dihydroxylbenzoylglycine. *J. Am. Chem. Soc.*, 80, 4645–4647 (1958).
41. Loper, J. E. Role of fluorescent siderophore production in biological control of *Pythium ultimum* by a *Pseudomonas fluorescens* strain. *Phytopathology*, 78, 166–172 (1988).
42. Raaijmakers, J. M., Leeman, M., van Oorschot, M. M. P., van der Sluis, I., Schippers, B., and Bakker, P. A. H. M. Dose-response relationships in biological control of Fusarium wilt of radish by *Pseudomonas* spp. *Phytopathology*, 85, 1075–1081 (1995).

43. Thomashow, L. S., and Weller, D. M. Role of a phenazine antibiotic from *Pseudomonas ftuorescens* in biological control of *Gaeumannomyces graminis* var. *tritici. J. Bacteriol.*, 170, 3499–3508 (1988).

44. Kita, N., Ohya, T., Uekusa, H., Nomura, K., Manago, M., and Shoda, M. Biological control of damping-off of tomato seedlings and cucumber *Phomopsis* root rot by *Bacillus subtilis* RB14-C. *Japan Agric. Res. Q.*, 39, 109–114 (2005).

45. Cook, R. J. Advances in plant health management in the twentieth century. *Annu. Rev. Phytopathol.*, 38, 95–116 (2000).

46. Hoitink, H. A. J., and Boehm, M. J. Biocontrol within the context of soil microbial communities: a substrate-dependent phenomenon. *Annu. Rev. Phytopathol.*, 37, 427–446 (1999).

47. Raupach, G. S., and Kloepper, J. W. Mixtures of plant growth-promoting rhizobacteria enhance biological control of multiple cucumber pathogens. *Phytopathology*, 88, 1158–1164 (1998).

48. Tuitert, G., Szczech, M., and Bollen, G. J. Suppression of *Rhizoctonia solani* in potting mixtures amended with compost made from organic household waste. *Phytopathology*, 88, 764–773 (1998).

49. Kondoh, M., Hirai, M., and Shoda, M. Integrated biological and chemical control of damping-off caused by *Rhizoctonia solani* using *Bacillus subtilis* RB14-C and flutolanil. *J. Biosci. Bioeng.*, 91, 173–177 (2001).

50. Merriman, P. R., Price, R. D., and Baker, K. F. The effect of inoculation of seed with antagonists of *Rhizoctonia solani* on the growth of wheat. *Aust. J. Agric. Res.*, 25, 213–218 (1974).

51. Guetsky, R., Shtienberg, D., Elad, Y., and Dinoor, A. Combining biocontrol agents to reduce the variability of biological control. *Phytopathology*, 91, 621–627 (2001).

52. Ebben, M. H., and Last, F. T. A sclerotial fungus pathogenic to cucumber roots. *Plant Pathol.*, 16, 96 (1967).

53. Moody, A. R., and Gindrat, D. Biological control of cucumber black root rot by *Gliocladium roseum. Phytopatholoy*, 67, 1159–1162 (1977).

4 Stability of *B. subtilis* and Its Derivative Strains in Soil

4.1 SURVIVAL OF *B. SUBTILIS* NB22 AND ITS TRANSFORMANT IN SOIL (1)

4.1.1 MATERIALS AND METHODS

4.1.1.1 Bacterial Strain and Plasmids

Bacillus subtilis NB22 was originally isolated in our laboratory from compost as a bacterium that shows a suppressive effect against various plant pathogens (Chapter 2) and was proven to be an antifungal-antibiotic iturin producer (Chapter 2). A spontaneous streptomycin-resistant mutant, *B. subtilis* NB22-l, which was obtained from NB22, was used. The growth rate and iturin productivity of NB22-l were similar to those of the parental strain NB22.

Plasmid pC194 (2, 3), which has a resistance determinant against chloramphenicol (Cm), was used initially. This is one of the most popular vector plasmids for *B. subtilis* since its nucleotide sequence is known and its DNA replication mode has been well analyzed (2-4). This plasmid was kept stable for 100 generations in *B. subtilis* NB22-1 during cultivation in a complex liquid medium, but unexpected instability was observed in soil (see details in Section 4.1.2.3). Therefore, another well-known plasmid, pUB110 (4-6), which has a kanamycin (Km)-resistant gene, was also used in comparison with those of pC194.

4.1.1.2 Media

L medium contained (per liter) 10 g of Polypepton (Nihon Pharmaceutical, Tokyo), 5 g yeast extract, and 5 g NaCl, adjusted to pH 7.2. No. 3 medium, which is the production medium of the antibiotic iturin, contained (per liter) 10 g Polypepton, 10 g glucose, 1 g KH_2PO_4, and 0.5 g $MgSO_4 \cdot 7H_2O$ (pH 6.8). Schaeffer's sporulation medium (7) contained (per liter) 8 g Difco nutrient broth powder, 0.25 g $MgSO_4 \cdot 7H_2O$, 1.0 g KCl, 0.001 mM $FeSO_4$, 1 mM $Ca(NO_3)_2$, and 0.01 mM $MnCl_2$. No. 3 agar medium (no. 3 medium plus 15 g agar) containing streptomycin (100 μg/ml) was used to count the number of viable cells of NB22-1 recovered from the soil. No. 3 agar medium containing Cm (5 μg /ml) or Km (10 μg/ml) was used to select the plasmid carriers of pC194 and pUB110, respectively.

4.1.1.3 Preparation of Plasmid DNA

The rapid alkaline lysis procedure (8) was used for detecting plasmid DNA, with modifications as previously described (9). For large-scale preparation, plasmids

extracted by the alkaline lysis method were further purified by CsCl-ethidium bromide density gradient ultracentrifugation, and electrophoretic analysis of DNA in
agarose gel was done by following standard methods (10).

4.1.1.4   Transformation of *B. subtilis* NB22-1

Transformation of *B. subtilis* NB22-1 with plasmids pC 194 and pUB 110 was carried out by a newly developed alkali metal ion treatment, as described in Chapter 5
(9). A logarithmic culture of *B. subtilis* NB22-1 in L medium was incubated in 4 M
KCl solution, and a portion of this was mixed with plasmid DNA in the presence of
35% polyethylene glycol 6000 followed by heat-shock treatment at 42°C for 5 min
and gene expression in L medium. Transformants were selected on L agar medium
(L medium plus 15 g agar) containing Cm or Km.

4.1.1.5   Soil Treatments

The soil used in this study was Hiratsuka coarse sandy loam, which has been used
for tomato cultivation in the Agricultural Research Institute of Kanagawa Prefecture,
Japan. The soil was sieved through an 8-mesh (about 2 mm) screen, air-dried for a
few days, and kept in sterilized plastic bags at room temperature. The main characteristics of the soil thus prepared are as follows: texture, coarse sandy loam; C (%),
1.52; N (%), 0.14; cation exchange capacity (meq/100 g dry soil), 22.0; moisture content (%), 14.7; maximum water-holding capacity (g/100 g dry soil), 95.2; bulk density
(g/100 ml), 89.8; pH, 5.7. The measurement of these properties followed the methods previously (11). All experiments were performed in non-sterile soil. However,
when sterile soil was used, it was sealed into a sterilizable polypropylene bag, and
autoclaved for 60 min (121°C) four times at 12 h intervals. The moisture content
was maintained at 60% of the maximum water-holding capacity by the addition of
sterilized water. In order to enhance survivability of the bacterium in soil, the soil
was nutrient-amended with slightly modified Spizizen's minimal medium, which
had been developed for *B. subtilis* cultivation (12). The final concentration of each
added component per dry soil was as follows: 0.5% glucose, 0.02% $MgSO_4 \cdot 7H_2O$,
0.6% KH_2PO_4, 1.4% K_2HPO_4, 0.2% $(NH_4)_2SO_4$, 0.1% sodium citrate, and 0.01%
yeast extract. The C/N ratio of the soil before and after the amendment was 10.9 and
9.6, respectively, indicating a slight modification of this soil property.

4.1.1.6   Inoculation of *B. subtilis* NB22-1 into Soil

Vegetative cells of *B. subtilis* NB22-1 without spores were prepared as follows. Cells
cultured overnight in no. 3 medium (30°C) were inoculated into 100 ml of fresh
no. 3 medium (1%) and the culture recultivated overnight. When this procedure
was repeated four times, all cells were judged to be vegetative in the liquid culture,
because no living cells were detected after 15 min heat treatment at 80°C. Spores of
the bacterium were prepared in 100 ml of Schaeffer's sporulation medium by cultivating for 24 h at 30°C. Vegetative cells or spores were collected by centrifugation
(7600 × *g*, 10 min at 4°C), washed in 0.85% NaCl solution, and recentrifuged under
the same conditions. The washed cells or spores were resuspended in 10 ml of sterile
distilled water, and the suspension was mixed well with the soil at a ratio of about
1 ml of cell or spore suspension to 1 g of dry soil. A 50 g soil sample was placed in a

200 ml Erlenmeyer flask with a silicone sponge closure and kept at 15°C or 25°C in the dark. The moisture content was adjusted to 60% of the maximum water-holding capacity of the soil before each sampling.

4.1.1.7 Recovery of Bacterium from Soil and Counting of Viable Cells

The soils were sampled periodically after inoculation of the bacterium. On each sampling, the soils from different parts of each flask were sampled, mixed together and 2 g was suspended in 8 ml of 0.85% NaCl solution in 50 ml Erlenmeyer flasks, followed by 15 min shaking (140 strokes per minute at room temperature). The suspension was diluted serially in the NaCl solution, and plated onto no. 3 agar containing streptomycin (100 μg/ml). The plates were incubated at 30°C and viable cells (total cell number) counted after 12 h. To determine the number of spores, 1 ml of the suspension was heated for 15 min at 80°C, and the samples were serially diluted and spread onto the same plates as described earlier. All data are expressed as colony-forming unit (cfu)/g dry soil.

Discrimination between the inoculated NB22-1 and streptomycin-resistant bacteria indigenous to soil was easy because NB22-1 grew on no. 3 agar medium plates faster than did any other bacteria indigenous to soil; thus after 12 h the colonies of this bacterium became visible first. Within 24 h the surfaces of the colonies of NB22-1 changed from transparent to turbid white. Such morphological characteristics and streptomycin resistance enabled this bacterium to be clearly distinguished from other bacteria.

4.1.1.8 Determination of Plasmid Stability

The stability of plasmids in the host strain was determined as follows: 100 colonies that appeared on no. 3 agar medium with streptomycin and were judged to be NB22-1 by surface appearance as mentioned earlier were randomly selected and transferred by replica plating onto no. 3 agar medium with and without each selective antibiotic. The fraction of bacteria expressing resistance and therefore carrying plasmid among 100 colonies was taken as the indicator of the plasmid stability. The existence of plasmid pC194 or pUB110 in the antibiotic-resistant colonies was confirmed by extraction of the plasmid from 12 colonies chosen at random.

4.1.1.9 Plasmid Stability in Liquid Culture

Vegetative cells prepared in no. 3 medium as described earlier were inoculated in Schaeffer's sporulation medium at 1% and cultivated at 30°C followed by periodical sampling. Viable counts of total cell numbers and heat-resistant cell (spore) numbers were determined at each sampling; plasmid stabilities were also determined as described earlier.

4.1.1.10 Qualitative Assay and Identification of Iturin Produced by the Isolated Bacteria

A phytopathogenic fungus, *Fusarium oxysporum* f.sp. *lycopersici* race J1 SUF119 (Chapter 2), was grown on potato-dextrose agar (potato-dextrose agar 39 g, distilled water 1 liter, pH 5.6) at 30°C for 5 days and suspended in sterile distilled water. A portion of this suspension was mixed into melted L agar medium to make an L agar plate containing *F. oxysporum*. After spotting each of the colonies isolated from soil onto this plate, the suppressiveness of each colony was investigated by observing clear inhibitory zones of mycelial growth. Those colonies that produced inhibitory zones were considered to be iturin producers. Iturin was assayed by reversed-phase HPLC as previously described in Chapter 2.

4.1.1.11   Statistical Analysis

Soil samples were prepared in triplicate for each experiment. In each cfu deter-mination for total cell number and spore number, at least three different dilutions were spread in duplicate onto the selective plates, and an average value was calcu-lated. Statistical significance was evaluated using Student's t-test in Stat View on a Macintosh computer. The maximum error was less than 10%; errors were included in each symbol of the figures mainly because logarithmic units were used.

4.1.2   Results

4.1.2.1   Survival of *B. subtilis* NB22-1 in Nonsterile Hiratsuka Soil

The survival patterns of vegetative cells of *B. subtilis* NB22-1 at 15°C in non-sterile Hiratsuka soil and sterile soil is illustrated in Figure 4.1a. The total cell counts on the no. 3 agar medium decreased during the first 10 days and stabilized thereafter at a level of 10^4 cfu/g dry soil. The number of total cells and spores were similar, indicating that all *B. subtilis* cells were stabilized as spores. The level of *B. subtilis* population persisted at this level for at least 50 days.

The colonies of the bacteria were judged as NB22-1 by the appearance of the surface of the colonies as mentioned earlier. To confirm the accuracy of this observa-tion, 100 colonies appearing on the no. 3 agar plates were randomly chosen, and rep-lica-plated onto L agar plates containing *F. oxysporum*. As all the colonies showed clear inhibitory zones against *F. oxysporum* on the plates, they were judged to be *B. subtilis* NB22-1 (an iturin producer). The production of iturin was further confirmed by HPLC analysis of methanol extracts of the cell-free culture broth prepared by the growth of randomly chosen colonies in no. 3 medium (data not shown).

4.1.2.2   Effects of Nutrients and Temperature on Stabilization of NB22-1

The effect of the addition of nutrients to the soil at 15°C is illustrated in Figure 4.1b. The result at 25°C without nutritional amendment is shown in Figure 4.1c. In both cases, vegetative cells declined steadily from 10^7 to 10^3 cfu/g dry soil, and the spores reached the level of 10^3 cfu/g dry soil after several days of incubation.

Those levels were maintained during the 50-day incubation period. The declining pattern of the numbers of *B. subtilis* NB22-1 under those conditions showed no significant difference from that observed in Figure 4.1a. However, a stable population size of NB22-1 was observed when simultaneous raising of the temperature to 25°C and nutritional amendment were carried out (Figure 4.1d). This stabilization of NB22-1 was due to rapid spore formation by the vegetative cells at the level of 10^7 during the first 5-day incubation. A similar pattern was observed when the temperature was raised to 30°C with nutritional addition (data not shown).

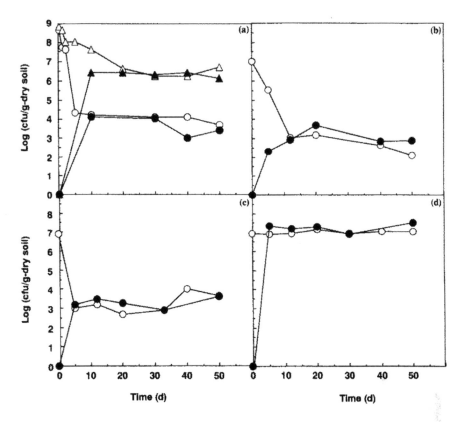

FIGURE 4.1 Survival of vegetative cells of *B. subtilis* NB22-1 in non-sterile Hiratsuka soil: (a) at 15°C without nutritional amendment; (b) at 15°C with nutritional amendment; (c) at 25°C without nutritional amendment; and (d) at 25°C with nutritional amendment. o, total cell number; •, spore number. Part (a) also shows data for numbers of total cells (Δ) and spores ($\blacktriangle$) in sterile soil at 15°C without nutritional amendment.

4.1.2.3 Survival of *B. subtilis* NB22-1 Bearing Plasmid pC194 or pUB110 in the Soil

B. subtilis NB22-1 was transformed with plasmid pC194 or pUB110 by the alkali metal ion treatment method developed by us because the conventional competent cell method was not applicable to this wild strain (9). The detail of this method is described in Chapter 5. The survival patterns of vegetative *B. subtilis* NB22-1 (pC194) and NB22-1 (pUB110) in non-sterile Hiratsuka soil at 15°C are shown in Figure 4.2. The cell number of both strains decreased quickly from the initial concentration of 10^8 cfu/g dry soil to the lower limit of detection in this soil system several days after inoculation. At days 2 and 5, both transformants were present at significantly lower concentrations than the host strain (Figure 4.2), although initial concentrations were not statistically different. When the incubation temperature was raised to 25°C and nutritional amendment carried out as the case of the host strain in Figure 4.1, rapid sporulation for both transformants occurred (Figure 4.3a and b).

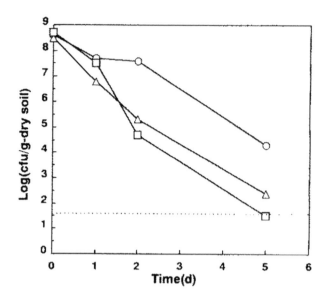

FIGURE 4.2   Survival of the vegetative cells of *B. subtilis* NB22-1 (o); *B. subtilis* NB22-1(pC194) (Δ); and *B. subtilis* NB22-1 (pUB110) (□). Both transformants were introduced into Hiratsuka soil and incubated at 15°C without nutritional amendment. The dotted line indicates the lower limit of detection in this soil system.

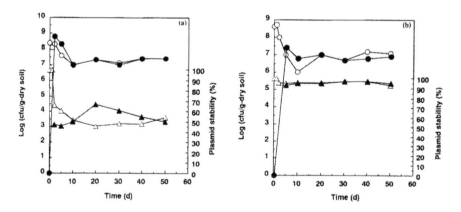

FIGURE 4.3   Survival of the vegetative cells of (a) *B. subtilis* NB22-1 (pC194) and (b) the vegetative cells of *B. subtilis* NB22-1 (pUB110). Both transformants were introduced into Hiratsuka soil and incubated at 25°C with nutritional amendment. o, total cell number; •, spore number; Δ, plasmid stability of total cells; ▲, plasmid stability of spores.

Since counts of total cells and spores showed no significant difference (P > 0.05), the population at the level of 10^7 cfu/g dry soil was considered to be maintained as spores during the 50-day incubation period. As for plasmid stability, pC194 was cured significantly to the level of about 50% during sporulation, which remained stable in the following period. In sharp contrast to pC194, pUB110 was very stable at more than 90% not only in vegetative cells for the initial few days but also in sporu-lated cells for the subsequent experimental period.

4.1.2.4 Survival of *B. subtilis* NB22-1 and Its Transformants Introduced into the Soil as Spores

Spores of *B. subtilis* NB22-1 and its transformants with the plasmids pC194 and pUB110 prepared in the sporulation medium were mixed with the soil and survival of each was investigated. With or without plasmid, all strains were found to be sta-ble at the initial level of 10^8 cfu/g dry soil for 50 days (Figure 4.4a–c). The values of plasmid stability were about 50% and 90% for pC194 and pUB110, respectively (Figure 4.4b and c), and no significant changes of the plasmid stability observed dur-ing the incubation period. As plasmid stability was almost the same as that in liquid culture (see later), the plasmid-holding ratio of the spores for both transformants was maintained in soil after the spore suspension was inoculated into the soil.

4.1.2.5 Stability of Plasmids in Liquid Culture

B. subtilis NB22-1 (pC194) and NB22-1 (pUB110) were cultivated in the sporulation medium at 30°C, and the stability of each plasmid was investigated periodically in batch cultures. Sporulation of the two recombinants and the changes of the plasmid stability are shown in Figure 4.5a and b, respectively. Although plasmid pC194 was maintained stably at 100% in *B. subtilis* NB22-1 in no. 3 medium with or without chloramphenicol at 30°C (data not shown), the plasmid was quickly cured to about 50% in the sporulation medium (Figure 4.5a). This curing was attributed to sporula-tion. In sharp contrast to this plasmid, pUB110 showed high stability, i.e., over 96% even in the sporulation medium (Figure 4.5b).

4.1.3 DISCUSSION

B. subtilis NB22, which was newly isolated in this study, expresses broad suppress-ibility against phytopathogenic microorganisms by producing the lipopeptide anti-biotics iturin A and surfactin and is expected to be applied as a biocontrol agent (in Chapters 2 and 3). Current concerns about the release of genetically engineered microorganisms (GEMs) into soil have arisen and many studies have been carried out to assess the beneficial or adverse effects resulting from the release of GEMs or to identify the factors affecting their survival and multiplication in the environment (13–20).

In order to use the iturin-producing *B. subtilis* safely and effectively in the envi-ronment, it is important to know its basic characteristics in soil systems. Many such biochemical or genetic studies have been performed with *B. subtilis*. The plasmid stabilities of *B. subtilis* in liquid culture have been intensively (21–26). Their infor-mation, however, was mostly limited to the standard strain *B. subtilis* Marburg 168 or its derivatives. To use the wild *B. subtilis* NB22 effectively in the field, it is very important to assess the basic ecological behavior of the strain or its transformants with plasmid DNA in soil. *B. subtilis* has been proven safe over many years, because it is a non-pathogenic bacterium and has been consumed in large quantities as a Japanese food, natto. Control of phytopathogens using this bacterium appears to be

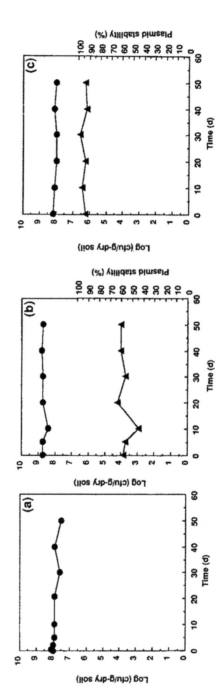

FIGURE 4.4 Survival of (a) spores of *B. subtilis* NB22-1; (b) spores of *B. subtilis* NB22-1 (pC194); and (c) spores of *B. subtilis* NB22-1 (pUB1 10). All spores were introduced into Hiratsuka soil and incubated at 15°C without nutritional amendment. ●, spore number; ▲, plasmid stability.

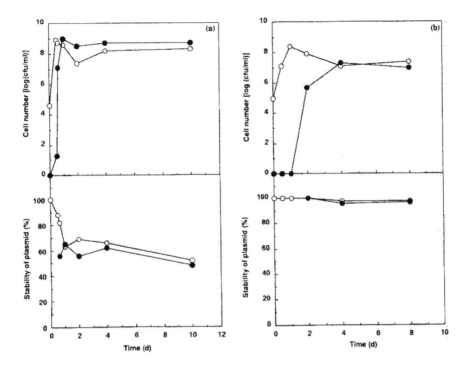

FIGURE 4.5 Survival and stability of plasmid of (a) *B. subtilis* NB22-1 (pC194); and (b) NB22-1 (pUB110) in liquid sporulation medium. o, total cell; •, spores.

safe and ecologically sound. However, due to relatively low-density distribution in soil (27, 28) and poor survival of *B. subtilis* introduced into soils (29–31), conditions required for higher and longer stability of the host bacterium and its recombinant in soil should be established for biological control of plant diseases.

Survival of *B. subtilis* NB22-1 (a spontaneous antibiotic-resistant mutant of NB22) and its transformants with plasmid DNA in a soil system was investigated for 50 days, a period sufficient for biocontrol of plant diseases reported (Chapter 3). Changes in the stability of the plasmids were also studied. The host strain NB22-1 rapidly declined in number in the initial days and stabilized at 10^4 cfu/g dry soil for the remainder of the experimental period at 15°C (Figure 4.1a). This stable survival of NB22-1 was the result of spores formed during incubation in the soil, indicating that induction of sporulation was an important factor to stabilize *B. subtilis* in the soil. Initial rapid decline in the population size of vegetative cells was thought to be partly due to an autolytic reaction triggered by nutritional starvation in the soil. In fact, lysis of NB22 had been clearly observed in the stationary phase of liquid culture when nutritional starvation occurred (data not shown). The similar declining pattern of the survival was observed even in the sterile soil, but the decreasing rate in the cell number was slower in a sterile soil (Figure 4.1a). This indicates that a biotic effect by microorganisms indigenous to the soil affected the survival of the cells in non-sterile soil. However, the final density of *B. subtilis* NB22-1 in sterile soil was the same as that in the non-sterile one, suggesting that an abiotic stress might be a crucial factor to determine the survival level of spores of NB22-1 in soil.

Enrichment of soil with nutrients and increase of the temperature to 25°C were effective to stabilize *B. subtilis* NB22-1 at a high level for 50 days, presumably because the two factors stimulated spore formation by NB22-1. Supplementing a soil with 1.0% glucose was reported to be effective to prevent the marked loss of viability of *B. subtilis* for a short experimental period of 8 days (29). Their observation was also considered to be associated with spore formation.

A rapid death rate and extinction of vegetative cells of the transformants was observed in soil (Figure 4.2). This was due to the poorer survival of the transformants in soil or less active sporulation or a higher susceptibility to starvation than the host cells. However, high persistence of the transformants was observed by nutritional addition at 25°C (Figure 4.3), indicating that sporulation is also a key factor for the recombinants of *B. subtilis* to be stabilized in soil.

Plasmids pC194 and pUB110 demonstrated different stabilities, not only in liquid sporulation medium (Figure 4.5) but also in soil (Figure 4.3). Although the precise mechanism of the instability of pC194 is not clear, the instability may be affected by factors such as the size or copy number of the plasmid, especially in a nutrition-deficient condition. As curing of plasmid pC194 was observed even in the presence of chloramphenicol and cured cells could not grow in the medium with the antibiotic, curing will occur after cell division and most probably at the stage of septum formation in sporulation. As the copy numbers of pUB110 and pC194 are likely to be around 50 and 15 per chromosomal equivalent, respectively (32, 33), the probability of distribution of each plasmid into the spore cells will be higher in pUB110 and, thus, this will lead to higher stability of the plasmid. The similar results were obtained in four different soils (34).

When liquid cultivation of NB22-1 (pC194) was repeated in no. 3 medium so as to maintain vegetative cells in a logarithmic growth phase, pC194 was maintained stably at 100% over 100 generations (data not shown). However, the reduction in stability of plasmid pC194 to about 50% was seen in nutrient-deficient condition such as a Schaeffer's sporulation medium or in soil where sporulation occurs easily. When genetically manipulated *B. subtilis* are used in conditions that only allow sporulation to occur in an oligotrophic environment, selection of vector plasmid is important to maintain higher plasmid stability. In this respect, plasmid pUB110 is preferable as a vector for NB22-1.

When spores of NB22-1 and its transformants were introduced into the soil, they were maintained at a high level throughout the experimental period (Figure 4.4). This long persistence of spores shows that spores are tolerant not only of abiotic stresses but also of biotic stress such as predation, parasitism, or lytic enzyme reactions by the indigenous microbial community.

Although *B. subtilis* has been reported to both decline in number and to occur as spores in soils (18, 29, 30), this work clearly showed that the spores, when either induced in, or introduced into, a soil guarantee a high and stable survival of the bacterium for a long period, which will lead to a suppressive effect of the bacterium on plant diseases. The occurrence of such a suppressive effect against plant pathogens in soil by the bacterium is thought to be triggered by an antifungal peptide, iturin, which is produced by vegetative cells germinated from spores. The growth of *B. subtilis* and germination of spores are reported to be influenced by acidity of the forest soil (35) or growth of fungal hyphae (27). As *B. subtilis* NB22 suppressed phytopathogenic fungi and bacterium of tomato in soil (Chapter 3), spores of this bacterium may respond to the growth of phytopathogenic fungi in a soil system. Further investigation of characteristics of vegetative cells and spores of *B. subtilis* in soil, either of the host or of a genetically engineered organism, in the presence of a plant rhizosphere has to be carried out in relation to detection of the antifungal peptide iturin in soil.

4.2 PERSISTENCE OF *B. SUBTILIS* RB14 AND ITS DERIVATIVE STRAINS IN SOIL WITH RESPECT TO THE *LPA-14* GENE (36)

For a biocontrol agent to fulfill its purpose, the organism has to be released into the environment. It is thus important to understand the characteristics of microorganisms introduced into soils. Few reports have appeared on the survival of *B. subtilis* in soil because *B. subtilis* is not regarded as a typical soil bacterium (37, 38). It is also of interest to know whether the gene related to lipopeptide antibiotic production contributes to the survival of RB14 in soil.

In the present section, RB14, RΔ1, the *lpa-14* deleted mutant of RB14, and RΔ1(pC115), a recombinant in which the productivity of iturin A and surfactin is restored by plasmid pC115 carrying *lpa-14*, were introduced into soil and their persistence was compared at 30°C and 15°C.

4.2.1 Materials and Methods

4.2.1.1 Bacterial Strains and Plasmid

The bacterial strains and plasmid used are listed in Table 4.1. *B. subtilis* RB14 was originally isolated from compost and characterized as a bacterium suppressive

TABLE 4.1
Bacterial Strains and Plasmid Used

Strain or Plasmid	Phenotype Plasmid Marker[a]	Reference
B. subtilis		
RB14	IT+ SF+	(40)
RB14-C	I T+ SF+ Sm^r	(36)
RΔ1	IT– SF– Em^r	(41)
RΔ1(pC115)	IT+ SF+ Em^r Cm^r	(39)
Plasmid		
pC115	Cm^r, 4.2 kb, *lpa-14*+	(39)

[a] IT, iturin A production; SF, surfactin production; Sm, streptomycin; Em, erythromycin; Cm, chloramphenicol.

against several phytopathogenic microorganisms as shown in Chapter 2. *B. subtilis* RB14-C, a spontaneous streptomycin-resistant mutant of RB14, was used because of its easy recovery from soil. The growth rate, iturin A and surfactin productivities, and antifungal activity of RB14-C were confirmed to be similar to those of the parental strain RB14 (Chapter 2).

B. subtilis RΔ1 is an iturin A and surfactin nonproducer derived from RB14. The characteristics of RΔ 1 are described in Chapter 5.

The plasmid pC115 carried the *lpa-14* fragment, which was cloned in this study (39). The productivities of iturin A and surfactin were restored in RΔ1 by transforming it with pC115. Although the recovered level of iturin A production by this transformant, RΔ1(pC115), was almost the same as that of RB14, the productivity of surfactin was only about one-eighth (36).

4.2.1.2   Media

L medium, containing (per liter) 10 g of Polypepton (Nihon Pharmaceutical Co., Tokyo), 5 g of yeast extract, and 5 g of NaCl (pH 7.0), was used. To cultivate strains or select transformants, the antibiotics chloramphenicol (Cm), erythromycin (Em), or streptomycin (Sm) were added at concentrations of 5, 5, and 100 µg/ml, respectively. The medium was solidified with 2.0% agar when necessary.

4.2.1.3   Soil Treatments

A low humic andosol obtained from a field at the Kanagawa Horticultural Experimental Station, Japan (42), was used. The soil was sieved through a size 8 mesh (about 2 mm) screen and air-dried for a few days. The main characteristics of the soil thus prepared were texture, low-humic andosol; moisture content (%), 12.7; maximum water-holding capacity (g/100 g dry soil), 137; pH, 5.9; bulk density (g/100 ml), 52.2. Measurement of these properties followed the methods previously described in Chapter 3. This soil was mixed with vermiculite in a ratio of 4: 1 (w/w) and the mixture was nutrient-amended to give final N, P_2O_5, and K_2O concentrations of 70, 240, and 70 mg per 100 g dry soil, respectively. This prepared soil mixture was kept in plastic bags at room temperature. When a sterile soil was to be used, soil mixture was put into a sterilizable polypropylene bag and autoclaved for 60 min at 121°C four times with 12 h intervals between each sterilization. The moisture was maintained at 60% of the maximum water-holding capacity by addition of sterilized water.

4.2.1.4   Introduction of Bacteria to Soil and Incubation Conditions

Cells of RB14-C, RΔ1, or RΔ1 (pC115) cultured for 16 h at 30°C in L medium with antibiotics were inoculated at 1% into 300 ml of fresh L medium with antibiotics and cultivated for 24 h at 30°C. The cells were collected by centrifugation (8000 × *g*, 10 min, 4°C), washed in 0.85% NaCl solution (pH 7.0), and then centrifuged again under the same conditions. Washed cells were suspended in 25 ml sterile distilled water.

Soil prepared as described earlier was put into 200 ml Erlenmeyer flasks (about 60 g per flask) and 5 ml of each cell suspension was added to the soil. The moisture content in each flask was then adjusted to 60% of the maximum water-holding capacity of the soil. The soil was mixed well, and after fitting silicone sponge closures, flasks were kept at 15°C or 30°C in the dark. The moisture content was adjusted before each sampling.

4.2.1.5   Recovery of Bacteria from Soil and Counting of Viable Cell Numbers

The soils were sampled periodically after inoculation of the bacteria. On each sampling, a total 3 g of soil collected from different parts of each flask and mixed together was suspended in 8 ml of 0.85% NaCl solution (pH 7.0) in a 50 ml Erlenmeyer flask, followed by 15 min shaking at 140 strokes/min at room temperature. The suspension was diluted serially in the NaCl solution and plated onto L agar medium containing appropriate antibiotics. The plates were incubated at 37°C and the total cell number, which is the sum of both vegetative cells and spores, was counted after 12 h. To determine the number of spores, 1 ml of the suspension was heated for 15 min at 80°C, and the samples were serially diluted and spread onto the same plates as described earlier. All data are expressed as colony-forming unit (cfu)/g dry soil.

4.2.1.6   Determination of Plasmid Stability

The stability of pC115 in RΔ1 (pC115) was determined as follows: Of the colonies which appeared on L agar medium with erythromycin and were judged from their surface appearance to be RΔ1(pC115), 100 were randomly selected and transferred by replica plating onto L agar medium with and without chloramphenicol. The fraction of bacteria expressing chloramphenicol resistance, and therefore carrying pC115, among the 100 colonies was taken as an indicator of pC115 stability. The existence of plasmid pC115 in the chloramphenicol-resistant colonies was confirmed by extraction of the plasmid from 10 colonies chosen randomly.

4.2.2   Results

4.2.2.1   Survival of *B. subtilis* RB14-C at 30°C

The survival pattern of *B. subtilis* RB14-C in non-sterile and sterile soils at 30°C is shown in Figure 4.6. As the difference between the total cell number and spore number indicates the number of vegetative cells, it can be seen that most of the inoculated cells were vegetative ones. The initial decline in the total cell number reflects the death of vegetative cells. On the other hand, spores increased after inoculation, with the total number of cells and the number of spores becoming almost the same after 2 days. The population was then maintained at that level during the remainder of the observation period (30 days). Thus, all the cells were stabilized as spores, their numbers being in the order of 10^7 cfu/g dry soil. There was no difference in the survival pattern between cells in non-sterile and sterile soil, showing that indigenous microorganisms existing in the soil did not affect the persistence of RB14-C.

4.2.2.2   Survival of *B. subtilis* RΔ1 at 30°C

As shown in Figure 4.7, a slight decrease in the total number of RΔ1 cells was observed. The number of spores increased rapidly and became almost equal to the total cell number after 2 days. Thereafter, the population persisted at that level for the 30 days. The cells of RΔ1 were also stabilized as spores in the same manner

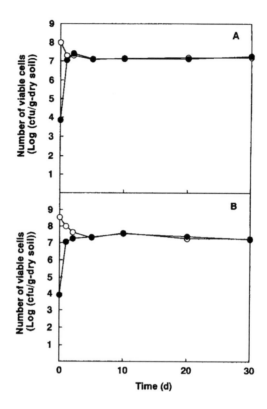

FIGURE 4.6   Survival of *B. subtilis* RB14-C cells at 30°C in (A) non-sterile and (B) sterile soils. ○, total cell number (vegetative cells + spores); ●, spore number.

and at the same population level as RB14-C. As no difference was observed in the survival pattern between cells in non-sterile and sterile soil, RΔ1, lacking the *lpa-14* gene, was also not affected by existence of microorganisms indigenous to the soil at 30°C.

4.2.2.3   Survival of *B. subtilis* RΔ1(pC115) at 30°C

The total cell number of RΔl(pC115) decreased immediately after inoculation (Figure 4.8), and as in the cases of RB14-C and RΔ1, the number of spores increased rapidly. All the cells had sporulated by day 2, and thereafter the population level was maintained for the 30 days. The stabilized population of RΔl(pC115) was as high as that of RB14-C and RΔ1. The tendency was similar in non-sterile and sterile soils. The stability of plasmid pC115 in RΔ1(pC115) is also illustrated in Figure 4.8. The plasmid was maintained at 70–80% in spores of RΔl(pC115) 10 days after inoculation.

4.2.2.4   Survival of *B. subtilis* RB14-C at 15°C

At 15°C, a significant difference was observed in the changes of the total number of cells and the number of spores of RB14-C compared with those at 30°C (Figure 4.6).

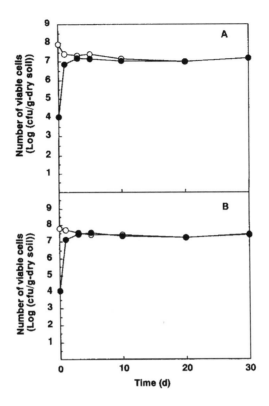

FIGURE 4.7 Survival of cells of *B. subtilis* RΔ1, an *lpa-14*-defective strain of RB14, at 30°C in (A) non-sterile and (B) sterile soils. o, total cell number; •, spore number.

There was a slower decline in the total cell number and a slower increase in the spore number at the lower temperature (Figure 4.9). After about 10 days, the cell number leveled off as spores and this amount lasted for the 30 days. The level of stabilization was as high as at 30°C and there was no marked difference between the number of cells in non-sterile and sterile soil.

4.2.2.5 Survival of *B. subtilis* RΔ1at 15°C

In non-sterile soil, the total number of RΔ1 cells decreased more rapidly (Figure 4.10A) than in the parental strain, RB14-C (Figure 4.9A), and became equivalent to the spore number after 10 days. This number was maintained thereafter for the 30 days. In sharp contrast to the rapid sporulation of RΔ1 at 30°C (Figure 4.7A), at 15°C sporulation hardly occurred, and the number of spores did not increase during the 30 days. The death rate of RΔ1 vegetative cells (Figure 4.10B) was markedly lower in sterile than in non-sterile soil (Figure 4.10A). During the period the vegetative cell number was stable, sporulation progressed to increase the spore number to about 10-fold that in sterile soil. The stabilized RΔ1 population in sterile soil was therefore higher than that in non-sterile soil. RΔ1 appears to be susceptible to dying out in non-sterile soil, partly due to slower sporulation at low temperature.

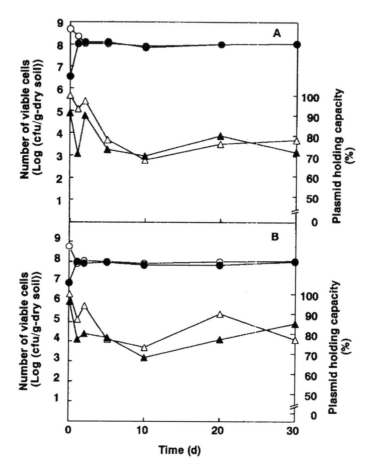

FIGURE 4.8 Survival and plasmid stability of cells of *B. subtilis* RΔ1(pC115), a transformant of RΔ1 with pCl15, at 30°C in (A) non-sterile and (B) sterile soils. o, total cell number; •, spore number; Δ, plasmid stability of all cells; ▲, plasmid stability of spores.

4.2.2.6   Survival of *B. subtilis* RΔ1(pC115) at 15°C

The total number of RΔl(pC115) cells decreased gradually after inoculation (Figure 4.11), whereas the number of spores had increased about 10-fold after 2 days. Thereafter, the number stabilized at this level. The total number of cells and the number of spores were almost the same after 10 days. This number was maintained thereafter for the 30 days. RΔl(pC115) cells thus also stabilized as spores. Compared with RΔ1 or RB14-C, a clear increase in the number of spores of RΔl(pC115) was observed, but the sporulation level of RΔl(pC115) was considerably lower than that of RB14-C.

As there was no difference between the survival pattern of RΔl(pC115) cells in non-sterile and sterile soil, other microorganisms in the soil did not affect the persistence of RΔl(pCl15) at 15°C. The stability of pC115 in RΔl(pC115) was as high (about 80%) as at 30°C for 30 days both in the total cell number and in spores.

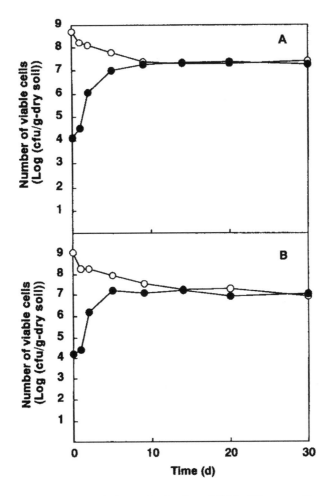

FIGURE 4.9 Survival of *B. subtilis* RB14-C cells at 15°C in (A) non-sterile and (B) sterile soils. ●, total cell number; ●, spore number.

4.2.3 DISCUSSION

Many bacteria have been found to suppress plant diseases caused by soilborne pathogens, with most of their disease-suppressing ability being attributed to the production of antimicrobial metabolites. Genes coding for the biosynthesis and regulation of some of these metabolites have been cloned (43–49). Many attempts have been made to construct genetic recombinants that possess higher antibiotic activities than the parental strains, either by manipulation at the gene regulation level or by transferring the genes to new hosts (46–49). Since in such cases, structural and regulatory genes for production of antimicrobial metabolites are targets of genetic engineering, it is important to know whether such genes affect the survival of recombinants in soil.

The gene *lpa-14*, which was cloned from RB14, is responsible for producing the antibiotics iturin A and surfactin. Although the precise function of *lpa-14* is not

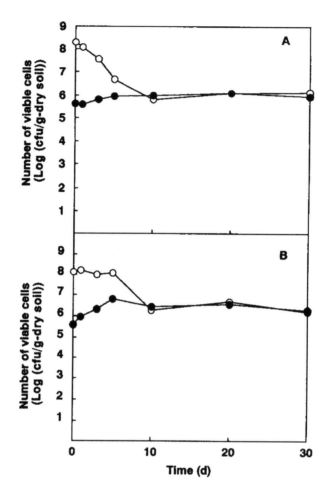

FIGURE 4.10   Survival of *B. subtilis* RΔ1 cells at 15°C in (A) non-sterile and (B) sterile soils. o, total cell number; •, spore number.

still clear, survival of the mutant RΔ1, from which *lpa-14* was deleted, and of the recombinant RΔ1(pCl15), in which productivity of the antibiotics was restored, were compared with that of parental strain, RB14-C.

When the soil temperature was 30°C, all three strains exhibited similar survival patterns. After a quick decline in the number of vegetative cells and rapid sporulation following inoculation, all the viable cells were detected as spores by day 2 and the cell number was maintained at this level thereafter until the end of the observation period of 30 days. Cells of all the strains were found to be stabilized as spores at a level of 10^7 to 10^8 cfu/g dry soil. At 30°C, there was little difference between the findings for non-sterile and sterile soil among the three strains, indicating that indigenous microorganisms in the soil did not affect the persistence of these strains at this temperature. The presence or absence of *lpa-14* was not clearly associated with the persistence of the bacteria in soil at 30°C, indicating that production of the antibiotics

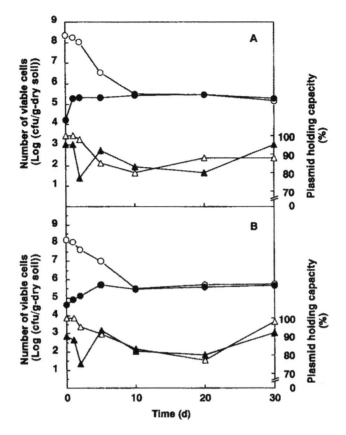

FIGURE 4.11 Survival and plasmid stability of *B. subtilis* RΔ1(pC115) cells at 15°C in (A) non-sterile and (B) sterile soils. o, total cell number; •, spore number; Δ, plasmid stability of all cells; ▲, plasmid stability of spores.

iturin A and surfactin seemed to have no significant role in the survival of *B. subtilis* RB14 in soil at 30°C. However, at 15°C, differences among the three strains were observed both in the rate of sporulation and the rate of decline in the vegetative cell number. In the case of RB14-C, the rate of sporulation at 15°C was obviously slower than that at 30°C, but the number of spores increased by about a thousand-fold during the initial 10 days after inoculation (Figure 4.9). The sporulation rate of RΔ1, however, declined markedly at 15°C and the number of spores hardly increased over the 30 days (Figure 4.10). This suggests that deletion of the DNA fragment that includes *lpa-14* is related to the decline in the sporulation ability of RΔ1.

The sporulation ability of RΔ1(pC115) were not restored to the level of RB14-C at 15°C (Figure 4.11). The reason for this is not yet clear. One possibility is that the gene *srfA*, which is related not only to surfactin production but also to efficient sporulation (50), was found to be located next to *lpa-14* (data not shown). The destructive *lpa-14* was constructed in RΔ1 with the integration plasmid holding *lpa-14* and a fragment of *srfA*, causing partial destruction of *srfA* in RΔ1. As plasmid pC115 carries only *lpa-14*, it cannot make up for the defective *srfA*, and hence the sporulation ability of RΔl(pC115) was not perfectly restored. This would also explain the experimental finding that the surfactin production of RΔl(pC115) was reduced to only one-eighth that of RB14-C (39).

At 15°C, the death rate of RΔl vegetative cells differed in the two soils used, being faster in non-sterile (Figure 4.10A) than in sterile soil (Figure 4.10B). This was partly because RΔl was unable to compete as well with indigenous microflora in the soil due to the loss of its ability to produce lipopeptide antibiotics. Mutants, which are defective in antibiotic production, generally suffer from a diminished ability to compete with indigenous microflora, as described in the case of fluorescent pseudomonads (51), and the production of antibiotics is therefore critical for mesophilic *B. subtilis* to survive in soil at relatively lower temperatures. RΔl may have experienced this phenomenon even at 30°C (Figure 4.7A and B), but the rate of sporulation at that temperature was so rapid that no clear difference between the two soils was observed. The rate of decline in the number of vegetative cells of RB14-C was the slowest among the three strains, mainly because RB14-C is a coproducer of iturin A and surfactin at relatively higher concentrations compared with RΔl and RΔl(pC115) and is thus able to compete with indigenous soil microorganisms. With respect to the faster death rate of RΔl at 15°C in non-sterile soil, other interactions between RΔl and indigenous microorganisms in the soil, such as competition for nutrients or antagonism, may be involved because fairly constant numbers of bacteria and fungi were reported year-round in a 2–15 cm layer of soil in which the maximum temperature was 20°C (52).

Control of plant diseases using *B. subtilis* appears to be safe, but a low density (51) or rather poor survival of this bacterium in soil (37, 53) have been reported. However, Tokuda et al. (1) found that inoculation of spores of *B. subtilis* was crucial for its stability in soil. In the present work, the persistence of RB14 in soil was affected by the death rate of the vegetative cells and the rate of sporulation, the bacterium being stabilized finally as spores. The gene *lpa-14* confers on RB14 the ability to compete with indigenous soil microorganisms by the production of the antibiotics, thus giving it much higher viability in soil. *lpa-14* also induced rapid sporulation of RB14. When the soil temperature was 15°C, the rate of sporulation declined markedly compared with that at 30°C; the deletion of *lpa-14* thus has an adverse effect on the persistence of the bacterium in soil, especially at lower temperatures. As an average temperature of soil in mainland Japan is 15°C, the results obtained in this study will be useful when considering the introduction of RB14 into soil in the future.

In Section 4.1, the survival in soil of *B. subtilis* bearing plasmid pC194 was investigated and found that the plasmid itself was quite unstable, having a stability of only about 50%. This instability was attributed to frequent curing during sporulation. In a previous study, it was found that 50% of plasmid pC194 was cured during sporulation in liquid culture (54). Although the plasmid pC115 was constructed with pC194 as a vector, the stability of pC115 was as high as about 80% in the present study (Figure 4.8 and Figure 4.11). The reason for the stability of pC115 is not clear, but several unknown factors that determine the stability of the plasmid during sporulation appears to be implicated in pC115.

Future research with these three strains will involve plant tests in soil to determine the role of *lpa-14* in the suppression of plant diseases and the survival of the strains in the root environment. Few reports have been published in this area for *B. subtilis* compared with the large amount of work that has been done with *Pseudomonas fluorescens* (43, 55).

REFERENCES

1. Tokuda, Y., Ano, T., and Shoda, M. Survival of *Bacillus subtilis* NB22 and its transformant in soil. *Appl. Soil Ecol.*, 2, 85–94 (1995).
2. Ehrlich, S. D. Replication and expression of plasmids from *Staphylococcus aureus* in *Bacillus subtilis*. *Proc. Natl Acad. Sci. U.S.A.*, 74, 1680–1682 (1977).
3. Horinouchi, S., and Weisblum, B. Nucleotide sequence and functional map of pC194, a plasmid that specifies inducible chloramphenicol resistance. *J. Bacteriol.*, 150, 815–825 (1982).
4. Gruss, A., and Ehrlich, S. D. The family of highly interrelated single-stranded deoxyribonucleic acid plasmids. *Microbiol. Rev.*, 53, 231–241 (1989).
5. Gryczan, T. J., Contente, S., and Dubnau, D. Characterization of *Staphylococcus aureus* plasmids introduced by transformation into *Bacillus subtilis*. *J. Bacteriol.*, 134, 318–329 (1978).
6. McKenzie, T., Hoshino, T., Tanaka, T., and Sueoka, N. The nucleotide sequence of pUB110: some salient features in relation to replication and its regulation. *Plasmid*, 15, 93–103 (1986).
7. Schaeffer, P., Millet, J., and Aubert, J. P. Catabolic repression of bacterial sporulation. *Proc. Natl Acad. Sci. U.S.A.*, 54, 704–711 (1965).
8. Birnboim, H. C., and Doly, J. A rapid alkaline extraction procedure for screening recombinant plasmid DNA. *Nucleic Acids Res.*, 7, 1513–1523 (1979).
9. Ano, T., Kobayashi, A., and Shoda, M. Transformation of *Bacillus subtilis* with the treatment by alkali cations. *Biotechnol. Lett.*, 12, 99–104 (1990).
10. Sambrook, J., Fritsch, E. F., and Maniatis, T. *Molecular Cloning: A Laboratory Manual*, 2nd edn. Cold Spring Harbor Laboratory, Cold Spring Harbor, NY, pp. 6.3–6.19 (1989).
11. Katayama, A., Hirai, M., Shoda, M., and Kubota, H. Factors affecting the stabilization period of sewage sludge in soil with reference to the gel chromatographic pattern. *Soil Sci. Plant Nutr.*, 32, 383–395 (1986).
12. Anagnostopoulos, C., and Spizizen, J. Requirements for transformation in *Bacillus subtilis*. *J. Bacteriol.*, 81, 741–746 (1961).
13. Bentjen, S. A., Fredrickson, J. K., Voris, P. V., and Li, S. W. Intact soil-core microcosms for evaluating the fate and ecological impact of the release of genetically engineered microorganisms. *Appl. Environ. Microbiol.*, 55, 198–202 (1989).
14. Bleakley, B. H., and Crawford, D. L The effect of varying moisture and nutrient levels on the transfer of a conjugative plasmid between *Streptomyces* species in soil. *Can. J. Microbiol.*, 35, 544–549 (1989).
15. Graham, J. B., and Istock, C. A. Genetic exchange in *Bacillus subtilis* in soil. *Mol. Gen. Genet.*, 166, 287–290 (1987).
16. Ramos, J. L., Doque, E., and Ramos-Gonzalez, M. Survival in soils of an herbicide-resistant *Pseudomonas putida* strain bearing a recombinant TOL plasmid. *Appl. Environ. Microbiol.*, 57, 260–266 (1991).
17. Sun, L., Bazin, M. J., and Lynch, J. M. Plasmid dynamics in a model soil column. *Mol. Ecol.*, 2, 9–15 (1993).

18. Van Elsas, J. D., Govaert, J. M., and van Veen, J. A. Transfer of plasmid pFT30 between bacilli in soil as influenced by bacterial population dynamics and soil condition. *Soil Biol. Biochem.*, 19, 639–647 (1987).

19. Wang, Z., and Crawford, D. L. Survival and effects of wild-type, mutant, and recombinant *Streptomyces* in a soil ecosystem. *Can. J. Microbiol.*, 35, 535–543 (1989).

20. Wellington, E. M. H., Cresswell, N., and Saunders, V. A. Growth and survival of *Streptomycete* inoculants and extent of plasmid transfer in sterile and nonsterile soil. *Appl. Environ. Microbiol.*, 56, 1413–1419 (1990).

21. Alonso, J. C., Viret, J., and Tailor, R. H. Plasmid maintenance in *Bacillus subtilis* recombinant-deficient mutants. *Mol. Gen. Genet.*, 208, 349–352 (1987).

22. Bron, S., Luxen, E., and Swart, P. Instability of recombinant pUB110 plasmids in *Bacillus subtilis*: Plasmid-encoded stability function and effects of DNA inserts. *Plasmid*, 19, 231–241 (1988).

23. Fleming, G., Dawson, M. T., and Patching, J. W. The isolation of strains of *Bacillus subtilis* showing improved plasmid stability characteristics by means of selective chemostat culture. *J. Gen. Microbiol.*, 134, 2095–2101 (1988).

24. Harington, A., Watson, T. G., Louw, M. E., Rodel, J. E., and Thomson, J. A. Stability during fermentation of a recombinant α-amylase plasmid in *Bacillus subtilis*. *Appl. Environ. Microbiol.*, 27, 521–527 (1988).

25. Pinches, A., Louw, M. E., and Watson, T. G. Growth, plasmid stability and α-amylase production in batch fermentations using a recombinant *Bacillus subtilis* strain. *Biotechnol. Lett.*, 7, 621–626 (1985).

26. Wei, D., Parulekar, S. J., Stark, B. C., and Weigand, W. A. Plasmid stability and α-amylase production in batch and continuous cultures of *Bacillus subtilis* TN106[pAT5]. *Biotechnol. Bioeng.*, 33, 1010–1020 (1989).

27. Siala, A., and Gray T. R. G. Growth of *Bacillus subtilis* and spore germination in soil observed by a fluorescent-antibody technique. *J. Gen. Microbiol.*, 81, 191–198 (1974).

28. Siala, A., Hill, I. R., and Gray, T. R. G. Population of spore-forming bacteria in an acid forest soil with special reference to *Bacillus subtilis*. *J. Gen. Microbiol.*, 81, 191–198 (1974).

29. Acea, M. J., Moore, C. R., and Alexander, M. Survival and growth of bacteria introduced into soil. *Soil Biol. Biochem.*, 20, 509–515 (1988).

30. Liang, L. N., Sinclair, J. L., Mallory, L. M., and Alexander, M. Fate in model ecosystems of microbial species of potential use in genetic engineering. *Appl. Environ. Microbiol.*, 44, 708–714 (1982).

31. Van Elsas, J. D., Dijkstra, A. F., Govaert, J. M., and van Veen, J. A. Survival of *Pseudomonas fluorescens* and *Bacillus subtilis* introduced into two soils of different texture in field microplots. *FEMS Microbiol. Ecol.*, 38, 151–160 (1986).

32. Alonso, J. C., and Trautner, T. A. A gene controlling segregation of the *Bacillus subtilis* plasmid pC194. *Mol. Gen. Genet.*, 198, 427–431 (1985).

33. Keggins, K. M., Lovett, P. S., and Duvall, E. J. Molecular cloning of genetically active fragments of *Bacillus subtilis* DNA in *Bacillus subtilis* and properties of the vector plasmid pUB 1 10. *Proc. Natl Acad. Sci. U.S.A.*, 75, 1423–1427 (1978).

34. Tokuda, Y., Ano, T., and Shoda, M. Survival of *Bacillus subtilis* NB22, an antifungal-antibiotic iturin producer, and its transformant in soil-systems. *J. Ferment. Bioeng.*, 75, 107–111 (1993).

35. Siala, A., Hill, I. R., and Gray, T. R. G. Populations of spore-forming bacteria in an acid forest soil, with special reference to *Bacillus subtilis*. *J. Gen. Microbiol.*, 81, 183–190 (1974).

36. Asaka, O., Ano. T., and Shoda, M. Persistence of *Bacillus subtilis* RB14 and its derivative strains in soil with respect to the *lpa-14* gene. *J. Ferment. Bioeng.*, 81, 1–6 (1996).

37. Liang, L. N., Sinclair, J. L., Mallory, L. M., and Alexander, M. Fate in model ecosystems of microbial species of potential use in genetic engineering. *Appl. Environ. Microbiol.*, 44, 708–714 (1982).
38. Amner, W., McCarthy, A. J., and Edwards, C. Quantitative assessment of factors affecting the recovery of indigenous and released thermophilic bacteria from compost. *Appl. Environ. Microbiol.*, 54, 3107–3112 (1986).
39. Huang, C. C., Ano, T., and Shoda, M. Nucleotide sequence and characteristics of the gene, *lpa-14*, responsible for biosynthesis of the lipopeptide antibiotics iturin A and surfactin from *Bacillus subtilis* RB14. *J. Ferment. Bioeng.*, 76, 445–450 (1993).
40. Hiraoka, H., Asaka, O., Ano, T., and Shoda, M. Characteristics of *Bacillus subtilis* RB14, coproducer of peptide antibiotics iturin A and surfactin. *J. Gen. Appl. Microbiol.*, 38, 635–640 (1992).
41. Hiraoka, H., Ano, T., and Shoda, M. Molecular cloning of a gene responsible for the biosynthesis of the lipopeptide antibiotics iturin and surfactin. *J. Ferment. Bioeng.*, 74, 323–326 (1992).
42. Phae, C. G., Shoda, M., Kita, N., Nakano, M., and Ushiyama, K. Biological control of crown and root rot and bacterial wilt of tomato by *Bacillus subtilis* NB22. *Ann. Phytopathol. Soc. Jpn.*, 58, 329–339 (1992).
43. Howie, W. J., and Suslow, T. V. Role of antibiotic biosynthesis in the inhibition of *Pythium ultimum* in the cotton spermosphere and rhizosphere by *Pseudomonas fluorescens*. *Mol. Plant Microbe Interact.*, 4, 393–399 (1991).
44. Gutterson, N., Ziegle, J. S., Warren, G. J., and Layton, T. J. Genetic determinants for catabolite induction of antibiotic biosynthesis in *Pseudomonas fluorescens* Hv37a. *J. Bacteriol.*, 170, 380–385 (1988).
45. Thomashow, L. S., and Weller, D. M. Role of a phenazine antibiotic from *Pseudomonas fluorescens* in biocontrol of *Gaeumannomyces graminis* var. *tritici*. *J. Bacteriol.*, 170, 3499–3508 (1988).
46. Sundheim, L., Poplawsky, A. R., and Ellingboe, A. H. Molecular cloning of two chitinase genes from *Serratia marcescens* and their expression in *Pseudomonas* species. *Physiol. Mol. Plant Pathol.*, 33, 483–491 (1988).
47. Voisard, C., Keel, C., Haas, D., and Defago, G. Cyanide production by *Pseudomonas fluorescens* helps suppress black root rot of tobacco under gnotobiotic conditions. *EMBO J.*, 8, 351–358 (1989).
48. Fenton, A. M., Stephens, P. M., Crowley, J., O'Callaghan, M., and O'Gara, F. Exploitation of gene(s) involved in 2,4-diacetylphloroglucinol biosynthesis to confer a new biocontrol capability to a *Pseudomonas* strain. *Appl. Environ. Microbiol.*, 58, 3873–3878 (1992).
49. Maurhofer, M., Keel, C., Schnider, U., Voisard, C., Haas, D., and Defago, G. Influence of enhanced antibiotic production in *Pseudomonas fluorescence* strain CHAO on its disease suppressive capacity. *Phytopathology*, 82, 190–195 (1992).
50. Nakano, M. M., Magnuson, R., Myers, A., Curry, J., Grossman, A. D., and Zuber, P. *srfA* is an operon required for surfactin production, competence development, and efficient sporulation in *Bacillus subtilis*. *J. Bacteriol.*, 173, 1770–1778 (1991).
51. Mazzola, M., Cook, R. J., Thomashow, L. S., Weller, D. M., and Pierson III, L. S. Contribution of phenazine antibiotic biosynthesis to the ecological competence of fluorescent pseudomonads in soil habitats. *Appl. Environ. Microbiol.*, 58, 2616–2624 (1992).
52. Higashida, S., and Takao, K. Seasonal fluctuation patterns of microbial numbers in the surface soil of a grassland. *Soil Sci. Plant Nutr.*, 31, 113–121 (1985).
53. van Elsas, J. D., Dijkstra, A. F., Govaert, J. M., and van Veen, J. A. Survival of *Pseudomonas fluorescens* and *Bacillus subtilis* introduced into soils of different texture in field microplots. *FEMS Microbiol. Ecol.*, 38, 151–160 (1986).

54. Asaka, O., Tokuda, Y., Ano, T., and Shoda, M. Plasmid instability in *Bacillus subtilis* during sporulation. *Biosci. Biotechnol. Biochem.*, 57, 336–337 (1993).
55. Natsch, A., Keel, C., Pfirter, H. A., Haas, D., and Defago, G. Contribution of the global regulator gene *gacA* to persistence and dissemination of *Pseudomonas fluorescens* biocontrol strain CHAO introduced into soil microcosms. *Appl. Environ. Microbiol.*, 60, 2553–2560 (1994).

5 Development of Transformation Methods of *B. subtilis* and Cloning of Genes Responsible for Biosynthesis of Lipopeptide Antibiotics

5.1 DEVELOPMENT OF TRANSFORMATION METHODS OF *B. SUBTILIS* (1–4)

Several methods have been established to accomplish plasmid transformation of *Bacillus subtilis*. The physiological competent cell method can be applied only to restricted strains of *B. subtilis*, such as Marburg 168, and monomer plasmids do not undergo transformation with this method (5). The protoplast transformation method is highly efficient, but it takes several days before acquiring transformed cells after laborious procedures (6). A new alkali cation method that is applicable to wild strains of *B. subtilis* is proposed.

5.1.1 TRANSFORMATION METHODS

The simplified alkali cation transformation method includes the following steps: (i) cell harvesting, (ii) alkali cation treatment, (iii) plasmid addition, (iv) addition of polyethylene glycol (PEG) 6000, (v) PEG dilution and washing, and (vi) gene expression.

The optimized transformation procedure is as follows:

i. An overnight culture of *B. subtilis* MI113 in L medium (10 g of Polypepton, 5 g of yeast extract, and 5 g of NaCl in 1 liter; pH 7.2) was inoculated into fresh L medium at 1% and incubated aerobically until the optical density at 660 nm (OD_{660}) reached 0.7–0.8 in the middle or late exponential phase. A 0.33 ml aliquot of the culture was transferred into a polypropylene tube (nominal volume: 1.5 ml), pelleted by centrifugation at $10,000 \times g$ for 2 min.

ii. The pellet was resuspended in 1 ml of 410 mM KCl solution, after being incubated statically for about 30 min at 30°C.

iii. An aliquot of this suspension (50 μl) was mixed with plasmid DNA (less than 5 μl).

iv. Then 50 μl of 70% PEG6000 solution were added to this mixture.

v. After mixing several times, PEG was diluted by the addition of 1 ml of LC medium (L medium supplemented with 100 mM $CaCl_2$), and was centrifuged at 10,000 × g for 2 min. The pelleted cells were suspended in 0.5 ml of L medium in the tube.

vi. The suspension was maintained statically for 2 h at 37°C to permit expression of the antibiotic-resistance gene. Transformants were selected by plating the cells onto L-agar plates containing chloramphenicol (5 μg/ml). The numbers of transformants and viable cells were counted after overnight incubation at 37°C.

Among the alkali cations, K^+ and Cs^+ were more effective at transformation than the other ions (Li^+, Na^+, Rb^+) (2). KCl was selected from the points of view of safety and cost because its effectiveness at transformation was virtually the same as that of CsCl. The recipient *B. subtilis* MI113 was treated with various concentrations of KCl solution for 10 min and transformed by pC194 in the presence of PEG by the procedure described above. The frequency of transformation was highest when 410 mM KCl was used (1).

The transformation efficiency increased proportionally with incubation time up to 45 min, during which a gradual decrease in cell turbidity was observed. Cell lysis proceeded during the KCl treatment, resulting in lower cell numbers at the end of gene expression (1). The possible relationship between the induction of competency and cell lysis is described later. Even at a treatment time of 15 min, several hundred transformants/μg DNA appeared. Plasmids recovered from the drug-resistant colonies by a rapid alkaline lysis method (7) were confirmed to be identical to the original plasmid transformed by agarose gel analysis. No transformants were obtained in the absence of pC194 DNA.

The effect of cell concentration during KCl treatment was investigated by diluting or concentrating the cell density at ratios from 0.1 to 10, where the cell concentration at the end of the main culture was taken as unity. The highest transformation efficiency was obtained with a ratio of 1/3. Lower or higher cell concentrations than this value were less effective or inhibitory.

Addition of $MgCl_2$ or $CaCl_2$ in the PEG-diluting solution at a concentration of 100 mM increased the number of transformants about 10-fold compared to the control (Table 5.1), and these divalent cations are known to inhibit lysis immediately, even after lysis has been triggered in *B. subtilis* cells (8).

The activation of autolysin(s) is known to be induced by monovalent cations such as K^+, Na^+, Cs^+, Rb^+, and Li^+ (8). The involvement of autolysin(s) was tested by the transformation of an autolysin-deficient mutant using this method. *B. subtilis* FJ2 *trp lyt* (9) deficient in the autolytic enzymes N-acetylmuramyl-1-alanine amidase and endo-3-N-acetylglucosaminidase, was treated by this transformation method. No transformants were obtained from this mutant strain, while its parent strain *B. subtilis* 168 was efficiently transformed with the plasmid DNA using the same method as shown in Table 5.2. This observation concerning the relationship between the

TABLE 5.1

Effect of Divalent Cations on Transformation

Added Cations	Time Added	Number of Transformants[a] (/ml)	Viable Cells[b] (cfu/ml)
$MgCl_2$ (100 mM)	Before PEG treatment	4.0×10	N.D.[c]
	After PEG treatment	6.1×10^3	N.D.
$CaCl_2$ (100 mM)	Before PEG treatment	0	N.D.
	After PEG treatment	6.1×10^4	1.2×10^7
None		8.6×10^2	8.1×10^6

[a] 0.65 μg of DNA was added.
[b] Viable cells were counted as colony forming units (cfu).
[c] N.D., not determined.

activation of autolysis and the induction of competency may be a clue for analyzing the mechanism of this KCL transformation method.

5.1.2 COMPARISON BETWEEN KCL METHOD AND COMPETENT CELL METHOD

Transformation frequencies obtained by the KCL method and by competent cell transformation methods were compared for DNA with different configurations as shown in Table 5.3. In the KCL treatment method, the monomeric form of pC194 transformed the host cells at almost the same frequency as did unfractionated plasmid DNA. This donor DNA did not transform naturally competent cells. Linear plasmid pC194, cleaved with *Hin*dIII at a unique site, could transform KCL-treated cells but not competent cells. However, the frequency was lower than with circular DNA. Experiments using chromosomal DNA, on the other hand, gave a high frequency of transformation of competent cells but no transformants with KCL-treated cells.

TABLE 5.2

Transformation of *B. subtilis* FJ2 Deficient in Autolytic Enzymes

	B. subtilis 168 (Parental Strain)	*B. subtilis* FJ2 (Mutant of 168)
OD_{660} at the end of main culture	0.43	0.65
Number of transformant[a] (/ml)	2.2×10^3	0
Viable cells[b] (cells/ml)	1.3×10^6	6.7×10^6

[a] 0.65 μg of DNA was used.
[b] Viable cells were counted as colony forming units (cfu).

TABLE 5.3

Effect of DNA Configuration on Transformation Efficiency Expressed as Number of Transformants per 1 µg DNA in Two Methods

DNA	KCl Method	Competent Cell Method	Selected Markers
Unfractionated pC194	5.8×10^4	6.4×10^3	Cm^r
Monomeric p194	8.6×10^4	0	Cm^r
Linear pC194	1.2×10^3	0	Cm^r
Chromosomal DNA of *B. subtilis* MI115	0	$1.0 \times 10^{4 \, a}$	*trp*$^+$
	0	$7.4 \times 10^{4 \, b}$	arg$^+$
Chromosomal DNA of *B. subtilis* 168	0	$8.5 \times 10^{4 \, c}$	arg$^+$

Notes: Minimal medium was supplemented with amino acids (50 µg/ml each):
[a]   arginine + leucine.
[b]   tryptophan + leucine.
[c]   tryptophan.

5.1.3   Electroporation (4)

Recent developments of the electroporation technique have allowed the transformation of several Gram-negative and Gram-positive bacteria, but there is still room for improvement in efficiency for *B. subtilis*. Several factors to optimize transformation efficiency of *B. subtilis* NB22 with plasmid pC194 were investigated.

5.1.3.1   Methods

B. subtilis NB22 was grown overnight at 37°C in L broth (Polypepton, 10 g/l, yeast extract, 5g/l, NaCl, 5g/l, pH 7.2). 0.5 ml of the culture was inoculated into 50 ml fresh L broth. After being grown at 37°C for 3 h with shaking at 110 rpm, the cells were collected by centrifugation ($2000 \times g$, 10 min) washed twice with cold (4°C) sterile 1 mM Hepes (N-2-hydroxyethylpiperazine-N′-ethanesulfonic acid) buffer (pH 7.0). Then the cells were washed once in cold transformation solution (25% PEG and 0.1 M mannitol), and resuspended in the solution at a cell density of 10^{10}–10^{11} cells/ml, and stored at 4°C for 10 min. Plasmid DNA (less than 5 µl) was mixed with a 40 µl cell suspension and 20 µl of this mixture was transferred to a 0.15 cm cuvette, followed by exposure to a single electric pulse (2.5 kV peak voltage, 2 µF capacitance, and 4 kΩ resistance), using a Cell Porator linked to the booster (BRL Life Technologies, Inc., Gaithersburg, Maryland, USA), which generates field strength about up to 16 kV/cm with the cuvette. In 2 or 3 min after pulsing, 10 µl of this suspension was put into 0.5 ml of L broth and stood still for 3 h at 37°C. Then the cell suspension was plated on selective media, containing L agar plus 5 µg/ml of chloramphenicol or 50 µg/ml of kanamycin for NB22 and 5 µg/ml of chloramphenicol or kanamycin for MI113. The viable cell number was determined on L agar plates without antibiotic.

5.1.3.2   Results

5.1.3.2.1   Introduction of PEG

Only a few transformants of pC194 DNA were obtained in the transformation of intact cells of *B. subtilis* NB22 by following the method described for *E. coli* (10), where 10% glycerol in the electroporation buffer was used. Then, polyethylene glycol (PEG) (Wako Pure Chemical Industries, Ltd., Osaka) was examined. Among PEG with various molecular weights tested, PEG 6000 was most effective and the number of transformants was maximum at 25% PEG.

5.1.3.2.2   Effect of Mannitol Concentration

Addition of mannitol in the transformation solution affected the transformation efficiency of *B. subtilis* NB22, and the highest efficiency of transformation was observed at the concentration of 0.1 M. Since the high electric field pulse is known to produce reactive oxygen species that cause DNA cleavage (11), the mannitol effect observed here may partly be explained by the action of a radical scavenger rather than that of an osmotic stabilizer.

5.1.3.2.3   Effect of the Electric Field Strength

Maximum transformation efficiency of $10^6/\mu g$ DNA was achieved at a voltage of 14.9 kV/cm. PEG may function to protect the cells against lethal damage by electric fields.

5.1.3.2.4   Effect of Cell and Plasmid Concentrations

With a higher concentration than 10^8 cells/ml, transformation efficiency was directly proportional to the number of cells to reach maximum transformation at 10^{11} cells/ml. The effect of plasmid DNA concentration of pC194 was then examined on the transformation efficiency under the cell concentration, 1.1×10^{11} cells/ml. The number of transformants increased significantly in a range of 0.01 to 1 μg of plasmid DNA and saturated at 10 μg.

5.1.3.2.5   Comparison of Transformation by KCl Method and Electroporation

The four iturin-producing *B. subtilis* that were isolated in this study and six plasmids were used for the comparison of the KCl method and electroporation as shown in Table 5.4. The KCl method gave transformants with almost equivalent efficiency in NB22 and YB8. However, the efficiency for UB24 and RB14 was significantly lower. All the iturin producers were successfully electroporated with newly isolated plasmids (3) as well as a reference plasmid pC194.

5.2   MOLECULAR CLONING OF A GENE RESPONSIBLE FOR THE SYNTHESIS OF ITURIN AND SURFACTIN (12)

5.2.1   CLONING PROCEDURE AND RESULTS

B. subtilis RB14 is a newly isolated strain and coproduces the lipopeptides iturin and surfactin (13, 14) as shown in Chapter 2, Section 2.8.

　　B. subtilis MI113, a derivative of the standard *B. subtilis* 168 (15–18) was used as a host for the shotgun cloning of the gene(s) responsible for the production of iturin

TABLE 5.4

Transformation of Iturin Producers
with Plasmid

		Transformants/μg DNA	
Strain	Plasmid	KCl Method	Electroporation
NB22	pCS6	2.0×10	8.8×10^5
NB22	pCS59	2.6×10^2	9.0×10^5
NB22	pES42	0	7.8×10^4
NB22	pES59	1.3×10^2	2.2×10^6
NB22	pTS59	1.6×10^2	1.3×10^6
NB22	pC194	4.1×10^3	1.6×10^6
YB8	pCS6	3.6×10^2	2.1×10^4
YB8	pCS59	2.4×10^3	2.2×10^4
YB8	pES42	6	9.7×10^3
YB8	pES59	1.2×10^3	9.3×10^4
YB8	pTS59	6.6×10	3.6×10^4
YB8	pC194	4.0×10^3	1.0×10^5
UB24	pCS6	0	2.5×10^4
UB24	pCS59	0	2.7×10^4
UB24	pES42	0	1.3×10^4
UB24	pES59	2.2×10^2	2.3×10^5
UB24	pTS59	2.7×10^2	1.2×10^5
UB24	pC194	2.1×10^3	3.9×10^5
RB14	pCS6	0	1.4×10^4
RB14	pCS59	0	5.1×10^3
RB14	pES42	0	4.3×10^3
RB14	pES59	5.0×10	2.2×10^3
RB14	pTS59	0	1.8×10^5
RB14	pC194	1.7×10^3	6.8×10^4

and surfactin, because it produced neither of the lipopeptides. Among 40,000 transformants obtained from the ligation mixture of the chromosomal DNA of RB14 and plasmid pTB522 (19), two positive colonies were obtained on tributyrin L agar plates (TB plates), on which 20 μl of tributyrin was spread. The plasmids isolated from the transformants were designated as pIB111 and pIB121. As these plasmids contain four *Hind*III fragments of the same sizes, the smaller plasmid pIB111 was chosen for further analysis.

When the methanol extract prepared from the culture broth of *B. subtilis* MI113 (pIB111) in no. 3 medium was analyzed by thin-layer chromatography and HPLC, only surfactin was detected. This was confirmed by the following analyses. The IR spectrum of the substance in KBr showed the characteristic infrared spectrum of surfactin, as previously reported (20). The amino acid composition of the hydrolysate of the substance was determined to be Asp, Glu, Val, Leu with a molar ratio of

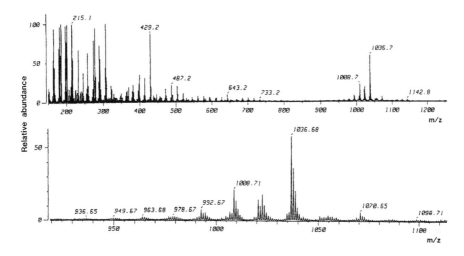

FIGURE 5.1 FAB-MS spectrum of the substance from the culture broth of *B. subtilis* MI113 (pIB111).

1:1:1:4, which was the same value as that expected from surfactin (20). The molecular weight determined by fast atom bombardment mass spectrometry (FAB-MS) represented the same molecular weight (1036) as that of surfactin, as a (M+H)⁺ ion peak at *m/z* 1036.68 (Figure 5.1).

All the aforementioned results indicated that the strain *B. subtilis* (pIB111) produced the biosurfactant surfactin. A reduction in size of pIB111 was achieved by replacing the vector plasmid pTB522 (10.5 kbp) by pC194 (2.9 kbp) (21) through *Hind*III ligation treatment. Two kinds of plasmids were obtained. One was designated pC111, which contained four *Hind*III fragments of pIB111, and the other was pC112 with two *Hind*III fragments (2.7 and 0.59 kbp) from pIB111. These two *Hind*III fragments were essential for the production of surfactin.

An integration plasmid was constructed to determine the function of the cloned *lpa* (lipopeptide antibiotic production) gene in the original host strain *B. subtilis* RB14. An 800-bp *Pst*I fragment was removed from pC112 (Figure 5.2), and the resulting plasmid pC112Δ could not offer surfactin-production capability to strain MI113. This plasmid pC112Δ was ligated with pE194 with a temperature-sensitive replication unit (22) at the unique *Pst*I site to make pEC112Δ. An integration plasmid pE112Δwas constructed by removing pC194 from pEC112Δ by *Hind*III-ligation treatment, as shown in Figure 5.2. The resulting plasmid was introduced into *B. subtilis* RB14 by electroporation (4). Erythromycin-resistant transformants were selected at 30°C. As a control to strain RB14 (pE112Δ), RB14 was transformed by the vector plasmid to obtain RB14 (pE194).

After cultivating strains RB14 (pE112Δ) and RB14 (pE194) in L medium containing 2 μg of erythromycin at 30°C to the stationary phase, they were diluted and plated on L agar plates with or without erythromycin, followed by incubation at 48°C for 12 h to obtain a surfactin-production-negative strain. No Emʳ colonies emerged from the culture of RB14 (pE194), but 96 Emʳ colonies were obtained from that of RB14 (pE112Δ) at a dilution rate 10⁻⁴, although almost the same number of viable cells was counted in

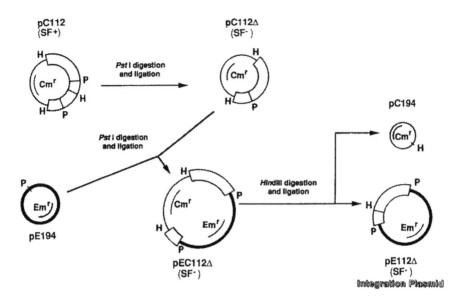

FIGURE 5.2   Construction of an integration plasmid pE112Δ. H and P designate *Hind*III and *Pst*I, respectively. The thick lines indicate plasmid pEl94 with a temperature-sensitive replicon. SF+ and SF− in parentheses mean surfactin production positive and negative, respectively.

RB14 (pE112Δ) and RB14 (pE194) on L agar plates. Integration ratios were about 17% and less than 3×10^{-5} in RB14 (pE112Δ) and RB14 (pE194), respectively. These data indicate the integration occurred by homologous recombination between the cloned fragment of plasmid pE112Δ and the corresponding chromosomal DNA of RB14, rather than by the random integration of the vector plasmid pE194 at high temperature.

Among 300 Em^r colonies, three (designated as RΔ1, RΔ2, and RΔ3) were negative on TB plate assay. The disappearance of plasmid pE112Δ in RΔ1, RΔ2, and RΔ3 was confirmed by alkaline lysis extraction, and the stability of the surfactin-negative property was tested by successive cultivation of the colonies under the nonselective condition at 30°C. After about 30 generations, no revertant appeared from all the colonies tested. *B. subtilis* RΔ1 was chosen for further analysis. As shown in Figure 5.3B, the nonproduction of surfactin by strain RΔ1 was confirmed by HPLC analysis. No growth inhibition zone against *Fusarium oxysporum* was observed *in vitro* using strain RΔ1 transformed with vector plasmid pTB522 (Figure 5.4B), and no iturin peaks were detected by HPLC analysis, as shown in Figure 5.5B. Strain RΔ1 was thus confirmed to be defective in the production of both surfactin and iturin.

Transformants of strain RΔ1 by the surfactin-producing plasmid pIB111 clearly restored the suppressiveness against *F. oxysporum*, as shown in Figure 5.4C, whereas no restoration was observed from the vector plasmid (Figure 5.4B). HPLC analyses showed that the transformant RΔ1(pIB111) produced both surfactin and iturin (Figure 5.3C and Figure 5.5C). All these results indicated that the *lpa* gene was responsible for the production of both iturin and surfactin. Although neither of the lipopeptides surfactin and iturin was detected from the supernatant of the culture

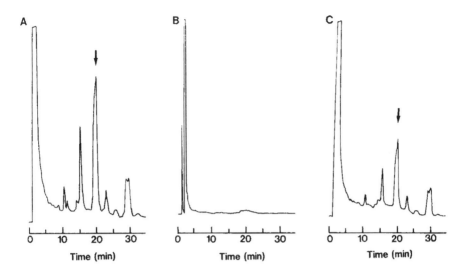

FIGURE 5.3 HPLC analysis of the production of surfactin. The separation patterns of methanol extracts from (A) *B. subtilis* RB14, (B) *B. subtilis* RΔI, and (C) *B. subtilis* RΔl(pIB111l). The arrows indicate the retention time of authentic surfactin. HPLC was operated by a reverse-phase column Inertsil ODS-2 (GL Sciences Inc., Tokyo, 4.6 Φ× 250 mm) at a flow rate of 1.5 ml/min and monitored at 205 nm with the solvent acetonitrile–acetic acid (1%) (80:20, v/v).

broth of the strain RΔl, the possibility of the defective excretion of the antibiotics from the cells into the culture medium was tested on the cells of RΔl(pTB522), RB14, and RΔl(pIB111) by extraction with chloroform/methanal (2: 1, v/v) according to the method of Besson et al. (26). Neither surfactin nor iturin was detected intracellularly in the cells of RΔl(pTB522), whereas small amounts of these substances were found in the cells of RB14 and RΔl(pIB111) (data not shown). Thus, strain RΔ1 was defective in the synthesis of iturin and surfactin, not an excretion-defective mutant.

A genetic locus responsible for surfactin production (*sfp*) was transferred from the original surfactin producer *B. subtilis* ATCC21332 to JH642, a derivative of strain 168,

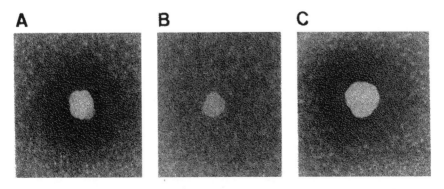

FIGURE 5.4 Iturin production assayed on agar plate containing *F. oxysporum*. (A) *B. subtilis* RB14(pTB522); (B) *B. subtilis* RΔ1 (pTB522); (C) *B. subtilis* RΔ1(pIB111).

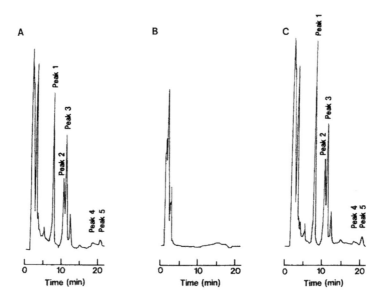

FIGURE 5.5   HPLC analysis of the production of iturin. The separation patterns of methanol extracts from (A) *B. subtilis* RB14, (B) *B. subtilis* RΔ1, and (C) *B. subtilis* RΔ1(pIB111). Iturin was analyzed by the same column used for the identification of surfactin (Figure 5.3). Peaks 1 to 5 correspond to the components of iturin with n-C_{14}-β-amino acid, *anteiso*-C_{15}-β-amino acid, *iso*-C_{15}-β-amino acid, n-C_{16}-β-amino acid, and *iso*-C_{16}-β-amino acid, respectively.

and this newly constructed strain, OKB105, produced surfactin (18). This *sfp* is encoded within a 1 kbp fragment, which complements *sfp⁰* (*sfp* allele of nonproducer strains). *lpa* is suggested to be a similar gene to *sfp*, because MI113, a derivative of strain 168 (*sfp⁰*), was complemented to become a surfactin producer by the introduction of this *lpa* gene.

The deduced amino acid sequence of *sfp* had no significant homology to other known proteins, and the function of this protein is not yet known (17, 18). Iturin and surfactin have an analogous structure that consists of a seven amino acid cyclic peptide that is linked to the moiety of the aliphatic chain, but the linkage modes and the amino acids components are quite different in the two compounds (23–26), suggesting that Lpa and Sfp are the regulatory proteins or the common factor required for the biosynthesis of the lipopeptides, such as a translocation element of the fatty acid moiety to the hydrophilic amino acid peptide part. As the introduction of *lpa* into strain RΔ1 transformed the strain into a coproducer of iturin and surfactin, a clue to the function of the gene products of *lpa* and *sfp* will be given by the further analysis of the genes.

5.3   CHARACTERISTICS OF THE GENE *LPA-14* (LIPOPEPTIDE ANTIBIOTIC-PRODUCTION OF RB14) RESPONSIBLE FOR BIOSYNTHESIS OF ITURIN A AND SURFACTIN (27)

In this section, the *lpa-14* cloned in the previous section is analyzed in detail and the nucleotide sequence of the gene was determined and compared to other similar genes.

5.3.1 Materials and Methods

5.3.1.1 Bacterial Strains and Plasmids

Bacterial strains and plasmids are listed in Table 5.5.

5.3.1.2 Media

L medium was used as shown in Chapter 3.

5.3.1.3 Transformation of *B. subtilis* MI113

Competent cells of *B. subtilis* MI113 were prepared by the method developed by Anagnostopoulos and Spizizen (51).

5.3.1.4 Transformation of the Iturin Producer and Its Derivative

Transformation of *B. subtilis* RB14 and its derivative RΔ1 with plasmid DNA was carried out by electroporation using a Cell Porator linked to a booster (BRL Life Technologies, Inc., Gaithersburg, Maryland, USA) (4).

5.3.1.5 Preparation of Plasmid DNA

The rapid alkaline lysis procedure (7) used for screening clones was performed.

TABLE 5.5
Strains and Plasmids

Strain or Plasmid	Genotype, Phenotype, Plasmid Marker	Source or Reference
B. subtilis		
MI113	*arg-15 trpC2 hsmM hsrM*	(12)
MI113(pC194)	Cm^r	This work
MI113(pTB522-2 kb)	SF$^+$ Tc^r	(30)
MI113(pC112)	SF$^+$ Cm^r	(12)
RB14	SF$^+$ IT$^+$	(58)
RΔ1	SF$^-$ IT$^-$ Em^r	(12)
RΔ1(pC194)	SF$^-$ IT$^-$ Em^r Cm^r	This work
RΔ1(pC115)	SF$^+$ IT$^+$ Em^r Cm^r	This work
RΔ1(pCSF1)	SF$^+$ IT$^+$ Em^r Cm^r	This work
B. pumilus A-1	SF$^+$	(30)
Plasmids		
pC194	Cm^r, 2.9 kb	(36)
pTB522-2 kb	SF$^+$, Tc^r, 12.5 kb	(30)
pC112	SF$^+$, reduced plasmid from pC111	(12)
pC113	SF$^+$, reduced plasmid from pC112	This work
pC115	SF$^+$, reduced plasmid from pC113	This work
pC112Δ	SF$^-$, *Pst*1 fragment was removed from pC112	(12)
pCSF1	SF$^+$, *psf-1* was cloned from pTB522-2 kb	This work

Notes: SF, surfactin production; IT, iturin production; Tc, tetracycline; Cm, chloramphenicol; Em, erythromycin.

For large-scale preparation, plasmids extracted by the alkali lysis method were further purified by CsCl–ethidium bromide density gradient ultracentrifugation (28).

5.3.1.6   DNA Sequencing and Analysis

DNA sequencing of double-stranded DNA cloned in pUC19 was performed by an ABI 373A auto-sequencer at Takara Shuzo Co. and was reconfirmed by a Pharmacia A.L.F. DNA sequencer at the Gene Experimental Laboratory of Tokyo Institute of Technology by the dideoxy-chain-termination method of Sanger et al. (29 using a T7 sequencing kit.

5.3.1.7   Detection of Biosurfactant Activity of the Lipopeptides Surfactin and Iturin A

The presence of lipopeptides was detected on tributyrin L agar plates (TB plates), using 20 μl of tributyrin spread onto an L agar plate by a glass rod. When a halo was made around a colony spotted on a TB plate, a biosurfactant was judged to be produced.

5.3.1.8   Fungal Growth Inhibition Test by Bacteria

A phytopathogenic fungus, *F. oxysporum* f. sp. *lycopersici* race J1 SUF119, was grown on a potato-dextrose agar medium (potato-dextrose agar 39 g, distilled water 1 liter, pH 5.6) at 30°C for 5 days and suspended in sterile distilled water. A portion of this suspension was mixed into L agar medium before it was solidified. After spotting the bacterium to be tested at the center of this plate, its suppressiveness was investigated by observing the zones inhibitory to the growth of the fungus.

5.3.1.9   HPLC Analysis of Iturin A and Surfactin

Iturin A was assayed by reversed-phase HPLC as previously described in Chapter 2. Surfactin was analyzed by a reversed-phase HPLC system on the same column used for the identification of iturin.

5.3.2   Results and Discussion

5.3.2.1   Subcloning of pC112

A 10 kb pC112, which was obtained in the previous section (2), was further reduced as shown in Figure 5.6 to determine the region indispensable for surfactin production by using *B. subtilis* MI113, a nonproducer of surfactin, as a host strain. Plasmid pC112 was cut by *Hpa*I and the 0.5 kb *Hpa*I fragment was removed and ligated to make pC113. As pC113 clearly transformed MI113 into a surfactin producer, the region required for surfactin production was estimated to be to the right or left of the *Hpa*I site of pC113 in Figure 5.6. A 2.0 kb fragment was removed from pC113 by the double digestion of *Hpa*I and *Pvu*II followed by blunt end ligation. The resulting plasmid, pC115, also transformed MI113 into a surfactin producer. Surfactin productivity in no. 3 medium by these surfactin-positive plasmids, pC112, pC113, and pC115, showed no significant variation (data not shown). Thus, we considered that the reduction in the size of the fragment was accomplished without the loss of any important part required for surfactin-production encoded on pC112.

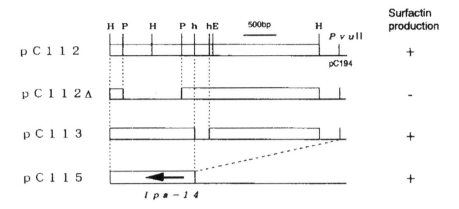

FIGURE 5.6 Physical map of the *lpa-14* region and localization of the *lpa-14* gene. The open boxes indicate the fragment cloned onto the vector plasmid pC194. The thin lines between the open boxes mean the part removed from pC112. The arrow indicates the orientation of the *lpa-14* open reading frame. +, Surfactin positive production; −, surfactin negative production in *B. subtilis* MI113. Restriction endonuclease sites are as follows: E, *EcoR*1; H, *Hind*III; h, *Hpa*l; P, *Pst*l.

5.3.2.2 Analysis of the Region Essential for Surfactin Production

The nucleotide sequence of the 1.1 kb region from *Hind*III to *Hpa*l in pC115 (Figure 5.6) was determined. Three overlapping fragments: *Hind*lll to *Hind*lll, *Pst*l to *Pst*l, and *Hind*III to *EcoR*I in pC113 from left to right in Figure 5.6 were subcloned into pUC19, respectively. The nucleotide sequence determined from these subclones is shown in Figure 5.7. The sequence of this region revealed the existence of a large open reading frame (ORF) consisting of 224 amino acids; this ORF showed high homology with that of *sfp* (18), which is known to be a regulatory gene for surfactin production (Figure 5.8).

B. subtilis RΔ1 is an iturin–surfactin nonproducer that was derived from *B. subtilis* RB14, a coproducer of the lipopeptides. A 10 kb insert comprising four *Hind*lll fragments originally isolated from *B. subtilis* RB14 could restore the productivity of both iturin A and surfactin in *B. subtilis* RΔ1 (12). The possibility that the region required for the production of iturin A and surfactin might be differently located on this fragment was first considered. The smallest plasmid, pC115, containing the ORF found earlier, and plasmid pC112Δ without the ORF, which had been constructed by removing the *Pst*l fragment in pC112 (Figure 5.6), were respectively introduced into RΔ1. While the clear recovery of the production of both lipopeptides was observed with pC115, no production of these substances by pC112Δ was observed (Table 5.6). From these results, the possibility described earlier was eliminated. It was confirmed that the gene product from this ORF was indispensable for the production of the lipopeptides iturin A and surfactin in *B. subtilis* RB14, and the gene containing the ORF was named *lpa-14* (lipopeptide antibiotic production of RB14).

5.3.2.3 Characteristics of *Bacillus pumilus* A-1 and *psf-1*

A gene named *psf-1*, which is necessary for surfactin production in *B. pumilus* A-1 as well as in MI113, was previously cloned and sequenced (30). As relatively high amino acid homology (41%) was observed between Psf-1 and Lpa-14 (Figure 5.8), the productivity of iturin A or other antifungal substance(s) of *Bacillus pumilus* A-1 was tested (Figures 5.9 and 5.10). The production of iturin A of strain A-1 was tested after cultivating it in no. 3 medium or spotting it on the center of a plate containing *F. oxysporum*, as described in Section 5.3.1.7. No antifungal zone in the bioassay and no peak of iturin A in HPLC analysis were detected from the cells or the culture broth of strain A-1, as shown in Figure 5.9B and Figure 5.10B. The gene *psf-1* was cloned

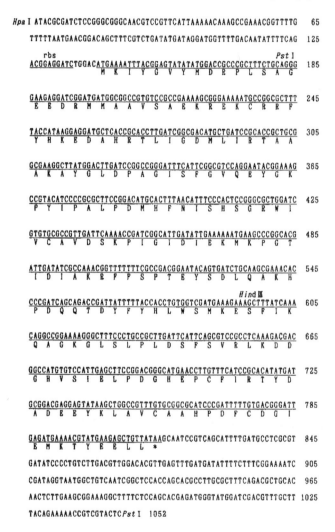

FIGURE 5.7   Nucleotide sequence of *lpa-14*. The *Hind*III, *Hpa*I, and *Pst*I sites are shown. The putative ribosome-binding site (rbs) is identified above the sequence and also underlined. The letters below the nucleotide sequence indicate the deduced amino acid sequence of LPA-14, using the single-letter notation. *, Stop codon. The nucleotide sequence data reported in this chapter will appear in the GSDB, DDBJ, EMBL, and NCBI nucleotide sequence databases with the following accession number D21876.

```
LPA14    1 :  MKIYGVYMDRPLSAGEEDRMMAAVSAEKREKCRRFYHKEDAHRTLIG
SFP      1 :  MKIYGIYMDRPLSQEENERFMTFISPEKREKCRRFYHKEDAHRTLLG
PSF-1    1 :  MKIFAIQLQPLDDKNARKQIEQLKPFVSFEKRAAAERFRFLIDARRTLLG
ORFX     1 :          IDRHVFNFLSSNVSKEKQQAFVRYVNVKDAYRSLLG

LPA14   48 :  DMLIRTAAAKAYGLDPAGISFGVQEYGKPYIPALPDMHFNISHSGRMIVC
SFP     48 :  DVLVRSVISRQYQLDKSDIRFSTQEYGKPCIPDLPDAHFNISHSGRMVIG
PSF-1   51 :  EVLIRHIIHEMYALPMEQIIFETEGNGKPVVRQIPSFHFNLSHSGDMVVG
ORFX    37 :  ELLIRKYLIQVLNIPNENILFRKNEYGKPFV--DFDIHFNISHSDEMVYC

LPA14   98 :  AVDSKPIGIDIEKMKPGTIDIAKRFFSPTEYSDLQAKHPDQQTDYFYHLM
SFP     98 :  AFDSQPIGIDIEKTKPISLEIAKRFFSKTEYSDLLAKDKDEQTDYFYHLM
PSF-1  101 :  AVDDAPVGIDIEEIKPIDLAIAERFFSADEYQDLLSQPAERQEAYFFHLM
ORFX    87 :  AISNHPVGIDIERISEIDIKIAEQFFHENEYIWLQSKAQNSQVSSFFELM

LPA14  148 :  SMKESFIKQAGKGLSLPLDSFSVRLKDDGHVSIELPDGHEPCFIRTYDAD
SFP    148 :  SMKESFIKQEGKGLSLPLDSFSVRLHQDGQVSIELPDSHSPCYIKTYEVD
PSF-1  151 :  SMKEAFIKLTGKGISYGLSSFTARLSEDGQATLRLPDHEAPCVVQTYSLD
ORFX   137 :  TIKESYIKAIGKGMYIPINSFWIDKNQTQTVIYKQNKKEPVTIYEPELFE

LPA14  198 :  EEYKLAVCAAHPDFCDGIEMKTYEELL
SFP    198 :  PGYKMAVCAAHPDFPEDITMVSYEELL
PSF-1  201 :  PAYQMAVCTRKPAAAEHVEILTCENMLSRLNNV
ORFX   187 :  -GYKCSCCSLFSSVTNLSITKLQVQELCNLFLD
```

FIGURE 5.8 Amino acid residue sequence comparison for Lpa-14 of *B. subtilis* RB14, Sfp of *B. subtilis* OKBI05, OrfX of *B. brevis*, and Psf-1 of *B. pumilus*. Regions of identity are shown as shaded areas. –, Amino acid deletion.

into pC194 from pTB522-2kb by *Hin*dIII and T4 ligase treatments, and the resulting plasmid, pCSFl, was introduced into *B. subtilis* RΔ1 by electroporation. The transformants exhibited the production of both iturin A and surfactin (Table 5.6), while the control strain RΔ1 with the vector plasmid pC194 showed neither production of the peptides (Table 5.6) nor inhibitory effect against *F. oxysporum* (Figure 5.9E). The recovery of the lipopeptides in the strain is interesting in that *psf-1* isolated from *B. pumilus*, a non-iturin producer, has the function of producing the two peptide antibiotics in *B. subtilis*, indicating the common features and uniqueness of peptide antibiotics of *Bacillus* species.

5.3.2.4 Productivity of Surfactin in MI113 and Transformants of RΔ1

Surfactin production of MI113(pC112) in no. 3 medium showed about twofold higher productivity than that of the parental strain, RB14 (31). The vector plasmid of pC112

TABLE 5.6
Recovery of Iturin A and Surfactin

Strain	Iturin A (ppm)	Surfactin (ppm)
B. subtilis RB14	140	335
B. subtilis RΔ1/pC194	0	0
B. subtilis RΔ1/pC112Δ	0	0
B. subtilis RΔ1/pC115	152	41
B. subtilis RΔ1/pCSF1	128	71

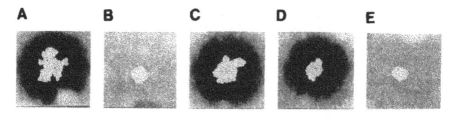

FIGURE 5.9 Iturin production assayed on agar plates containing *F. oxysporum* prepared as described in Section 5.3.1.7. (A) *B. subtilis* RB14; (B) *B. pumilus* A-1; (C) *B. subtilis* RΔI (pC115); (D) *B. subtilis* RΔ1 (pCSF1); (E) *B. subtilis* RΔ1 (pC194).

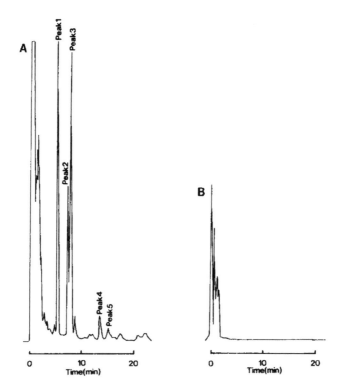

FIGURE 5.10 HPLC analysis of the production of iturin. The separation patterns of methanol extracts from (A) *B. subtilis* RB14 and (B) *B. pumilus* A-1, respectively. Peaks 1 to 5 correspond to the components of iturin as described in Chapter 2.

is pC194, which has a copy number of about 15 per chromosomal equivalent (32), and *lpa-14* was cloned onto pUB110, which has a much higher copy number of about 50 per chromosomal equivalent (33). However, the amounts of surfactin produced in the medium from both plasmids were almost the same (data not shown), and no gene dosage effect of this gene was observed.

The productivity of iturin A and surfactin was clearly restored in *B. subtilis* RΔ1 by plasmid pC115 or pCSF1, as shown in Table 5.6. Although the levels of iturin

A production by these transformants were almost the same of that of the parental strain, RB14, the recovery levels of surfactin in the transformants were only 10% to 20% that of RB14. The true reason for the latter poor restoration levels in R∆1 is not known. A regulatory gene for surfactin production in *B. subtilis* JH642, *sfp*, is reported when cloned onto a multicopy plasmid to repress the *lacZ* transcriptional fusion of the *srfA* operon, which encodes enzymes that catalyze surfactin synthesis (18). The poor recovery observed here may be partly explained by a similar effect of Lpa-14, because it has 72% homology with Sfp (Figure 5.8). Besides this, some factors involved in iturin A production by RB14 may be associated with the phenomenon. The effect of *lpa-14* on the production of lipopeptides will be clarified when the gene is introduced into a single defective strain of RB14, i.e., a nonproducer of either iturin A or surfactin. Isolation of such a mutant should be conducted in the next step.

5.3.2.5   Homology among *lpa-14*, *sfp*, *psf-1*, and *orfX*

A gene responsible for the production of iturin A and surfactin was analyzed in a derivative of the original wild strain as well as in a derivative of strain 168. The gene, *lpa-14*, permitted only surfactin production in the derivative of strain 168, and it had a high homology with *sfp*, a genetic locus responsible for surfactin production, which had been transferred from the original surfactin producer *B. subtilis* ATCC21332 to JH642, a derivative of strain 168. As *sfp* is located in the *B. subtilis* genome at a site closely linked to *srf A*, which encodes the surfactin synthetase enzymes (15), *ituA* (a putative gene encoding the iturin A synthetase enzyme[s]) might be coded just upstream of *lpa-14* in *B. subtilis* RB14 by the same analogy (34). Conversely, the possibility that *B. subtilis* MI113, a derivative of strain 168, might be a defective strain of *ituA* is also suggested. Analysis of the flanking region of *lpa-14* in *B. subtilis* RB14 and the shotgun cloning of *ituA* in MI113 in the presence of *lpa-14* should be conducted in the next step.

The deduced amino acid sequence of *sfp* had no significant homology to other known proteins, and the function of this protein is not yet known (17, 18). Only a relatively high homology (47%) was found between the sequence of *sfp* and an unknown open reading frame, *orfX* (35) in the upstream region of the *grs* operon for gramicidin S biosynthesis. Although enzyme activities such as ATP–PPi exchange (amino acid activating) activity, and thioesterase II, esterase, and surfactin degrading activity of the Psf-1 protein have not yet been detected (30), the relatively high homology among the *sfp*, *orfX*, *psf-1*, and *lpa-14* gene products is a clue to analyzing the role of those new gene products in the complex mechanism of peptide synthesis. If they share the same function(s) in their peptidic antibiotics, surfactin, gramicidin S, and iturin A, they are suggested to function in the upstream region in the synthetic metabolic pathways of the peptides because there is no apparent homology among the substances.

An analogous structure consisting of a seven-amino acid cyclic peptide linked to the fatty acid part is found in iturin A and surfactin, but the linkage modes and amino acid components are quite different in the two compounds (20, 21, 24). However, two synergistic effects of surfactin and iturin A have been reported: very extensive hemolysis induced by the presence of surfactin and iturin A (36), and increased antifungal activity of iturin A in the presence of surfactin as shown in Figure 2.16. The natural occurrence of these two lipopeptides in the same strain of *B. subtilis*, their synergistic effects, and the coregulation of their synthesis by the same gene, *lpa-14*, might provide useful information for developing more effective biological control agents.

5.4   A GENE, *LPA-14*, INVOLVED IN PRODUCTION OF AN IRON-CHELATING SIDEROPHORE (37)

The gene *lpa-14* showed high sequence homology with *sfp* of *B. subtilis* OKB105, *gsp* of *B. brevis* ATCC9999, *psf-1* of *B. pumilus* A-1, and *entD* of *E. coli* (27), an iron-chelating metabolite classified as a siderophore is synthesized and secreted in response to an iron-deficient medium by microorganisms. So far, 2,3-di-hydroxy-benzoylglycine (2,3-DHBG), shown in Figure 5.11, is the only known siderophore produced by the gram-positive *B. subtilis*. The gram-negative *E. coli* produces a siderophore, enterobactin, and this system has been well studied (38). The *E. coli entD* mutant was complemented for enterobactin production by the *sfp⁰* gene of the *B. subtilis* 168 strain (18), which differs from *sfp* by five base substitutions and one base insertion. These mutations change a 224-amino acid peptide in SFP to a 165-amino acid peptide of SFP⁰ (18). This indicates that there is functional interchangeability between *sfp⁰* and *entD*. Furthermore, we found that *lpa-14* showed 72% sequence homology with *sfp* (27). From these data the involvement of *lpa-14* in siderophore production in *B. subtilis* can be speculated.

5.4.1   Methods and Results

5.4.1.1   Growth in Iron-Deficient Medium

B. subtilis RΔ1 is an *lpa-14* defective mutant that was derived from RB14 (12). Plasmid pC115 has a fragment encoding *lpa-14*, and RΔ1/pC115 indicates a transformant of RΔ1 with pC115. The characteristics of these strains were described in detail (12, 27). RB14, RΔ1, and RΔ1/pC115 were precultured overnight in Luria-Bertani (LB) medium containing 10 g of Polypepton (Nihon Pharmaceutical Co., Tokyo), 5 g of yeast extract, and 5 g of sodium chloride per liter (pH 7.2), and washed with 1 mM Hepes buffer. Then, the suspension of each strain was used to inoculate synthetic iron-deficient medium (1 g of potassium sulfate, 3 g of dipotassium phosphate, 3 g of ammonium acetate, 20 g of sucrose, 1 g of citric acid, 2 mg of thiamin, 0.01%mg of copper(II) sulfate, 0.154 mg of manganese sulfate, 8.79 mg of zinc sulfate, and 0.81 g of magnesium sulfate per liter of water, pH 7.0) and the iron-deficient medium

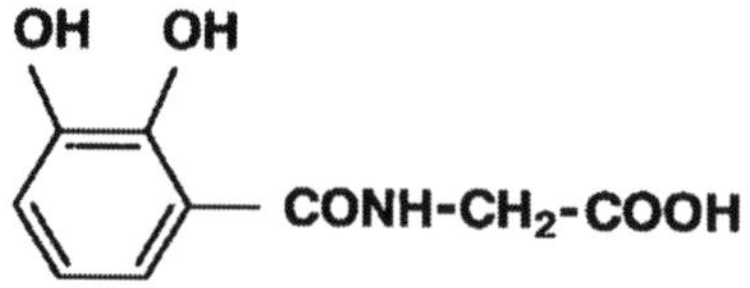

FIGURE 5.11   Structure of 2,3-dihydroxylbenzoylglycine.

containing an extremely low concentration of iron, which was estimated to be around 0.1 μM (39). The growth curves of RB14, RΔ1, and RΔ1/pC115 were monitored by measuring the OD at 660 nm (UV-1200, Shimadzu, Kyoto) as shown in Figure 5.12, and RB14 and RΔ1/pC115 grew well under iron-stressed conditions by producing a siderophore, but RΔ1 showed no growth. When 10 μM $FeCl_3$ was added to the iron-deficient medium, growth of RΔ1 was observed as shown in Figure 5.13, suggesting that *lpa-14* in RB14 is related to iron uptake. So far, 2,3-dihydroxybenzoylglycine

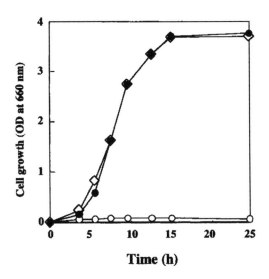

FIGURE 5.12   The growth of RB14 (◇), RΔI (o), and RΔ1/pCI15 (●) in iron-deficient medium.

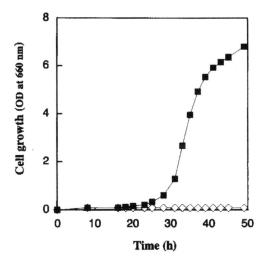

FIGURE 5.13   The growth of RΔ1 in iron-deficient medium (◇) and in iron-deficient medium fortified with 10 μM of $FeCl_3$ (■).

(2,3-DHBG), shown in Figure 5.11, is the only known siderophore produced by the gram-positive *B. subtilis*.

5.4.1.2   Purification and Identification of 2,3-DHBG Produced by RB14

Purification of 2,3-DHBG was conducted using a 5-day-old culture broth of RB14 in an iron-deficient medium using almost the same method as described in a previous paper (37), except that crystallization of the siderophore in hot water was repeated several times in this study. The purified material was used for the following analyses. Spectrometric analysis of the purified product was performed using fast atom bombardment mass spectrometers (FAB-MS, models JMX-DX303 and JMA-DA5100) and showed a peak at the molecular weight of 2,3-DHBG ($212[M +H]^+$).

^{1}H-NMR (JEOL JNM-EX-90) analysis of the product demonstrated a peak at 12.26 ppm for COOH, at 9.1 ppm for ArCONHR-(Ar:phenyl), at 6.60–7.35 ppm for C_6H_3, and at 4.06 ppm for CH_2COOH. The Fourier transform infrared spectrum chart showed absorption at 1739 cm^{-1} resulting from the C–O stretching mode for COOH, at 3352 cm^{-1} from the N–H stretching mode for CONH, at 1648 cm^{-1} from the stretching mode for CONH, and at 738 cm^{-1} from C6H3. The melting point was determined to be between 200°C and 206°C. The R_f value as determined by TLC (PSC-Fertigplatten Kieselgel 60 F254, Merck & Co., Inc., New Jersey, USA; solvent system [volume ratio]: t-butyl alcohol, 10; methyl ethyl ketone, 10; water, 5; diethyl amine, 1) was 0.36, while previously a value of 0.39 was reported (39). By comparing the data obtained here with those in a previous paper (39), the purified substance was judged to be 2,3-DHBG.

5.4.1.3   Quantitative Analyses of 2,3-DHBG by HPLC

To determine the concentration of 2,3-DHBG produced by RB14, an HPLC system was established by modification of a previously reported system (40). After the purification of 2,3-DHBG from the culture broth as mentioned earlier, the sample extracted into ethyl acetate was dissolved in 100% methanol and then filtered through a 0.2 μm PTFE membrane (JP020, Advantec Ltd., Tokyo). The filtrate was subjected to HPLC with the elution solvent consisting of a mixture of methanol–0.1% phosphoric acid (1:1 [v/v]) (41). The elution time of 2,3-DHBG was 3.5 min, and the concentration produced by RB14 grown in an iron-deficient medium for 5 days was determined to be approximately 70 ppm. As RΔ1 showed almost no growth in the iron-deficient medium after 5 days, no peak corresponding to 2,3-DHBG was observed in the HPLC trace of the filtrate derived from this culture. The fact that RΔ1/pC115 also produced about 70 ppm 2,3-DHBG indicates that *lpa-14* is associated with the production of the siderophore 2,3-DHBG.

5.4.1.4   Growth of RΔ1 in Iron-Deficient Medium Containing
Chemically Synthesized 2,3-DHBG

To clarify that the lack of growth of RΔ1 in the iron-deficient medium is related to the defective production of 2,3-DHBG, 2,3-DHBG was chemically synthesized and added to the iron-deficient medium. Although the synthesis of 2,3-DHBG was reported in previous papers (39), a simpler procedure to achieve higher purity and yields of 2,3-DHBG than previously report (39) was developed because 2,3-DHBG of high purity is critical for observing the effects on the growth of RΔ1. 2,3-Dihydroxybenzoic acid (2,3-DHB) (5 mmol), glycine ethyl ester hydrochloride (6 mmol), and water soluble 1-ethyl-3(3-dimethyl-aminopropyl) carbodiimide hydrochloride (6 mmol), a dehydrant, were dissolved in 25 ml of chloroform. Then, 0.83 ml of triethylamine (6 mmol) was slowly added, and the mixture was stirred overnight at room temperature under a N_2 atmosphere. The reaction mixture was dried by evaporation, and the residue was acidified with 2 N HCl (pH 3) and then extracted three times with ethyl acetate. The organic phase was washed with H_2O and 0.5 N $NaHCO_3$. The crude 2,3-DHBG ethyl ester thus obtained was eluted with ethyl acetate:n-hexane (1:1[v/v]) as an elution solvent through an EDTA-treated silica gel column. Recrystallization of the eluted solution produced a crystal of 2,3-DHBG ethyl ester in 25% yield. The crystal was dissolved in 2 N NaOH and stirred at room temperature for 1 h. The reaction mixture was acidified with dilute HCl (pH 4), extracted three times with ethyl acetate, and washed with 10% citrate and H_2O. After removing the solvent, the residue was recrystallized from diethyl ether–hexane. Several repeats of the recrystallization process gave pure white 2,3-DHBG in 10% yield. The structure and purity of the synthesized sample were confirmed using the previously mentioned methods. When chemically synthesized 2,3-DHBG was added to the iron-deficient medium at 10 ppm, 50 ppm, 100 ppm, and 200 ppm. The growth rate of RΔ1 increased with each increase in the concentration of 2,3-DHBG as shown in Figure 5.14. The growth of RB14, which produced about 70 ppm 2,3-DHBG, was between that of RΔ1 in medium with 50 ppm and 100 ppm of added 2,3-DHBG, indicating that the growth of RΔ1 in the presence of 2,3-DHBG was almost equivalent to that of RB14 in the iron-deficient medium. This confirms that the *lpa-14* gene is involved in the synthesis of 2,3-DHBG.

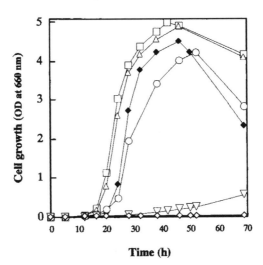

FIGURE 5.14   Growth of RB14 (◆) in iron-deficient medium and RΔ1 in iron-deficient medium containing 2,3-DHBG. 2,3-DHBG: 0 ppm (◇), 10 ppm (▽), 50 ppm (○), 100 ppm (Δ), 200 ppm (□).

This study showed that *lpa-14* functions not only in the synthesis of the lipopeptide antibiotics, iturin A, and surfactin, but also in siderophore synthesis in the RB14 strain.

Bacterial siderophores are known to help plant growth by supplying iron via iron-chelation termed the plant-growth-promoting rhizobacteria (PGPR) effect (42–44). The fact that *B. subtilis* RB14 produced not only antibiotics that suppress plant pathogens but also a siderophore, and that these products are commonly regulated by the gene *lpa-14*, indicate the possibility of enhancing biocontrol using such agents if *lpa-14* is appropriately manipulated.

5.5 CHARACTERISTICS OF THE GENE *LPA-8* (LIPOPEPTIDE ANTIBIOTIC-PRODUCTION OF YB8) RESPONSIBLE FOR BIOSYNTHESIS OF PLIPASTATIN AND SURFACTIN (45)

B. subtilis YB8 is one of the *B. subtilis* strains isolated for biocontrol as described in Chapter 2. The population of YB8 was present one to two orders higher than other isolated *Bacillus* in the compost. The suppressive spectrum of YB8 against plant pathogens was slightly different from that of strain NB22, and no iturin was detected in the culture supernatant. The detail of YB8 was analyzed.

5.5.1   MATERIALS AND METHODS

5.5.1.1   Bacteria, Plasmids, and Media

The bacterial strains and plasmids used in this work are listed in Table 5.7. L medium was shown in Chapter 3. ACS medium (48) contained sucrose (100 g), citric acid (11.7 g), Na_2SO_4 (4 g), yeast extract (5 g), $(NH_4)_2HPO_4$ (4.2 g), KCl (0.76 g), $MgCl_2 \cdot 6H_2O$ (0.420 g), $ZnCl_2$ (10.4 mg), $FeCl_3 \cdot 6H_2O$ (24.5 mg), $MnCl_2 \cdot 4H_2O$ (18.1 mg) in 1 liter, and was adjusted to pH 6.9 with NH_4OH, and was used for the production of the lipopeptides.

To select or cultivate the transformants, the antibiotics ampicillin (Ap), chloramphenicol (Cm), erythromycin (Em), kanamycin (Km), or tetracycline (Tc) were added at concentrations of 50, 5, 10, 10, and 20 μg/ml, respectively.

5.5.1.2   Purification of Surfactin and an Antifungal Substance

B. subtilis YB8 was cultivated in ACS medium for 5 days at 30°C. The acid precipitate of the culture supernatant was collected and extracted with methanol. The extract was separated by HPLC as previously (27), and a clear peak of surfactin was eluted at the same position as an authentic purchased sample (Wako, Osaka, Japan). The infrared (IR) spectrum, amino acid analysis, and molecular mass determination by FAB-MS of the peak sample also confirmed that the substance was surfactin (data not shown). The structure of surfactin (47) is shown in Figure 5.15.

Purification of the antifungal substance was carried out as follows by modifying the methods developed for purification of fengycin (48) or plipastatin (49). The same acid precipitate described earlier made from 4 liters of the culture supernatant was

TABLE 5.7
Bacterial Strains and Plasmids Used

Strain or Plasmid	Characteristics	Source or Reference
	Escherichia coli strain	
JM109	*relA1 supE44 endA1 hsdR17 gyrA96 mcrA mcrB⁺ thi Δ(lac-proAB)/F' [traD36 proAB⁺ lacI^q lacZΔM15]*	(28)
	Bacillus subtilis strains	
MI113	*arg-15 trpC2 hsmM hsrM*	(27)
MI113 (pTB81)	SF⁺ Tc^r	This work
MI113 (pC81)	SF⁺ Cm^r	This work
MI113 (pUB8)	SF⁺ Km^r	This work
YB8	SF⁺ PL⁺	(48)
YB8 (pE194)	SF⁺ PL⁺ Em^r	This work
YB8(pECΔ1)	SF⁺ PL⁺ Cm^r Em^r	This work
YΔI	SF⁻ PL⁻ Cm^r Em^r	This work
YΔI (pUB8)	SF⁺ PL⁺ Cm^r Em^r Km^r	This work
E. coli plasmids		
pUC19	Ap^r, *lac* promoter	(28)
pACΔ1	Ap^r Cm^r, *lpa-8* cloned in pUC19 and inactivated by Cm^r insertion	This work
	B. subtilis plasmids	
pC194	Cm^r, 2.9 kb	(21)
pE194	Em^r, 3.7 kb	(52)
pTB522	Tc^r, 10.5 kb	(59)
pUB110	Km^r, 4.5 kb	(33)
pTB81	SF⁺ Tc^r, 25 kb	This work
pC81	SF⁺ Cm^r, subcloned from pTB81	This work
pC81AP	SF⁺ Cm^r, subcloned from pC81	This work
pUB8	SF⁺ Km^r, *lpa-8* cloned into pUB110 from pC81AP	This work
pECΔ1	SF⁻, pACΔ1 and pE194 ligated at *PstI* sites	This work

Notes: SF, surfactin production; PL, plipastatin B1 production; Ap^r, ampicillin resistant; Cm^r, chloramphenicol resistant; Em^r, erythromycin resistant; Km^r, kanamycin resistant; Tc^r, tetracycline resistant.

extracted with 95% ethanol. The extract was mixed with 10% charcoal (w/v), stirred for 60 min, then filtered. The charcoal was stirred with 1 liter of chloroform–methanol–H_2O (65:25:4, by vol) for 60 min and then the mixture was filtered. The eluate was concentrated and dissolved in 4 vol of propanol (v/w). This solution was loaded on a propanol-filled column of silica gel (Silica gel 60; Merck) and was successively eluted with 1 column volume of propanol, 2 column volumes of 90% propanol, and

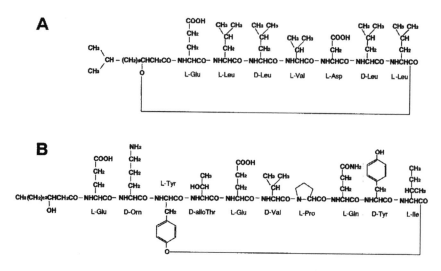

FIGURE 5.15   Structures of (A) surfactin and (B) plipastatin B1.

2.5 column volumes of 80% propanol. These fractions were bioassayed on an agar plate containing a phytopathogenic fungus, *F. oxysporum* f. sp. *lycopersici* race JI SUFI 19, which causes tomato crown and root rot (50).

The active eluate from the 80% propanol elution was further purified by reversed-phase preparative HPLC using a Prep-ODS column (GL Sciences, Tokyo, Japan; 2 cm diameter × 25 cm, flow rate 10 ml/min, detection at 205 nm) with a 3:4 (v/v) mixture of acetonitrile and 5 mM ammonium acetate. The main peak with the highest antifungal activity was passed through a Sephadex LH-20 (Pharmacia) column with 80% methanol; 3 mg of the purified antifungal substance was obtained.

5.5.1.3   HPLC Analysis of Plipastatin B1 and Surfactin

Plipastatin B1 and surfactin (Figure 5.15) were extracted as follows: after 5 days' cultivation of the bacterium in 40 ml of ACS medium, the culture was centrifuged at 10,000 × *g* for 10 min, and the supernatant was acidified to pH 2 with 12 N HCl. Then, the precipitate was collected by centrifugation and was extracted with methanol or 95% ethanol. The extracted solution was centrifuged at 10,000 × *g* for 10 min, and the supernatant was filtered through a 0.2 µm PTFE (polytetrafluoroethylene type) membrane (Advantec, Tokyo, Japan).

Plipastatin B1 was detected and quantified by reversed-phase HPLC as follows: the aforementioned filtrate was injected into an HPLC column (Inertsil ODS-2, 4.6 in diameter × 250 mm, GL Sciences). The column was eluted at a flow rate of 1.0 ml/min with acetonitrile–ammonium acetate (10 mM) 1:1 (v/v). The elution pattern was monitored at 205 nm. The purified sample was used as a standard.

Surfactin was analyzed by reversed-phase HPLC on the same column used for the identification of plipastatin B1. The column was eluted at a flow rate of 1.5 ml/min with acetonitrile–trifluoroacetic acid (3.8 mM; 80:20, v/v) and monitored at 205 nm, as previously described (27). Authentic surfactin was used as a standard sample.

5.5.1.4 Transformation of *B. subtilis*

Competent cells of *B. subtilis* MI113 grown in the transformation medium that contained the required amino acids (50 and 5μg/ml for the first and second transformation media, respectively) were prepared by the method (51). Transformation of wild-type *B. subtilis* YB8 and its derivative with plasmid DNA was performed by electroporation, as described in Section 5.1. Cells of a 150 ml late-exponential growth phase culture of *B. subtilis* were harvested and washed three times with cold 1 mM Hepes (pH 7.0) and were resuspended in a final volume of 400 μl. Aliquots (40 μl) were mixed with 5 μl of DNA and 40 μl of 50% PEG 6000 in distilled water. From this mixture, 20 μl was transferred to a chilled cuvette (0.15 cm electrode gap), and a single pulse of 16.0 kV/cm (2.4 kV, 4 kQ, 2 μF) was applied. The cell suspension was put into 0.5 ml of L medium and allowed to stand for 3 h at 37°C. Then, the cell culture was spread onto antibiotic-containing selective plates, which were then incubated overnight at 37°C.

5.5.1.5 Preparation of Plasmid DNA

Clones were screened using the rapid alkaline lysis procedure (7) with modifications as previously (27). For large-scale preparation, plasmids extracted by the alkaline lysis method were further purified by CsCl–ethidium bromide density gradient ultracentrifugation (28). Restriction endonuclease and T4 DNA ligase treatments and electrophoretic analysis of DNA in agarose gels were done by following standard methods (28). High-molecular-weight ligation products were recovered from an agarose gel after electrophoresis by the glass-binding method using the Geneclean II kit (Bio101, La Jolla, California, USA) and were used in the cloning experiment.

5.5.1.6 Detection of Biosurfactant Activity of the Lipopeptides

The presence of lipopeptides was detected on tributyrin L agar plates (TB plates; containing 20 μl of tributyrin spread onto an L agar plate with a glass rod) by observing biosurfactant activities of the colonies. When a halo appeared on the film of tributyrin after cells were placed on the TB plate, substances with biosurfactant activity were judged to have been produced.

5.5.1.7 Fungal Growth Inhibition by Bacteria or Cell-Free Culture Broth

A phytopathogenic fungus, *F. oxysporum* f. sp. *lycopersici* race JI SUFl 19 (50), was grown on potato dextrose agar medium at 30°C for 5 days and suspended in sterile distilled water. A portion of this suspension was mixed into L agar medium before solidification. Bacterial colonies or cell-free culture broth were spotted on this plate. Clear zones of growth inhibition of the fungus were observed when antibiotic(s) were produced by the bacteria.

5.5.1.8 Cloning of a Lipopeptide Antibiotic Gene (*lpa*)

Chromosomal DNA of *B. subtilis* YBS was partially digested with *Hin*dIII and ligated with *Hin*dIII-digested pTB522. Competent cells of *B. subtilis* MI113 were transformed with high-molecular-weight ligation products. Tetracycline-resistant colonies selected on plates were transferred with toothpicks onto TB plates containing 20 μg/ml of tetracycline. Colonies around which clear halos formed on TB plates were selected as candidates and were further analyzed.

5.5.1.9 DNA Sequencing and Analysis

Double-stranded DNA cloned in pUC19 was sequenced using a Pharmacia A.L.F. DNA sequencer, the dideoxy-chain-termination method (29), and the T7 sequencing kit (Pharmacia).

5.5.2 Results

5.5.2.1 Isolation and Identification of the Lipopeptides

The hydrolysate of the purified substance from *B. subtilis* YB8, analyzed using an amino acid analyzer, was comprised of Val (1), Thr (1), Glu (3), Pro (1), Ile (1), Tyr (2), and Orn (1) residues. The protonated molecular ion peak of this compound was displayed at *m/z* 1491.9 by FAB-MS analysis. The IR spectrum of the purified antifungal substance molded in KBr showed the characteristic absorption of peptide bonds (1654 and 1543 cm^{-1}) and a lactone carbonyl absorption (1753 cm^{-1}) of plipastatin, as previously (49). From these results, the antifungal substance was determined to be plipastatin B1, an inhibitor of phospholipase A2 (49), and its structure is shown in Figure 5.15B. *B. subtilis* YB8 also produces the potent biosurfactant surfactin and thus is the first strain reported to produce both plipastatin B1 and surfactin.

5.5.2.2 Cloning of a Gene Required for the Production of Lipopeptide Antibiotics

A clear halo formed almost instantly when purified plipastatin B1 or surfactin was put on a TB plate. *B. subtilis* YB8 also made a clear halo around its colony within 3 h of incubation at 37°C after replica plating, but *B. subtilis* MI113, a derivative of strain 168 that does not produce lipopeptides, did not make any halo. Thus, strain MI113 was used as a host for shotgun cloning of the gene(s) responsible for the production of plipastatin B1 and surfactin. Among 250 transformants obtained after transformation with the ligation mixture of the chromosomal DNA of strain YB8 and plasmid pTB522, three positive colonies were obtained on TB media. Plasmids were extracted from these colonies and analyzed by agarose gel electrophoresis. Sizes of plasmids with insertion were about 15, 40, and 50 kb, and the plasmids were designated as pTB81, pTB82, and TB83, respectively. The high cloning efficiency observed in this shotgun cloning was attributed to the partial enrichment of high-molecular-weight DNA as described earlier.

5.5.2.3 Identification of the Substances Produced by *B. subtilis* MI113 Containing Plasmid pTB8 I, pTB82, or pTB83

Methanol extracts prepared from cultures of *B. subtilis* MI113 (pTB81), MI113 (pTB82), and MI113 (pTB83) were analyzed by HPLC. Surfactin was detected in the three culture extracts. However, no plipastatin B1 was detected by HPLC analysis, and the three transformants formed no inhibitory zones on a plate containing *F. oxysporum* (data not shown). Therefore, it appeared that the transformants only produced surfactin.

5.5.2.4 Subcloning of pTB81

As the presence of each of the recombinant plasmids in strain MI113 led to the same phenotype, the smallest plasmid, pTB81, was chosen for further analysis. The plasmid was further reduced in size by replacing the vector plasmid pTB522 (10.5 kb) by pC194 (2.9 kb) by *Hin*dIII digestion and ligation. Plasmids of the chloramphenicol-resistant and halo-forming colonies on TB plates were selected; the smallest plasmid, designated pC81, contained three *Hin*dIII fragments (2, 0.7, and 0.3 kb). The plasmid was further reduced in size, forming pC81AP, which contained a 1.5 kb *Apa*I–*Pvu*II fragment that caused the production of surfactin. The smaller 1.3 kb *Stu*I–*Pvu*II fragment of pC81AP subcloned into vector pUB110 was designated as pUB8, which also led to surfactin activity.

5.5.2.5 Nucleotide Sequence Analysis of the Essential Region for TB-Plate-Positive Phenotype

The DNA of pC81AP essential for the production of surfactin was determined by sequencing nine subclones containing some part of this region. Figure 5.16 shows the determined nucleotide sequence. One ORF consisting of 224 amino acids, preceded by a sequence (rbs) homologous to the ribosome-binding sites of *B. subtilis* genes was found. The upstream region of this ORF had complete identity with that of *sfp*, as reported by Nakano et al. (18), who determined the transcriptional start site in this identical region and detected the promoter activity using the *sfp–lacZ* fusion method. This suggested that the identical promoter sequence contained in this region was used for expression of this ORF.

5.5.2.6 Construction of an Integration Plasmid pECΔ1

An integration plasmid was constructed to inactivate this ORF and to determine the function of this gene in the original host strain *B. subtilis* YB8. The 1.3 kb *Stu*I–*Pvu*II fragment of pC81AP was cloned into the *Hinc*II site of the *E. coli* plasmid pUC19 by blunt-end ligation. Two *Eco*RV fragments within the region were replaced by a *Dra*I fragment carrying the chloramphenicol-resistance gene of pC194. This newly constructed plasmid, designated pACΔ1, was ligated with pE194 at the *Pst*I sites of the plasmids to produce pECΔ1. Plasmid pE194 replicates at 30°C, but at temperatures above 45°C, replication ceases and the plasmid is lost from *B. subtilis* in the absence of selective pressure (52). pECΔ1 was introduced into *B. subtilis* YB8 by electroporation. Erythromycin-resistant transformants were selected at 30°C. Vector plasmid pE194 introduced into strain YB8 was used as a control.

5.5.2.7 Inactivation and Restoration of Lipopeptide-Production Activity in *B. subtilis* YB8

After cultivating one transformant in L medium at 30°C for 12 h, the culture was plated on L agar plates containing erythromycin, followed by incubation at 48°C to integrate plasmid pECΔ1 into the chromosome of strain YB8. Many Em^r colonies were obtained from the culture with pECΔ1, but no Em^r colonies emerged from the culture with pE194. These results indicate that the presence of the cloned fragment on the plasmid was necessary for chromosomal integration,

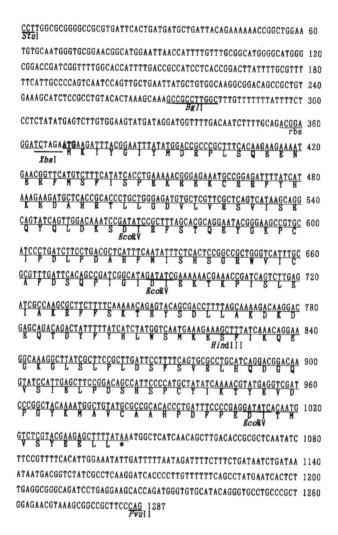

FIGURE 5.16 Nucleotide sequence and deduced amino acid sequence of *lpa-8* gene of *B. subtilis* YB8. The *Stu*I, *Bgl*I, *Xba*I, *Eco*RV, *Hind*III, and *Pvu*II sites are shown. The putative ribosome-binding site (*rbs*) is identified above the sequence and is underlined. The single-letter notation below the nucleotide sequence indicates the deduced amino acid sequence of LPA-8. *, Stop codon. The nucleotide sequence data reported here is available from the DDBJ, EMBL, and GenBank Nucleotide Sequence Databases under accession no. D50562.

and the random integration of the vector plasmid at high temperature occurred infrequently or not at all.

Among 230 Em^r colonies, one, designated YΔ1, which gave negative results in the TB plate assay, was used in the following experiments. HPLC analyses of a culture of strain YΔ1 showed that strain YΔ1 produced neither surfactin nor plipastatin B1, as shown in Figure 5.17B. Strain YΔ1 was transformed with pUB8, which carried the smallest fragment that contained the ORF; the resulting transformant strain YΔI

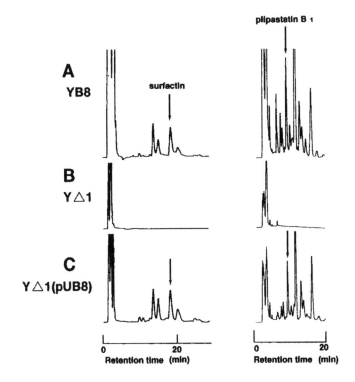

FIGURE 5.17 HPLC analysis of the production of surfactin and plipastatin B1 showing the separation patterns of methanol extracts of the acid–precipitate of the culture broth of (A) *B. subtilis* YB8, (B) *B. subtilis* YΔ1, and (C) *B. subtilis* YΔ1(pUB8). The arrows indicate the retention times of authentic surfactin and plipastatin B1.

(pUB8) produced both lipopeptides, as shown in Figure 5.17C. When the cell-free culture medium of strain YΔI (pUB8) was used in the antifungal activity assay, clear recovery of the activity was demonstrated, while that of strain YΔI showed no antifungal activity.

From these results my laboratory confirmed that the gene product from this ORF was indispensable for the production of the lipopeptides surfactin and plipastatin B1 in *B. subtilis* YB8 and named the gene *lpa-8* (lipopeptide antibiotic production of YB).

5.5.2.8 Homology between *lpa-8*, *sfp*, *lpa-14*, *psf-1*, *gsp*, *entD*, and *hetI*

The deduced amino acid sequence of Lpa-8 is similar to that of the following genes: *sfp* of *B. subtilis* (18), *lpa-14* of *B. subtilis* RB14 (27), *psf-1* of *B. pumilus* (30), *gsp* of *B. brevis* (53), *entD* of *E. coli* (54), and *hetI* of *Anabaena* sp. (46).

Lpa-8 and Sfp proteins were identical except for two amino acids: serine-22 and cysteine-97 in Lpa-8 are threonine and glycine in Sfp, respectively. A comparison between Lpa-8 and Lpa-14, Psf-1, Gsp, EntD, and HetI revealed 72%, 46%, 33%, 17%, and 24% similarity, respectively. Extensive regions of identity common to all proteins compared are shown in Figure 5.18.

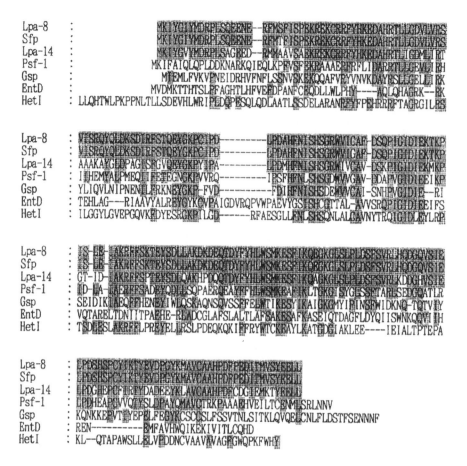

FIGURE 5.18 Amino acid residue sequence comparison of Lpa-8 of *B. subtilis* YB8, Sfp of *B. subtilis* OKB105, Lpa-14 of *B. subtilis* RB14, Psf-1 of *B. pumilus*, Gsp of *B. brevis*, EntD of *E. coli*, and HetI of *Anabaena* sp. Regions of identity are shown as shaded areas. Hyphens represent gaps introduced to optimize sequence alignments.

5.5.3 Discussion

B. subtilis YB8, isolated from a suppressive compost, was found to be a plipastatin B1 and surfactin producer. A gene, *lpa-8*, was cloned from strain YB8, and the nucleotide sequence was determined. When the gene was insertionally inactivated in strain YB8, the lipopeptides were no longer produced; however, the introduced intact *lpa-8* could restore the ability to produce the lipopeptides. From these results, *lpa-8* was identified as a gene involved in the production of the antibiotics plipastatin B1 and surfactin. This is the first identification of a gene that is involved in the production of the lipopeptide antibiotic plipastatin B1.

Lipopeptides represent a unique class of bioactive microbial secondary metabolites, and many of them show attractive therapeutical and biotechnological properties. Surfactin was isolated as an inhibitor of fibrin clotting and is known to lyse erythrocytes (20), and it is a potent surface-active reagent (13) and antibiotic (14).

Biochemical analyses of the surfactin synthetase multienzyme system (55, 56) and genetic analyses of the biosynthesis genes have recently been carried out (17, 18, 34).

Plipastatin B1 was originally isolated as a new inhibitor of phospholipase A2 (49), and later its antifungal activity against the growth of *Alternaria mali* (apple leaf spot pathogen), *Botrytis cinerea* (gray mold pathogen), and *Pyricularia oryzae* (rice blast pathogen), and considerable high protective value against *Helminthosporium* leaf spot was reported (57). As synergistic effects of surfactin and iturin A in extensive hemolysis induced by the two lipopeptides (36) and in increased antifungal activity of iturin A in the presence of surfactin (58) have been reported, a synergistic effect between plipastatin B1 and surfactin might exist. The deduced amino acid sequence of *lpa-8* showed sequence similarity to that of *sfp* (18), *lpa-14* (27), *psf-1* (30), *gsp* (53), and *entD* (54). Sfp, Lpa-14, and Psf-1 are required for the production of surfactin in *B. subtilis* or *B. pumilus*. Lpa-14 is also required for the production of iturin A (27). Although the function of Gsp is not yet known, Gsp is encoded by the *gsp* gene, which is located at the 5 end of the operon (*grs* operon) that encodes the proteins involved in the biosynthesis of the cyclic lipopeptide gramicidin S, and Gsp has functional homology with Sfp in producing surfactin (53). EntD is required for production of the peptidic siderophore enterobactin, which is a cyclic trimer of 2,3-dihydroxybenzoyl serine. The activity of Sfp is interchangeable with that of EntD (54). Since these proteins have functional homology and sequence similarity, they might represent a new class of proteins involved in (cyclic) peptide biosynthesis or secretion (54). As plipastatin Bl is also a cyclic peptide (Figure 5.15B) and requires this kind of protein, my data also seem to support the hypothesis.

Although significant sequence similarities were found among these proteins, their functions were not yet elucidated. Enzyme activities such as ATP–PP exchange (amino acid activating) activity, and thioesterase II-, esterase-, and surfactin-degrading activity of the Psf-1 protein could not be detected (30). No substance requiring *Het*l for its synthesis has been detected (46), and direct involvement of Gsp in gramicidin synthesis has not been shown. However, the requirement of Lpa-8 for the synthesis of the cyclic peptides surfactin and plipastatin B1 is obviously clear. Furthermore, some peaks in Figure 5.17A other than those of the lipopeptides corresponded also to the peptidic substances (data not shown), and this relationship between the gene and the substances will aid us in elucidating the function(s) of the protein. Mutational analysis of Lpa-8 and the effect on the production of the cyclic peptides are next steps for further analysis.

At present, we do not know the function of the cloned gene; however, since the linkages and amino acid components are quite different in surfactin and plipastatin Bl (Figure 5.15), the following three hypotheses can be proposed: (i) Lpa-8 is a regulatory protein that acts as a transcription initiation factor. (ii) Lpa-8 is involved in the biosynthesis of these compounds and is required for a common process in the biosynthesis, such as amino acid transfer, or (iii) Lpa-8 is required for peptide secretion or for the transport system of the peptides. Further experimental analyses of this new class of proteins might reveal a new mechanism of peptide secretion or regulation of the secondary metabolites.

The gene, *lpa-8*, led to surfactin production in the derivative of strain 168, and it was almost identical with *sfp*, a genetic locus responsible for surfactin production, which had been transferred from the original surfactin producer *B. subtilis* ATCC 21332 to strain JH642, a derivative of strain 168. As *sfp* is located in the *B. subtilis* genome at a site closely linked to *srfA*, which encodes the surfactin synthetase enzymes (34), *srfA* or *pliA* (a putative gene encoding the plipastatin B1 synthetase enzyme[s]) might be located in the vicinity of *lpa-8* in *B. subtilis* YB8 by the same analogy. Conversely, the possibility that *B. subtilis* MI113, a derivative of strain 168, might be a strain defective in *pliA* is also suggested. Pulsed-field gradient gel electrophoresis analysis of the chromosomal DNA fragment containing the flanking region of *lpa-8* in *B. subtilis* YB8 and the cloning of *pliA* in strain MI113 in the presence of *lpa-8* are the next research targets.

REFERENCES

1. Hiraoka, H., Ano, T., and Shoda, M. Rapid transformation of *Bacillus subtilis* using KCl-treatment. *J. Ferment. Bioeng.*, 74, 241–243 (1992).
2. Ano, T., Kobayashi, A., and Shoda, M. Transformation of *Bacillus subtilis* with the treatment by alkali cations. *Biotechnol. Lett.*, 12, 99–104 (1990).
3. Matsuno, Y., Hitomi, T., Ano, T., and Shoda, M. Transformation of *Bacillus subtilis*, antifungal-antibiotic iturin producers with isolated antibiotic resistance plasmids. *J. Gen. Appl. Microbiol.*, 38, 13–21 (1992).
4. Matsuno, Y., Ano, T., and Shoda, M. High-efficient transformation of *Bacillus subtilis* NB22, antifungal antibiotic iturin producer, by electroporation. *J. Ferment. Bioeng.*, 73, 261–264 (1992).
5. Canosi, U., Morelli, G., and Trautner, T. A. The relationship between molecular structure and transformation efficiency of some *S. aureus* plasmids isolated from *B. subtilis*. *Mol. Gen. Genet.*, 166, 259–267 (1978).
6. Chang, S., and Cohen, S. N. High frequency transformation of *Bacillus subtilis* protoplasts by plasmid DNA. *Mol. Gen. Genet.*, 168, 111–115 (1979).
7. Birnboim, H. C., and Doly, J. A rapid alkaline extraction procedure for screening recombinant plasmid DNA. *Nucleic Acids Res.*, 7, 1513–1523 (1979).
8. Svarachorn, A., Shinmyo, A., Tsuchido, T., and Takano, M. Autolysis of *Bacillus subtilis* induced by monovalent cations. *Appl. Microbiol. Biotechnol.*, 30, 299–304 (1989).
9. Fein, J. E., and Rogers, H. J. Autolytic enzyme-deficient mutants of *Bacillus subtilis* 168. *J. Bacteriol.*, 127, 1427–1442 (1976).
10. Dower, W. J., Miller, J. F., and Ragsdale, W. High efficiency transformation of *E. coli* by high voltage electroporation. *Nucleic Acids Res.*, 16, 6127–6145 (1988).
11. Tamiya, E., Nakajima, Y., Kamioka, H., Suzuki, M., and Karube, I. DNA cleavage based on high voltage electric pulse. *FEBS Lett.*, 234, 357–361 (1988).
12. Hiraoka, H., Ano, T., and Shoda, M. Molecular cloning of a gene responsible for the biosynthesis of the lipopeptide antibiotics iturin and surfactin. *J. Ferment. Bioeng.*, 74, 323–326 (1992).
13. Cooper, D. G., MacDonald, C. R., Duff, S. J. B., and Kosaric, N. Enhanced production of surfactin from *Bacillus subtilis* by continuous product removal and metal cation additions. *Appl. Environ. Microbiol.*, 42, 408–412 (1981).
14. Bernheimer, A. W., and Avigad, L. S. Nature and properties of a cytolytic agent produced by *Bacillus subtilis*. *J. Gen. Microbiol.*, 61, 361–369 (1970).
15. Nakano, M. M., Marahiel, M. A., and Zuber, P. Identification of a genetic locus required for biosynthesis of the lipopeptide antibiotic surfactin in *Bacillus subtilis*. *J. Bacteriol.*, 170, 5662–5668 (1988).

16. Nakano, M. M., and Zuber, P. Cloning and characterization of *srfB*, a regulatory gene involved in surfactin and competence in *Bacillus subtilis. J. Bacteriol.*, 171, 5347–5353 (1989).
17. Nakano, M. M., and Zuber, P. Molecular biology of antibiotic production in bacillus. *Crit. Rev. Biotechnol.*, 10, 223–240 (1990).
18. Nakano, M. M., Corbell, N., Besson, J., and Zuber, P. Isolation and characterization of *sfp*: a gene that functions in the production of the lipopeptide biosurfactant, surfactin, in *Bacillus subtilis. Mol. Gen. Genet.*, 232, 313–321 (1992).
19. Himeno, T., Imanaka, T., and Aiba, S. Effect of in vitro DNA rearrangement in the NH$_2$-terminal region of the penicillinase gene from *Bacillus licheniformis* on the mode of expression in *Bacillus subtilis. J. Gen. Microbiol.*, 131, 1753–1763 (1985).
20. Arima, K., Kakinuma, A., and Tamura, G. Surfactin, a crystalline peptidelipid surfactant produced by *Bacillus subtilis*: isolation, characterization and its inhibition of fibrin clot formation. *Biochem. Biophys. Res. Commun.*, 31, 488–494 (1968).
21. Ehrlich, S. D. Replication and expression of plasmids from *Staphylococcus aureus* in *Bacillus subtilis. Proc. Natl. Acad. Sci. U.S.A.*, 74, 1680–1682 (1977).
22. Scheer-Abramowitz, J., Gryczan, T. J., and Dubnau, D. Origin and mode of replication of plasmids pE194 and pUBl 10. *Plasmid*, 6, 67–77 (1981).
23. Isogai, I., Takayama, S., Murakoshi, S., and Suzuki, A. Structures of β-amino acids in antibiotics iturin A. *Tetrahedron Lett.*, 23, 3065–3068 (1982).
24. Peypoux, F., Guinand, M., Michel, G., Delcambe, I., Das, B. C., and Lederer, E. Structure of iturin A, a peptidolipid antibiotic from *Bacillus subtilis. Biochemistry*, 17, 3992–3996 (1978).
25. Kakinuma, A., Oushida, A., Shima, T., Sugino, H., Isono, M., Tamura, G., and Arima, K. Confirmation of the structure of surfactin by mass spectrometry. *Agric. Biol. Chem.*, 33, 1669–1671 (1969).
26. Besson, F., Chevanet, C., and Michel, G. Influence of the culture medium on the production of iturin A by *Bacillus subtilis. J. Gen. Microbiol.*, 133, 767–772 (1987).
27. Huang, C.-C., Ano, T., and Shoda, M. Nucleotide sequence and characteristics of the gene, *lpa-14*, responsible for biosynthesis of the lipopeptide antibiotics iturin A and surfactin from *Bacillus subtilis* RB14. *J. Ferment. Bioeng.*, 76, 443–450 (1993).
28. Sambrook, J., Fritsch, E. F., and Maniatis, T. *Molecular Cloning: A Laboratory Manual*, 2nd edn. Cold Spring Harbor Laboratory, Cold Spring Harbor, New York (1989).
29. Sanger, F., Nicklen, S., and Coulson, A. R. DNA sequencing with chain terminating inhibitors. *Proc. Natl. Acad. Sci. U.S.A.*, 74, 5463–5467 (1977).
30. Morikawa, M., Ito, M., and Imanaka, T. Isolation of a new surfactin producer *Bacillus pumilus* A-1, and cloning and nucleotide sequence of the regulator gene, *psf-1. J. Ferment. Bioeng.*, 74, 255–261 (1992).
31. Ohno, A., Ano, T., and Shoda, M. Production of a lipopeptide antibiotic surfactin with recombinant *Bacillus subtilis. Biotechnol. Lett.*, 14, 1165–1168 (1992).
32. Alonso, J. C., and Trautner, T. A. A gene controlling segregation of the *Bacillus subtilis* plasmid pC194. *Mol. Gen. Genet.*, 198, 427–431 (1985).
33. Keggins, K. M., Lovett, P. S., and Duvall, E. J. Molecular cloning of genetically active fragment of *Bacillus* DNA in *Bacillus subtilis* and properties of the vector plasmid pUBl 10. *Proc. Natl. Acad. Sci. U.S.A.*, 75, 1423–1427 (1978).
34. Cosmina, P., Rodriguez, F., de Ferra, F., Grandi, G., Pergo, M., Venema, G., and van Sinderen, D. Sequence and analysis of the genetic locus responsible for surfactin synthesis in *Bacillus subtilis. Mol. Microbiol.*, 8, 821–831 (1993).
35. Kratzsmar, J. M., Krause, M., and Marahiel, M. A. Gramicidin S biosynthesis operon containing the structural genes *grsA* and *grsB* has an open reading frame encoding a protein homologous to fatty acid thioesterases. *J. Bacteriol.*, 171, 5422–5429 (1989).

36. Magnet-Dana, R., Thimon, L., Peypoux, F., and Ptak, M. Surfactin/iturin A interactions may explain the synergistic effect of surfactin on the biological properties of iturin A. *Biochime*, 74, 1047–1051 (1992).

37. Huang, C.-C., Lian, Z. M., Hirai, M., Ano, T., and Shoda, M. *lpa-14*, a gene, involved in the production of lipopeptide antibiotics, regulates the production of a siderophore, 2,3-dihydoroxybenzoylglycine, in *Bacillus subtilis* RB14. *J. Ferment. Bioeng.*, 86, 603–607 (1998).

38. Armstrong, S. K., Pettis, G. S., Forrester, L. J., and Meintosh, M. A. The *Escherichia coli* enterobactin biosynthesis gene, *entD*: nucleotide sequence and membrane localization of its protein product. *Mol. Microbiol.*, 3, 757–766 (1989).

39. Ito, T., and Neilands, J. B. Products of "low iron fermentation" with *Bacillus subtilis*: isolation, characterization and synthesis of 2,3-dihydroxybenzoylglycine. *J. Am. Chem. Soc.*, 80, 4645–4647 (1958).

40. Bemer, I., Greiner, M., Metzger, J., Jung, G., and Winkelmann, G. Identification of enterobactin and linear dihydroxybenzoylserine compounds by HPLC and ion spray mass spectrometry. *Biol. Metals*, 4, 113–118 (1991).

41. O'Brien, I. G., and Gibson, F. The structure of entrochelin and related 2,3-dihydroxy-N-benzoylserine conjugates from *Escherichia coli*. *Biochim. Biophys. Acta*, 215, 393–402 (1970).

42. Crowley, D. E., Wang, Y. C., Reid, C. P. P., and Szaniszlo, P. J. Mechanisms of iron acquisition from siderophores by microorganisms and plants. *Plant Soil*, 130, 179–198 (1991).

43. Kloepper, J. W., Leong, J., Teintze, M., and Schroth, M. H. Enhanced plant growth by siderophores produced by plant growth promoting Rhizobacteria. *Nature*, 286, 885–886 (1980).

44. Leong, J. Siderophores: their biochemistry and possible role in the biocontrol of plant pathogen. *Annu. Rev. Phytopathol.*, 24, 187–209 (1986).

45. Tsuge, K., Ano, T., and Shoda, M. Isolation of a gene essential for biosynthesis of the lipopeptide antibiotics plipastatin B1 and surfactin in *Bacillus subtilis* YB8. *Arch. Microbiol.*, 165, 243–251 (1996).

46. Black, T. A., and Wolk, C. P. Analysis of a Her-mutation in *Anabaena* sp. strain PCC 7120 implicates a secondary metabolite in the regulation of heterocyst spacing. *J. Bacteriol.*, 176, 2282–2292 (1994).

47. Kakinuma, A., Oushida, A., Shima, T., Sugino, H., Isono, M., Tamura, G., and Arima, K. Confirmation of the structure of surfactin by mass spectrometry. *Agric. Biol. Chem.*, 33, 1669–1671 (1969).

48. Vanittanakom, N., Loeffler, W., Koch, U., and Jung, G. Fengycin—a novel antifungal lipopeptide antibiotic produced by *Bacillus subtilis* F-29-3. *J. Antibiotics*, 39, 888–901 (1986).

49. Umezawa, H., Aoyagi, T., Nishikiori, T., Okuyama, A., Yamagishi, Y., Hamada, M., and Takeuchi, T. Plipastatins: new inhibitors of phospholipase A2, produced by *Bacillus cereus* BMG302-tF67. *J. Antibiotics*, 39, 737–744 (1986).

50. Phae, C. G., Sasaki, M., Shoda, M., and Kubota, H. Characteristics of *Bacillus subtilis* isolated from composts suppressing phytopathogenic microorganisms. *Soil Sci. Plant Nutr.*, 36, 575–586 (1990).

51. Anagnostopoulos, C., and Spizizen, J. Requirements for transformation in *Bacillus subtilis*. *J. Bacteriol.*, 81, 741–746 (1961).

52. Scheer-Abramowitz, J., Gryczan, T. J., and Dubnau, D. Origin and mode of replication of plasmids pEI 94 and pUB110. *Plasmid*, 6, 67–77 (1981).

53. Borchert, S., Stachelhaus, T., and Marahiel, M. A. Induction of surfactin production in *Bacillus subtilis* by *gsp*, a gene located upstream of the gramicidin S operon in *Bacillus brevis*. *J. Bacteriol.*, 176, 2458–2462 (1994).

54. Grossman, T. H., Tuckman, M., Ellestad, S., and Osburne, M. S. Isolation and characterization of *Bacillus subtilis* genes involved in siderophore biosynthesis: relationship between *B. subtilis sfp* and *Escherichia coli entD* genes. *J. Bacteriol.*, 175, 6203–6211 (1993).

55. Menkhaus, M., Ullrich, C., Kluge, B., Yater, J., Vollenbroich, D., and Kamp, R. M. Structural and functional organization of the surfactin synthetase multienzyme system. *J. Biol.Chem.*, 268, 7678–7684 (1993).

56. Ullrich, C., Kluge, B., Palacz, Z., and Yater, J. Cell-free biosynthesis of surfactin, a cyclic lipopeptide produced by *Bacillus subtilis*. *Biochemistry*, 30, 6503–6508 (1991).

57. Yamada, S., Takayama, Y., Yamanaka, M., Ko, K., and Yamaguchi, I. Biological activity of antifungal substances produced by *Bacillus subtilis*. *J. Pest. Sci.*, 15, 95–96 (1990).

58. Hiraoka, H., Asaka, O., Ano, T., and Shoda, M. Characterization of *Bacillus subtilis* RB14, coproducer of peptide antibiotics iturin A and surfactin. *J. Gen. Appl. Microbiol.*, 38, 635–640 (1992).

59. Imanaka,T., Himeno, T., and Aiba, S. Effect of in vitro DNA rearrangement in the NH2-terminal region of penicillinase gene from Bacillus licheniformis on the mode of expression in *Bacillus subtilis*. *J. Gen. Microbiol.*, 131, 1753–1756 (1985).

6 Genetic Analysis of *B. subtilis* Related with Production of Three Peptide Substances

6.1 CLONING, SEQUENCING, AND CHARACTERIZATION OF THE ITURIN A OPERON (1)

Many *B. subtilis* strains produce a small peptide(s) with a long fatty moiety, the so-called lipopeptide antibiotics. On the basis of the structural relationships, the lipopeptides that have been identified in *B. subtilis* are generally classified into three groups: the surfactin group, the plipastatin–fengycin group, and the iturin group. The members of the surfactin and plipastatin–fengycin groups are composed of 1 β-hydroxy fatty acid and 7 and 10 α-amino acids, respectively, while the members of the iturin group consist of 1 β-amino fatty acid and 7 α-amino acids. The presence of the β-amino fatty acid is the most striking characteristic of the iturin A group and distinguishes this group from the other two groups. Mycosubtilin (2), which is a member of the iturin A group, has been sequenced and characterized. In particular, a study of the mycosubtilin operon of *B. subtilis* ATCC 6633 (2) showed that MycA, a novel template enzyme that has functional domain homology to β-ketoacyl synthetase and amino transferase and amino adenylation, was present, which implied that MycA is responsible for the incorporation of the β-amino fatty acid.

The amino acid compositions of iturin A and mycosubtilin are almost identical, except that the sixth and seventh amino acids are inverted, as shown in Figure 6.1. In Chapter 5, it was shown that a gene, *lpa-14*, was cloned that encodes the 4′-phosphopantheteinyl transferase required for maturation of the template enzyme of iturin A (3, 4). In this section, the features of the iturin A synthetase operon based on the nucleotide sequence and gene disruption were examined. By comparing the iturin A operon with the mycosubtilin operon, the difference between the two operons may be a result of intragenic swapping of amino acid adenylation domains.

6.1.1 Materials and Methods

6.1.1.1 Strains and Media

The strains and plasmids used in this study arc listed in Table 6.1. Luria-Bertani (LB) medium was used for cultivation of *Escherichia coli* and *B. subtilis*. When necessary, antibiotics were added at the following concentrations: ampicillin, 50 µg/ml;

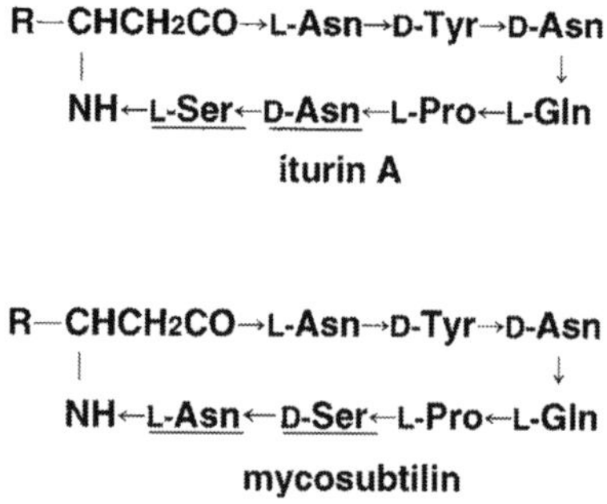

R—CHCH₂CO→L-Asn→D-Tyr→D-Asn

NH←L-Ser←D-Asn←L-Pro←L-Gln

iturin A

R—CHCH₂CO→L-Asn→D-Tyr→D-Asn

NH←L-Asn←D-Ser←L-Pro←L-Gln

mycosubtilin

FIGURE 6.1   Structures of iturin A and mycosubtilin. R indicates an alkyl moiety (generally C_{14} to C_{17}). The arrows represent peptide bonds in the -CO-NH- direction. The differences between the two lipopeptides are indicated by underlining.

chloramphenicol, 5 μg/ml; erythromycin, 10 μg/ml; tetracycline, 20 μg/ml; and neomycin, 20 μg/ml.

Iturin A production *in vitro* was indicated by the formation of a clear inhibitory zone on LB agar (LB medium with 1.5% agar) containing a spore suspension of a phytopathogenic fungus, *Fusarium oxysporum* f. sp. *lycopersici* race JI SUFI I9, as described in Chapter 2 (5). No. 3 medium was used for lipopeptide production in liquid cultures. No. 3S medium contained 10 g of Polypepton S (Nihon Pharmaceutical Co.) per liter instead of Polypepton because iturin A productivity was enhanced by Polypepton S.

6.1.1.2   Transformation and DNA Manipulation

B. subtilis RBl4 was transformed by electroporation as described in Chapter 5(7). *B. subtilis* MI113 was transformed by a competent cell method as described in Chapter 5(7). Chromosomal DNA of *B. subtilis* strains were prepared by using a method previously reported (12). Routine DNA manipulation and *E. coli* transformation were performed as previously described (7). Plaque and Southern hybridizations were performed by a digoxigenin enzyme-linked immunosorbent assay using a DNA labeling and detection kit (Roche) as recommended in the instruction manual.

6.1.1.3   Transposon Technique

Transposon-harboring plasmid pHV1249 (8) was introduced into RBl4 by electroporation, and random mutagenesis of mini-Tn*10* was performed by a previously described method (8). Cloning of the mini-Tn*10*-disrupted gene was carried out as follows. To construct a *B. subtilis* MI113 plasmid library of the RBl4 chromosome, chromosomal DNA of the transposon containing mutant was digested with *Hind*III, ligated at the *Hind*III site of plasmid pTB522, which could be replicated in *B. subtilis* (9), and then transformed into *B. subtilis* MI113 competent cells. The plasmid of the chloramphenicol-resistant transformant resulting from cloning of a chromosomal fragment containing mini-Tn*10* was obtained from the library and designated pTBI006. The *Hind*III insert of pTBI006 was transplanted into the *Hind*III site of pBR322 by transformation of *E. coli* JM109, resulting in pBRHd8k (Figure 6.2).

TABLE 6.1
Strains and Plasmids Used in This Study

Strain or Plasmid	Characteristic[a]	Reference
E. coli JM109	*recA1 endA1 gyrA96 thi hsdR17 supE44 Δ(lac-proAB) /F' [traD36 proAB+ lacI^q lacZΔM15*	(10)
B. subtilis Strains		
RB14	Wild type, IT$^+$ SF$^+$ PL$^+$	(5)
1006	RB14 *ituA*: : Tn*10*, I^{T-} SF$^+$ PL$^+$	This study
RΔIA1	RB14 *ituD*: : neo, IT$^-$ SF$^+$ PL$^+$	This study
R-PM1	RB14 P$_{repU}$$^-$ neo-*ituD*, IT$^+$ SF$^+$ PL$^+$	This study
MI113	Marburg 168 derivative, *arg-15 trpC2 hsmM hsrM*	(11)
E. coli Plasmids		
pUC19	Cloning vector, Ap^r	(10)
pBEST502	P$_{repU}$-*neo* cassette, Ap^r Nm^r	(6)
pSC24Pst	pUC19 carrying the 2-kb *Sau*3A1–*Pst*I fragment containing *ituD*, Ap^r	This study
pSC24PstNm	Nm^r gene cassette inserted into the internal *Sal*I–*Nsi*I region of *ituD* in pSC24Pst, Ap^r Nm^r	This study
pSC53EcoT	pUC19 carrying the 1.2 kb *Tth*HB8I–*Eco* T14I fragment containing P$_{ituD}$, Ap^r	This study
pKODP4P5	pUC19 carrying the 0.8 kb PCR product containing 5'-terminal region of *ituD* as well as the putative ribosome binding site, Ap^r	This study
pP4P5Nm	P$_{repU}$-*neo* inserted into the *Xba*I site located upstream of the putative ribosome binding site of *ituD* in pKODP4P5, Ap^r Nm^r	This study
pUCIPNm	pP4P5Nm carrying the 0.8 kb *Nde*I–*Tth*HB8I fragment containing 3'-terminal region of *yxjF*, Ap^r Nm^r	This study
B. subtilis Plasmids		
pHV1249	Mini-Tn*10* Ap^r Em^r Cm^r	(8)
pTB522	Cloning vector, Tc^r	(9)
pE194	Temperature sensitive for replication, Em^r	(11)
pE24PstNmα	Disruption plasmid, pE194 carrying the 2.0 kb insert from pSC24PstNm, *ituD*: :*neo* gene cassette, Em^r Nm^r	This study
pEIPNm2	pE194 carrying the 3.0-kb insert from pUCIPNm, pUCIPNm, P$_{ituD}$-*neo-ituD*, Em^r Nm^r	This study

[a] IT, iturin A production; SF, surfactin production; PL, plipastatin-like antifungal agent production; Ap^r, ampicillin resistant; Em^r, erythromycin resistant; Nm^r, neomycin resistant; Tc^r, tetracycline resistant; Cm^r, chloramphenicol resistant.

6.1.1.4　Phage Library Construction and Screening

The RB14 chromosome was digested partially with *Eco*RI and then was separated by electrophoresis to obtain homogeneous fragments that were 10 to 20 kb long. The fragments obtained were used for lambda DASH II library construction with a Lambda DASH 11/*Eco*RI vector kit (Stratagene) as recommended in the instruction manual. First, screening by plaque hybridization in which the 1.9 kb *Nde*I–*Pst*I fragment of pBRHd8k was used as a probe identified a positive phage with a 19.2 kb

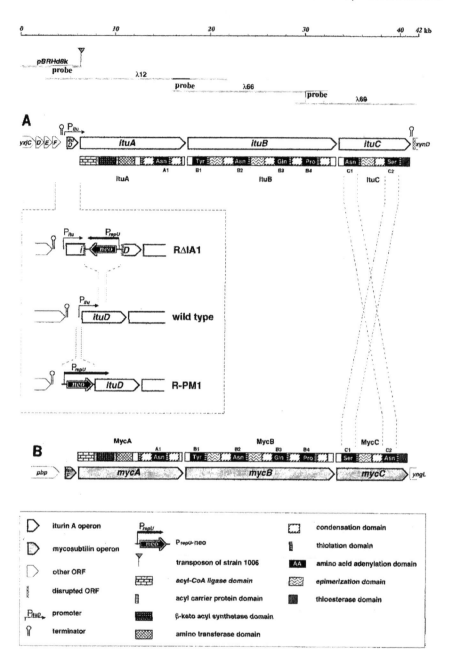

FIGURE 6.2   ORF organization of (A) iturin A and (B) mycosubtilin operons. The positions of the plasmid and phages sequenced are shown above the iturin A operon. The sequences of derivative strains RΔIAl and R-PMl associated with *ituD* are shown in a box. The intersecting dotted lines indicate the difference in the amino acid adenylation domain arrangement between the two operons. The mycosubtilin operon was drawn by referring to reference 2.

insert, designated λ12 (Figure 6.2). By using the internal 1.7 kb *Eco*RI–*Hind*III fragment of C12 as a probe, λ66, harboring a 16.0 kb insert, was obtained (Figure 6.2). The end of the 1.9 kb *Eco*RI fragment of λ66 was employed in the next screening, and λ69 with a 12.4 kb insert was obtained (Figure 6.2).

6.1.1.5 Disruption of *ituD* by the Plasmid Pop In–Pop Out Method

A 2 kb *Sau*3AI-*Pst*I fragment from pBRHd8k, which harbored the entire *ituD* gene, was inserted at the *Bam*HI–*Pst*I site of pUC19, generating pSC24Pst. The *Sal*I–*Nsi*I fragment of pSC24Pst was removed, and the *Sal*I–*Pst*I fragment of the neomycin resistance gene cassette (*neo*) from pBEST502 was inserted (6), resulting in pSC24PstNm. The *Hind*III–*Kpn*I fragment of pSC24PstNm, which harbored the disrupted *ituD* gene, was inserted at the *Pst*I site of pE194 by blunt-end ligation. The ligated mixture was transformed into MI113. A neomycin-resistant colony was selected and thus obtained pE24PstNmα. The *ituD* coding region in RB14 was disrupted by using the thermosensitive replication origin of pE194, as previously described (11). First, pE24PstNmα was transformed into RB14 by electroporation. Strain RB14(pE24PstNmα) was plated onto LB agar containing neomycin and erythromycin and then incubated at 48°C. The resulting strain, RB14::pE24PstNmα, was cultivated in LB medium without selective pressure at 30°C for 10 generations. The culture was diluted and plated onto LB agar to obtain single colonies. The neomycin resistance and erythromycin resistance of the resulting colonies were assayed. Finally, the disrupted mutant, RΔIA1, was isolated by screening for neomycin-resistant and erythromycin-sensitive colonies.

6.1.1.6 Primer Extension Analysis

Total mRNA was prepared from a stationary-phase culture of RB14 cultivated in no. 3 medium for 17 h. Four milliliters of the culture was centrifuged and washed with STE buffer (10 mM Tris-HCl [pH 8.0], 1 mM EDTA, 0.1 M NaCl). The pellet was then resuspended in 50 μl of TE buffer (10 mM Tris-HCl [pH 8.0], 1 mM EDTA), and 50 μl of TE buffer containing lysozyme (20 mg/ml) was added. The suspension was incubated at room temperature for 10 min. Further purification was performed with an RNeasy mini kit (Qiagen). For primer extension analysis, IRD41-labeled primer ITU-PREX (5′-ATCGCATCGCTCGCITCITCAAAC-3′) (Figure 6.3), which was complementary to the sequence from nucleotide 96 to nucleotide 119 downstream of the putative start codon of *ituD*, was purchased from Nisshinbo Co. (Tokyo, Japan). Total RNA was ethanol precipitated and dissolved in 20 μl of hybridization buffer [80 mM piperazine-N, N′-bis(2-ethanesulfonic acid) (PIPES) buffer (pH 6.4), 2 mM EDTA, 800 mM NaCl, 50% formamide]. The solution was supplemented with 1.8 μl of primer ITU-PREX (1 pmol/μl), denatured at 80°C for 15 min, and then cooled to 30°C with gentle shaking. After ethanol precipitation, the pellet was dissolved in extension buffer [4 μl of Moloney murine leukemia virus reverse transcriptase (Toyobo Inc., Osaka, Japan), 8 μl of 5 x buffer for Moloney murine leukemia virus reverse transcriptase, 16 μl of a solution containing each deoxynucleoside triphosphate at a concentration of 2.5 mM, 10 μl of H$_2$O, 2 μl of RNase inhibitor] and then incubated at 42°C for 1 h. The resulting cDNA was subjected to Li-cor dNA automated DNA sequencing.

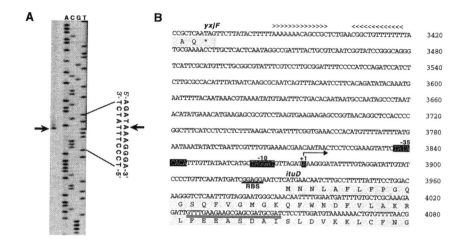

FIGURE 6.3   Determination of the transcription initiation site of *ituD* by primer extension analysis. (A) The position of the band corresponding to the *ituD*-specific primer extension product is indicated by arrows. (B) Nucleotide sequence of the *ituD* promoter region (P_{ituD}). The positions of the *ituD* transcription initiation site (highlighted, +1), putative −10 and −35 regions (highlighted), and putative ribosome binding site (RBS) (underlined) are indicated. Putative protein sequences of yxjF and ituD are shown below the DNA sequence (shaded). The sequence complementary to the oligonucleotide used for primer extension analysis (ITU-PREX) is double underlined. The putative rho-independent terminator downstream of *yxjF* is indicated by > and < above the sequence.

6.1.1.7   Promoter Exchange

A 0.8 kb fragment of the 5′ region of *ituD*, which contained up to the ribosome binding site but not the promoter, was obtained by polymerase chain reaction (PCR). PCR amplification with primers ITU P4-F (5′-CCCCTGT<u>TCTAGA</u>TGA TCGGAGGAATCTC-3′; underlining indicates an *Xba*I site, and italics indicates substituted bases) and ITUP5-R (5′-TG<u>CATCGA</u>TTCTGTCCATCTAACCG GCATC-3′; underlining indicates a *Cla*I site) was performed with the RB14 chromosome by using KOD DNA polymerase (Toyobo Inc.). The fragment obtained was double digested with *Xba*I and *Cla*I and then inserted between the *Xba*I and *Cla*I sites of pUC19. The resulting plasmid was confirmed to have the correct sequence and was designated pKODP4P5. P_{repU} accompanied by the neomycin resistance gene cassette (P_{repU}–*neo*) was excised from pBEST502 (6) by digestion with *Xba*I and then inserted into the *Xba*I site of pKODP4P5. A plasmid with P_{repU}–*neo*, whose direction of transcription is the same as that of *ituD*, was selected and designated pP4P5Nm. The 0.8 kb *Nde*I–*Tlh*HB8I fragment, which was upstream of the *ituD* transcription start site prepared from plasmid subclone pSC53EcoT of the phage library, was blunt ended and inserted into the *Sma*I site of pP4P5Nm, generating pUCIPNm. The *Hind*III–*Kpn* I fragment of pUCIPNm, which was P_{repU}–*neo* accompanied by the P_{itu} flanking region, was inserted into the *Pst*I site of pE194 by blunting and ligation and then transformed into MI113, which resulted in plasmid pEIPNm2. Replacement of P_{itu} of RB14 by P_{repU}–*neo* by the plasmid pop in–pop out method using pEIPNm2 was performed by the method described earlier for *ituD* disruption. Promoter replacement was confirmed by PCR (data not shown).

6.1.1.8   Quantitative Analysis of Iturin A and Surfactin

Iturin A and surfactin in the extracted solution of *B. subtilis* culture were quantified by reversed-phase high-performance liquid chromatography (HPLC), as previously described (12).

6.1.1.9   Nucleotide Sequence Analysis

Double-stranded DNA was cloned in pUCl9 sequenced with a Li-cor dNA 4000L DNA sequencer by using an IRD41 dye-labeled primer (Nisshinbo Co., Tokyo, Japan) and a Thermo Sequenase cycle sequencing kit (Amersham Pharmacia Biotech Co.). A partial sequence of the insert of the screened lambda DASH II phage was determined by Nippon Seifun Co., Atsugi, Japan. Motif retrieval for proteins was performed by using the PROSITE package. Harrplot analysis was performed with the Genetyx package (Software Development Co., Tokyo, Japan). Multiple-alignment analysis and phylogenetic analysis were performed by using Clustal W.

6.1.1.10   Nucleotide Sequence Accession Number

The nucleotide sequence of the iturin A operon of RB14 was deposited in the DDBJ, EMBL, and GenBank nucleotide sequence databases under accession number AB050629.

6.1.2   Results

6.1.2.1   Cloning, Sequencing, and Analysis of Iturin A Operon

To identify the genes responsible for iturin A production, transposon mutagenesis with mini-Tn*10* was used. About 5000 transposon-containing colonies were replicated on an LB agar plate containing *F. oxysporum* to assay for iturin production deficiency. Fifteen colonies exhibited no antifungal activity on the plate, and none of these colonies had the ability to produce iturin, as determined by HPLC. Ten of the colonies were selected at random, and chromosomal DNA were prepared, digested by *Hin*dlll, and subjected to Southern hybridization. All 10 colonies contained mini-Tn*10* in the same 8 kb *Hin*dIII fragment (data not shown). One colony, designated strain 1006, was selected for further study. No iturin A production by 1006 was observed (Figure 6.4). The 8 kb *Hin*dIII fragment with a transposon was screened from the plasmid library of the strain 1006 chromosome by using the chloramphenicol resistance marker of mini-Tn*10*.

To clone a complete iturin A operon, an RB14 chromosome library was constructed by using the lambda DASH II phage, and four contiguous inserts that collectively encompassed a 42 kb region were obtained. Figure 6.2 shows that this region contains nine open reading frames (ORFs). All nine ORFs are oriented in the same direction. The four ORFs located upstream in the region sequenced have high levels of homology to the *B. subtilis* 168 genes *yxjC* (77% identical), *yxjD* (83%), *yxjE* (83%), and *yxjF* (77%). These genes are thought to encode 3-oxoadipate coenzyme A (CoA) transferase (*yxjD* and *yxjE*) and gluconate 5-dehydrogenase (*yxjF*) (13), but their actual

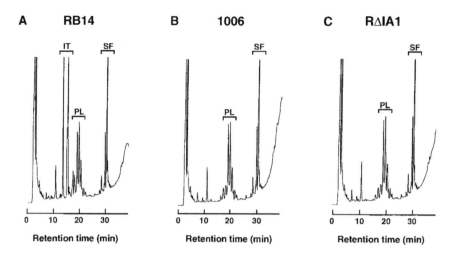

FIGURE 6.4 Qualitative HPLC analysis of the lipopeptides produced by (A) RB14, (B) 1006, and (C) RΔ1Al. Peaks corresponding to iturin A (IT) and surfactin (SF) are indicated. Plipastatin-like peaks (not identified) (PL) are also indicated.

functions have not been determined. There is a putative rho-independent terminator downstream of *yxjF* that appears to terminate transcription from both directions. The fifth ORF, designated *ituD*, is separated from *yxjF* by a 0.6 kb intercoding region. The 45 kDa ItuD protein has a high level of homology to FenF encoded by the mycosubtilin synthetase operon of *B. subtilis* ATCC 6633 (79%) (2) and FenF of *B. subtilis* F29-3 (89%) (14) and lower levels of homology to malonyl–CoA transacylase of *E. coli* (44%) (15) and *B. subtilis* 168 (37%) (16), suggesting that *ituD* encodes malonyl–CoA transacylase. The transposon-containing ORF, located 19 bp downstream of *ituD*, is designated *ituA*. The deduced amino acid sequence encoded by *ituA* corresponds to a 449 kDa protein that has a high level of homology to MycA (79%), which is the first subunit of mycosubtilin synthetase. The striking feature of MycA, the fact that three functional domains homologous to β-ketoacyl synthetase, amino transferase, and amino acid adenylation are combined, is conserved in ItuA. The gene 43 bp downstream of *ituA* is called *ituB*. ItuB is a 609 kDa peptide synthetase consisting of four amino acid adenylation domains, two of which are flanked by an epimerization domain. ItuB also exhibits 79% homology to MycB. The next gene is designated *ituC*, which encodes a peptide synthetase that has two adenylation domains, one epimerization domain, and a thioesterase domain that is probably responsible for peptide cyclization. Although ItuD, ItuA, and ItuB have high levels of homology to their counterparts in mycosubtilin synthetase, ItuC exhibits only 64% homology to MycC, which may reflect a structural difference between iturin A and mycosubtilin. There is a putative rho-independent terminator downstream of *ituC*. The final ORF exhibits homology to the *xynD* gene (17) of *B. subtilis* 168, which encodes endo-1,4-xylanase. As shown in Figure 6.5, there is a sequence that has a high level of homology to the *xynD* gene and is located between *ituC* and *xynD* in a direct-repeat manner (see later).

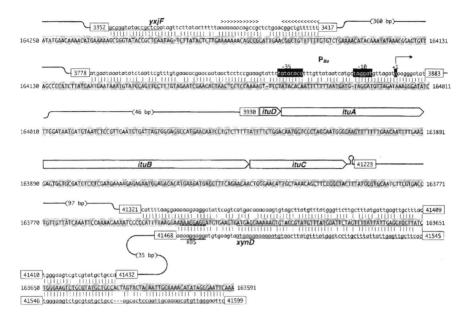

FIGURE 6.5 Alignment of iturin A operon of RB14 and *xynD* regions of strain 168. Shaded uppercase letters indicate the sequence of the *xynD* region of strain 168 (based on the complementary sequence from positions 164,250 to 163,691 of the accession no. Z99113 sequence). Lowercase letters indicate the iturin A operon sequence (accession no. AB050629) that has homology to the strain 168 sequence. Sequences encoding iturin A synthetase are indicated by arrow-shaped boxes. Identical nucleotides in the two sequences are indicated by vertical lines. The numbers in the boxes are the positions in the deposited sequence. The lines indicate linkage between discrete homologous segments of the iturin A operon. Underlining indicates coding regions of *yxjF* and *xynD*. –35 and –10 are promoter sites of *ituD*, and +1 is a transcription start site of *ituD*. The alignment was constructed based on the results of a BLAST search. RBS, ribosome binding site.

6.1.2.2 Disruption of the *ituD* Gene to Derive the Iturin A-Deficient Phenotype

To confirm that *ituD* is responsible for iturin A synthesis, *ituD* was disrupted (Figure 6.2). The resulting disruptant, RΔIA1, was inoculated into no. 3S medium and cultivated at 30°C for 60 h to assay for iturin A production. As shown in Figure 6.4, analytical HPLC of an βRΔIA1 culture extract resulted in no iturin A peak, while the production of the lipopeptide surfactin by RΔIA1 was the same as the production by the wild type, indicating that *ituD* disruption only resulted in an iturin A deficiency. Based on these results, it was concluded that *ituD* is essential specifically for iturin A synthesis.

6.1.2.3 Promoter Analysis of the *ituD* Gene

The transcription start site of *ituD* was determined by the primer extension method. Total RNA of RB14 was extracted from a 17 h culture in no. 3 medium. The total RNA obtained was hybridized with oligoDNA by using a fluorescent probe that was designed to hybridize with nucleotides 96 to 119 of *ituD* from the translation initiation codon. cDNA was synthesized with reverse transcriptase, and the resulting cDNA was then sequenced. As shown in Figure 6.3, the transcription start site of *ituD* was found to be an A residue 56 bp upstream of the first residue of the *ituD* initiation codon. Upstream of this start site, a TATACACA-16 bp-TAGGAT sequence that exhibited low levels of homology to the consensus -10 and -35 (TTGACA-17 bp TATAAT) sequences of σ^A (Figure 6.3) was found. This promoter was designated P_{ituD}. Other transcription start sites up to the putative rho-independent terminator of *yxjF* were sought, but no other such sites were found.

6.1.2.4   Promoter Replacement of the Iturin A Operon for High Iturin A Production

Because the mutant with disrupted *ituD* had an iturin A deficient phenotype and there is no coding region between *ituD* and *imC*, which implies that there is a promoter, it is possible that P_{itu} governs the expression of the iturin A operon. To confirm this, we replaced P_{itu} with a constitutively expressed promoter by using the neomycin resistance gene cassette, P_{repU}–*neo*, of pBEST502 (18). P_{repU}–*neo* is the chimera of the promoter of the replication protein gene (P_{repU}) and the neomycin resistance gene (*neo*) of plasmid pUB110 (19). Since P_{repU} of P_{repU}–*neo* lacks autoregulation of transcription by RepU (20), constitutive expression of P_{repU} is expected. Moreover, P_{repU} is sufficiently strong that neomycin resistance can be conferred by a single copy of neo in the chromosome, and there is no terminator to stop transcription starting from P_{repU}. The presence of the neomycin resistance gene cassette in a transformant, therefore, indicates that P_{repU} has been introduced and implies that constitutive and enforced expression of genes located downstream of P_{repU}–*neo* occurs in a polycistronic manner. The neomycin resistance gene cassette was ligated with a PCR-amplified fragment of the 5′ portion of *ituD* at the ribosome binding site of *ituD* and then attached to the upstream fragment of P_{itu} in front of the neomycin resistance gene. The resulting three-fragment array was ligated to the thermosensitive plasmid pE194 and transformed by electroporation, and then the promoter was replaced by the pop in-pop out method, as described earlier for *ituA* disruption. The resulting strain, R-PM1, contained the inserted P_{repU}–*neo* P instead of 250 bp of the P_{itu} sequence (Figure 6.2).

The strain generated and the wild type were inoculated into no. 3 medium and cultured at 30°C for 120 h to monitor iturin A production. The changes in the growth of strain R-PM1 and the pH of no. 3 medium over time were the same as those observed with the wild type (data not shown). However, as shown in Figure 6.6, the rate of iturin A production by R-PM1 was significantly greater than that by the wild type. The concentration of iturin A for R-PM1 after 72 h of cultivation was 330 µg/ml, whereas the concentration for the wild type was 110 µg/ml, indicating that production was threefold greater in R-PM1. Since the two strains did not differ significantly in terms of surfactin production (Figure 6.6), my colleagues and I concluded that the promoter exchange affected only iturin A production.

6.1.3   Discussion

In this study, the 42 kb region of *B. subtilis* RB14 that contains the complete iturin A synthetase operon was identified, which is more than 38 kb long. The iturin A

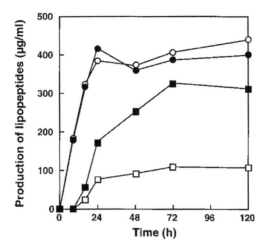

FIGURE 6.6 Time courses for production of two lipopeptides: iturin A produced by RBI 4 (□) and R-PMl (■) and surfactin produced by RB14 (○) and R-PMl (●).

operon is composed of four ORFs: *ituD*, *ituA*, *ituB*, and *ituC* (in that order). The iturin A operon closely resembles the mycosubtilin operon. ItuD, as well as FenF of *B. subtilis* ATCC 6633 (2) and F29-3 (14), exhibits homology to the malonyl CoA–transacylase FabD, which participates in fatty acid synthesis in *E. coli* (15) and *B. subtilis* 168 (16). It is thought that in some *Streptomyces* spp. malonyl CoA–transacylase may be responsible for not only fatty acid synthesis but also type II polyketide antibiotic synthesis (21, 22). To clarify whether *ituD* is involved in iturin A production and/or fatty acid synthesis, *ituD* was disrupted, which resulted in a specific iturin A deficiency. Since neither the growth of nor surfactin production by strain RΔ1Al, a strain in which *ituD* was disrupted, was inhibited, we concluded that *ituD* is indispensable for iturin A production. This is the first evidence that a gene related to *fenF* participates in synthesis of an iturin group antibiotic. On the other hand, ItuA exhibits homology to β-ketoacyl synthetase, amino transferase, and peptide synthetase in one molecule, and is probably responsible for synthesis of a β-amino fatty acid accompanied by ituD dipeptide (β-amino fatty acid-Asn) formation, as has been proposed for the mycosubtilin synthetase MycA (2). These features are consistent with the report that cerulenin, which inhibits β-ketoacyl synthetase, is more active in iturin A β-amino fatty acid synthesis than in fatty acid synthesis, which implies that β-ketoacyl synthesis occurs during iturin A production independent of fatty acid synthesis (23). Other genes, *ituB* and *ituC*, encode large peptide synthetases that build peptide chains on the precursor from ItuA. The level of homology between *ituC* and *mycC*, the counterpart of *ituC* in the mycosubtilin operon, is relatively low (64%) compared to the levels of homology for other groups (79%), reflecting the difference in the amino acid arrangements in ItuC and MycC. As shown in Figure 6.2, ItuC is predicted to be responsible for the D-Asn→L-Ser portion of iturin A, while MycC is thought to synthesize the D-Ser→L-Asn part of mycosubtilin. Therefore, it is not unreasonable to assume that swapping between two adenylation domains occurs.

To examine the possibility that intragenic swapping occurs in the adenylation domain, the sequences of *ituC* and *mycC* were compared. The results of a dot matrix analysis of the nucleotide sequences of *ituC* and the other relevant synthetase genes (*mycC*, *ituA*, and *ituB*), including *ituC* itself, are shown in Figure 6.7. Clearly, two regions of *ituC* exhibit discrete homology to *mycC*; the homology junction is roughly 10 amino acids upstream of L(TS)xEL (A1 motif sequence (24)) and the C terminus of GRxDxQVKIRGxRIELGEIE (A8 motif sequence (24)). Homologies are observed between asparagine adenylation domains (ItuC2 and MycC1) and between serine adenylation domains (ItuC1 and MycC2), implying that domain swapping occurs. Since there are three asparagine adenylation domains (ItuA1, ItuB2, and ItuC1) in iturin A synthetases (Figure 6.2), it is likely that the asparagine adenylation domain of ItuC1 exhibits the highest level of homology to ItuA1 or Itu2. However, as shown in Figure 6.8, phylogenetic analysis of 14 nucleotide sequences encoding the adenylation domain of iturin A and mycosubtilin revealed that the sequence most similar to the sequence encoding the asparagine adenylation domain of ItuC1 is the sequence encoding the asparagine adenylation domain of MycC2. All iturin A domains also show phylogenetic similarities to their counterparts in mycosubtilin domains. These results imply that iturin A and mycosubtilin have a common ancestor. Therefore, a model is proposed in which *ituC* or *mycC* swapped nucleotide sequences encoding adenylation domains after a common ancestor became established.

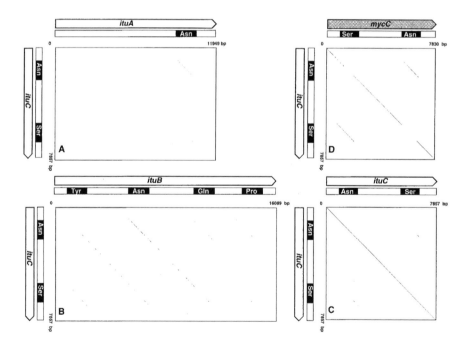

FIGURE 6.7   Dot matrices *ituC* and (A) *ituA*, (B) *ituB*, (C) *ituC*, and (D) *mycC*. Dots were placed at locations with identical nucleotides when more than 45 of 60 nucleotides were identical. Amino acid adenylation domains of functional modules of synthetases are indicated by black boxes.

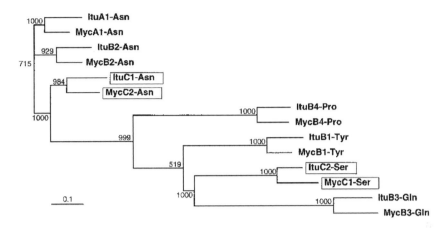

FIGURE 6.8 Phylogenetic tree constructed by the neighbor-joining method based on the sequences of the adenylation domains (from part of the A3 motif, PKG to the A6 motif, GELC[Y]) of iturin A synthetase and mycosubtilin synthetase. The numbers are bootstrap values based on 1000 replicates.

Although the overall sequence of the iturin A operon is highly homologous to that of the mycosubtilin operon, the flanking region of the iturin A operon is significantly different from that of the mycosubtilin operon. It has been shown that the mycosubtilin operon of ATCC 6633, which is not a plipastatin producer, is between *pbp* and *yng*L (2), while the plipastatin operon (167° to 171°), instead of the mycosubtilin operon, is present in strain 168 (11). In the case of RB14, the iturin A operon lies between *yxjF* and *xynD*. Since strain RBl4 is also a producer of a plipastatin-like compound (Figure 6.4), it is reasonable to conclude that the iturin A operon of RB14 is in a region other than the plipastatin region. However, in the case of strain 168, *yxjF* (341°) is located on the side opposite xynD (166°) in the chromosome (13). This can be explained by the fact that the SfiI digestion pattern of the RB14 genome, as determined by pulsed-field gel electrophoresis, was significantly different from that of strain 168 (data not shown), suggesting that the entire genomic structure of RB14 has little similarity to that of strain 168, including the location of *yxjF* and/or *xynD*.

However, unexpectedly, as shown in Figure 6.5, the promoter region of the iturin A operon has significant homology to the upstream sequence of *xynD* of strain 168. Since the downstream region of the iturin A operon has two tandem repeat DNA segments with significant homology to the *xynD* region of strain 168, it is highly probable that the coding region of the iturin A operon is transferred into the promoter of *xynD* and resides in this promoter, rather than that the promoter of *xynD* is simply transferred into the iturin A operon. The possibility that such a large antibiotic gene transfer occurs was also suggested for the tyrocidine operon, in which some peptide synthetase genes are integrated into the parental gramicidin S operon (25). In terms of horizontal transfer, it is possible that large DNA transfers occur frequently. The duplicated copy of *xynD* may indicate transient existence of a tandem repeat of early iturin A operons that were intermediates in the adenylation domain swapping mentioned earlier.

The attempt to introduce a *repU* promoter instead of an intrinsic promoter into RB14 resulted in increased production of iturin A (Figure 6.6), and the advantage of using the P_{repU}–*neo* cassette was identified. This method might be useful for generating bacteria that act as effective biological control agents. At present, we are able to compare very similar operons, such as surfactin operons of *B. subtilis* (26) and two lichenysin operons of *B. licheniformis* (27, 28), plipastatin (29, 30) and fengycin (31) operons, and iturin A and mycosubtilin (2) operons. These comparisons may provide good techniques for elucidating the boundary of the functional domain that directs engineered peptide synthesis (32–35).

6.2 CONVERSION OF *B. SUBTILIS* 168 INTO AN ITURIN PRODUCER BY TRANSFER OF ITURIN A OPERON (36)

Many *B. subtilis* strains simultaneously produce some lipopeptide antibiotics, whose peptide moiety is synthesized nonribosomally by large template enzyme complexes (e.g., *B. subtilis* RB14 (1)). Such lipopeptide antibiotics identified thus far have been divided into three groups according to their structure, as follows: the surfactin group (37), the plipastatin and fengycin group (37), and the iturin group (18, 39, 40). Iturin group lipopeptides are composed of seven α-amino acids and one β-amino acid with a long lipid moiety and are potent antifungal agents (39). Three distinct operons that belong to the iturin group have been cloned and sequenced thus far: the mycosubtilin operon of *B. subtilis* ATCC 6633 (2), the iturin A operon of *B. subtilis* RB14 (1), and the bacillomycin D operon of *Bacillus amyloliquefaciens* FZB42 (41). All of these operons are composed of a putative transcriptional unit with four genes: one small gene that encodes malonyl–coenzyme A transferase and is probably responsible for β-amino acid synthesis, and three large genes that encode large template enzymes for the synthesis of peptides with defined sequences and chiralities. Although the percent amino acid sequence identities between the operons range from approximately 75% to 85%, the flanking regions of the operons are quite different from each other, as shown in Figure 6.9. The flanking region of the mycosubtilin operon is identical to that of the plipastatin operon of *B. subtilis* 168 (2). On the other hand, the iturin A and bacillomycin D operons are flanked by sequences homologous to the *xynD* of strain 168 (1, 41) (Figure 6.9). These findings strongly indicate the dynamic features of the iturin group operons. Although 168 does not produce lipopeptides due to its mutation of the 4′-phosphopantetheine transferase gene *sfp*, which is responsible for the conversion of nascent antibiotic synthetase to its active form (42), it is a potential producer of two lipopeptide antibiotics, surfactin and plipastatin. Section 6.3 section will show that 168 is converted into a coproducer of surfactin and plipastatin by introducing *sfp* and the pleiotropic regulator gene *degQ*, which is also mutated in 168 (11). However, on the basis of whole-genome sequence data, 168 does not have iturin group operons (13).

To investigate the nature of the horizontal transfer of an antibiotic producer gene in terms of antibiotic production, the iturin A operon of RB14 was transferred to the non-iturin-producing strain 168, using a positive-selection system that employs the *cI* repressor gene of lambda phage as a reverse marker (43–45). In this study,

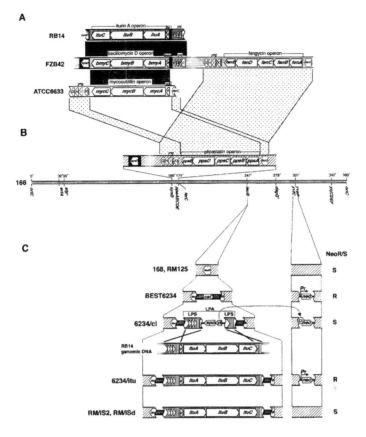

FIGURE 6.9 (A) Structure and flanking regions of the iturin A, mycosubtilin, and bacillomycin D operons. Shaded bridges between different strains connect regions homologous with the iturin A operon of RB14, while dotted bridges connect regions homologous with the plipastatin operon of 168. Solid black regions in the iturin A and bacillomycin D operons indicate regions homologous with the *xynD* gene of 168. Vertical lines, horizontal lines, and areas not lined represent *B. subtilis* RB14 DNA, *B. amyloliquefaciens* FZB42 DNA (41), and *B. subtilis* ATCC 6633 DNA (2), respectively. Fengycin is another name for plipastatin (41). (B) Whole-genome map of strain 168 (diagonal lines). The counterparts of the relevant genes in panel A are mapped. (C) Strain development. The positive-selection system is also indicated. The CI repressor in 6234/cl represses *Pr-neo* expression, resulting in neomycin susceptibility. However, once substitution of the cl gene by transformed DNA takes place, the transformant becomes neomycin resistant. Solid black regions show the tetracycline (labeled pBR) and ampicillin (labeled 322) resistance gene sides of the split pBR322 sequence. The genes *cat* and *spc* represent the chloramphenicol and spectinomycin resistance genes, respectively. The genes cl and *Pr-neo* indicate the CI repressor gene and the *Pr* promoter-driven neomycin resistance gene, respectively. BEST6234, 6234/cl, and 6234/itu also have a *Pr-neo* cassette in their genome; however, this cassette is absent in the RM/iS2 and RM/iSd series. The NeoR/S labels at the far right indicate neomycin resistance (R) or sensitivity (S). LPS, landing pad sequence; LPA, LPSs array.

although *sfp* is necessary, it shows that the iturin A operon is essentially the only operon required for conversion of strain 168 into an iturin A producer. It is also demonstrated that in the presence of *degQ*, an iturin A, production almost comparable to that of the donor strain was achieved.

6.2.1   MATERIALS AND METHODS

6.2.1.1   Bacterial Strains, Phage, Plasmids, and Media

The bacterial strains, phage, and plasmids used in this study are listed in Table 6.2.
   The LB medium and no. 3S medium used are described in Section 6.1.

6.2.1.2   Transformation and DNA Manipulation

The *B. subtilis* strain was transformed by the method of Anagnostopoulos and Spizizen (3, 46). Routine DNA manipulation, *E. coli* transformation, contour-clamped homogeneous electric-field (CHEF) pulsed-field gel electrophoresis, and Southern hybridization were performed as previously described (11). The CHEF conditions were as follows: field strength, 3 V/cm; pulse time, 45 s; and running time, 44 h.

6.2.1.3   Construction of LPA

A landing pad sequence array (LPA), in which the *cl* repressor gene and spectino-mycin resistance gene are flanked by two landing pad sequences (LPSs), was constructed in plasmid pBR322 in *E. coli* as follows. A 9 kb *Eco*RI fragment with a downstream edge of the iturin operon was obtained from recombinant phage λ69, derived from a lambda DASH II phage library of the RB14 chromosome (1), and cloned into the *Eco*RI site of pBR322 to obtain plasmid pBRE9k. An 8 kb *Hind*III fragment, with an upstream edge of the iturin A operon and a transposon, was obtained from pBRHd8k (1) and ligated to the largest fragment of the products of the *Eco*RV digestion of pBRE9k, generating the plasmid pBRE4H8. pBRE4H8 was digested with both AgeI and *Cpo*I, and then the largest fragment (10 kb) obtained was purified by electrophoresis. The obtained fragment was dephosphorylated by alkaline phosphatase from *E. coli* (Toyobo, Inc., Japan), blunted with a DNA blunt-ing kit (Takara Shuzo Co., Japan), and then ligated to the Small fragment of the *cl-spc* cassette from pCISP303B#6 (44), resulting in plasmid pBRE2H4*cl*, containing the LPA.

6.2.1.4   Preparation of a Recipient Strain Containing the LPA

The *Pr-neo* cassette introduction plasmid was constructed with a substitutional insertion of an *Eco*RI fragment of *Pr-neo* from pBEST515C (43) between two *Not*I sites in pNEXT4 (12). The resulting plasmid, pNEXT4PN-2, was used to trans-form *B. subtilis* RM125, a strain 168 derivative that lacks a restriction/modifica-tion system, generating BEST6225. The LPA carried by pBRE2H4cl was inserted into the *leuB* region in the BEST6225 genome as follows. Plasmid pBR322Cm (50), which harbors a chloramphenicol resistance gene (*cat*) cassette in the *Eco*RI site of pBR322, was linearized at a unique *Pvu*II site and then inserted into a unique *Bam*HI site in the leuB gene of plasmid RSF2124.B.leuB (49) by using a T4 DNA

TABLE 6.2
Bacteria, Phage, and Plasmids Used in This Study

Strain, Phage, or Plasmid	Characteristic(s)[a]	Reference or Source
E. coli Strain		
JA221	*F⁻ hsdR hsdM⁺ trp leu lacY recAl*	(12)
B. subtilis Strains		
RB14	IT⁺ SF⁺ PL⁺	(1)
168	*trpC2 sfp⁰ degQ⁰*, IT⁻ SF⁻ PL⁻	(13)
RM125	*leuB8 arg-15 ΔSPβ* R(*hsdR*-colA⁺ -*purB*⁺)202-5 *sfp⁰ degQ⁰*, IT⁻ SF⁻ PL⁻	(48)
BEST6225	RM125 Δ(*yvfC-yveP*)::Pr-*neo*, IT⁻ SF⁻ PL⁻	This study
BEST6234	RM125 Δ(*yvfC-yveP*)::Pr-*neo leuB*::pBR322::*cat*, Cm^r Nm^r IT⁻ SF⁻ PL⁻	This study
6234/cI	BEST6234; *leuB*::pBR322::[(*yxjCDEF*$_{RB14}$-*ituD*)-(*cI-spc*)-*ituC-xynD*$_{RB14}$)], Sp^r Cm^s Nm^s IT⁻ SF⁻ PL⁻	This study
6234/itu	BEST6234 (host) ×RB14 (DNA); *leuB*::pBR322::(*yxjCDEF*$_{RB14}$-*ituD⁺ A⁺ B⁺C⁺* -*xynD*$_{RB14}$), Nm^r Sp^s IT⁻ SF⁻ PL⁻	This study
RM/iS2	6234/itu::pMMN6(*sfp⁺ cat*) R[Δ(*yvfC-yveP*)::Pr-*neo*]RM125, Cm^r Nm^s IT⁺ SF⁺ PL⁺	This study
RM/iSd4, -12, -14, and -16	RM/iS2::pUC19HPIN$_{mrf}$(*degQ*$_{YB8}$⁺ *neo*), Cm^r Nm^r IT⁺ SF⁺ PL⁺, clone numbers 4, 12, 14 and 16	This study
RM/Sp6	RM125::pMMN6(*sfp⁺, cat*), Cm^r IT⁻ SF⁺ PL⁺	This study
Phage		
λ69	Lambda DASH II cloned downstream edge of iturin A operon	(1)
Plasmids		
pBR322	Cloning vector, Ap^r Tc^r	(10)
pNEXT4PN-2	Pr-*neo* cassette introduction plasmid, Ap^r Nm^r	This study
RSF2124.B.leuB	EcoRI fragment of *B. subtilis leuB* gene cloned into ColiE1	(49)
pBR322Cm	*cat* cassette cloned into EcoR1 site of pBR322, Ap^r Tc^r Cm^r	(50)
pBMAP103-322CA	RSF2124.B.leuB-inserted pBRCm in the *leuB* gene	This study
pCISP303B#6	*cI-spc* cassette cloned into pBR322, Ap^r Tc^r Sp^r	(44)
pBRE9k	9 kb EcoRI fragment from λ69 containing downstream edge of iturin A operon (i.e., *ituC-xynD*$_{RB14}$) cloned in EcoR1 site of pBR322, Ap^r Tc^r	This study
pBRHd8k	8 kb HinDIII fragment from strain 1006 containing promoter region of iturin A operon (i.e., *yxjCDEF*$_{RB14}$-*ituDA*) inserted into HindIII site of pBR322, Ap^r Cm^r	(1)
pBRE4H8	pBRE9K-inserted 8 kb HindIII fragment from pBRHd8k into EcoRV site by replacing small EcoRV fragments, Ap^r Cm^r	This study
pBRE2H4cI (LPA)	pBRE 4H8-inserted 2.5 kb Sma1 fragment containing *cI-spc* cassette from pCISP303B#6 into AgeI-CpoI site by replacing a small fragment, Ap^r Sp^r	This study
pMMN6	*E.coli* plasmid containing functional *sfp*, Ap^r Cm^r	(51)
pUC19HPINmrF	*degQ* of strain YB8 cloned into *E. coli* plasmid, Ap^r Nm^r	(11)

[a] IT, iturin A production; SF, surfactin production; PL, plipastatin production; Ap^r, ampicillin resistant; Tc^r, tetracycline resistant; Cm^r or Cm^s, chloramphenicol resistant or sensitive; Nm^r or Nm^s, neomycin resistant or sensitive; Sp^r or Sp^s, spectinomycin resistant or sensitive; *cat*, chloramphenicol resistance gene; *neo*, neomycin resistance gene; *spc*, spectinomycin resistance gene.

polymerase-based blunting kit. The resulting plasmid, pBMAP103-322CA, was used to transform BEST6225, generating BEST6234, which has pBR322Cm inserted in the genomic *leuB* region (Figure 6.9C). BEST6234 was transformed with pBRE-2H4*cI* and selected using spectinomycin. Transformants were assayed for chloramphenicol sensitivity by plate replication, and then a chloramphenicol-sensitive strain, which resulted from the doublecrossover recombination between pBR322Cm and pBRE2H4*cI* (Figure 6.9C), was selected. The obtained strain was designated 6234/cI and used as the recipient of the iturin A operon.

6.2.1.5   Iturin Operon Transfer

High-molecular-weight whole DNA of strain RB14 was prepared according to a method previously reported (2). One hundred microliters of competent cell culture of 6234/cI was mixed with 10 μl of 1 μg/μl RB14 chromosome. Following incubation at 37°C for 30 min, 300 μl of LB medium was added to the culture, which was incubated with gentle agitation at 37°C for 3 h to allow the expression of neomycin resistance. The culture was then plated on LB plates containing neomycin and incubated at 30°C overnight. Colonies that appeared on the plates were streaked on two LB plates, one containing spectinomycin and the other containing neomycin, for the screening of spectinomycin-sensitive colonies. Selected colonies were then picked up with a toothpick and inoculated in 25 μl of PCR solution (5 U of TaKaRa Ex *Taq* DNA polymerase [Takara Shuzo, Kyoto, Japan], 10 μl of Ex *Taq* buffer, and 8 μl of deoxynucleoside triphosphate solution [2.5 mM each]) with the primers ITUPl-F (5′-AGCTTAGGGAACAATTGTCATCGGGGCTTC-3′, positioned from nucleotide 15353 to 15383 of the iturin A operon sequence [DDBJ/EMBU GenBank accession no. AB050629]) and ITUP2-R (5′-TCAGATAGGCCGCC ATATCGGAATGATTCG-3′, complementary sequence positioned from nucleotide 17326 to 17355 of AB050629), which are able to detect a 2 kb region that includes the intergenic sequence between *ituA* and *ituB*. The colony PCR conditions were as follows: 96°C for 5 min; 30 cycles of 96°C for 30 s, 60°C for 30 s, and 72°C for 150 s.

6.2.1.6   Introduction of *sfp* and *degQ*

The *sfp*-harboring *E. coli* plasmid pMMN6 (51) was inserted into the genome of 6234/itu by Campbell-type insertion. In this transformation, genomic DNA of RM125 was simultaneously transferred to remove *Pr-neo* from the *yvfC–yveP* region for the following experiment. Thus, a chloramphenicol-resistant, neomycin-sensitive colony was selected and designated RM/iS2. This strain harbors pMMN6 in the *sfp⁰* region. We did not determine in which site the actual insertion in RM/iS2 occurred. The *degQ*$_{YB8}$⁻ containing *E. coli* plasmid pUC19HP1NmrF (11) was transformed into the RM/iS2 strain and selected for neomycin. Since pUC19HP1NmrF has three potential sites for Campbell-type insertion in the RM/iS2 genome (one is *degQ⁰* and the others are ampicillin resistance genes in genomic pBR322 and pMMN6), several transformants were selected and designated the RM/iSd series.

6.2.1.7   Quantitative Analysis of Iturin A, Plipastatin, and Surfactin

The culture (40 ml of no. 3S medium, 30°C) of the *B. subtilis* strain was acidified to pH 2.0 with 12 N HCl. Then, the precipitate formed was collected by centrifugation and extracted with methanol. Iturin A, plipastatin, and surfactin in the extracted solution were quantified by reversed-phase high-performance liquid chromatography (HPLC) using a two-eluent gradient as described previously (1, 11). For the detailed composition analysis of the fatty moiety of the β-amino acid of iturin A, the methanol extract was subjected to another reversed-phase HPLC using one eluent as described previously (5).

6.2.2 Results

6.2.2.1 Horizontal Transfer of Iturin A Synthetase Operon to Strain 168

To horizontally transfer the complete iturin A operon (42 kb) in one step, *B. subtilis* BEST6234, which was constructed from RM125 (48), a derivative of 168, was used. Briefly, strain BEST6234 has two functional units in its genome. One is a *Pr-neo* cassette, which is a neomycin resistance gene (*neo*) whose original promoter was replaced with the *Pr* promoter from lambda phage to control *neo* expression by a cI gene of lambda phage. This cassette was substitutionally inserted between the Not1 sites of *yvfC* and *yveP*. The other is pBR322Cm, containing a chloramphenicol resistance gene, inserted into the *Bam*HI site in the *leuB* coding sequence (Figure 6.9). Prior to the horizontal transfer of the iturin A operon, two landing pad sequences (2 kb and 4 kb), corresponding to the two edges of the region to be transferred and bracketing a cassette of the *cI* repressor gene with the spectinomycin resistance gene (*spc*) between them, were assembled in pBR322 in *E. coli* and then localized in the genomic pBR322 sequence in the *leuB* gene of BEST6234, as shown in Figure 6.9. The resulting strain, 6234/cI, was cultivated to develop competent cells, supplemented with whole chromosomal DNA of RB14, and then plated onto neomycin-containing LB plates. About 100 neomycin-resistant transformants were then screened for sensitivity to spectinomycin, and 20 transformants showing a neomycin-resistant and spectinomycin-susceptible phenotype were selected. These transformants were subjected to a colony PCR assay using specific primers for the internal region of the iturin A operon. Two of them yielded a 2 kb PCR product, and one of these was chosen for further experiments and named 6234/itu. To confirm the transfer of the iturin A operon to the *leuB* region of the recipient strain, the whole chromosome of 6234/itu was digested with Not1 or SfiI, fractionated by pulsed-field gel electrophoresis, and then subjected to Southern hybridization analysis using pBRE4H8 as the probe. As shown in Figure 6.10, strong signals from 6234/itu were observed because pBRE4H8 has 12 kb long iturin A operon sequences plus the pBR322 sequence. On the other hand, BEST6234 showed weak bands, which are due to hybridization between the genomic copy of pBR322 and the pBR322 sequences of pBRE4H8. The sizes of the observed bands are consistent with the expected values from the whole-genome sequence (Figure 6.10). 6234/itu was thus confirmed to have integrated the 42 kb region containing the iturin A operon into the *leuB* region.

6.2.2.2 Conversion of Iturin A Operon-Transferred Strain into an Iturin A Producer

In Chapter 5, it was demonstrated that the 4′-phosphopantetheinyl transferase *sfp* gene is essential for iturin A production as well as for surfactin production in the RB14 strain (3, 4).

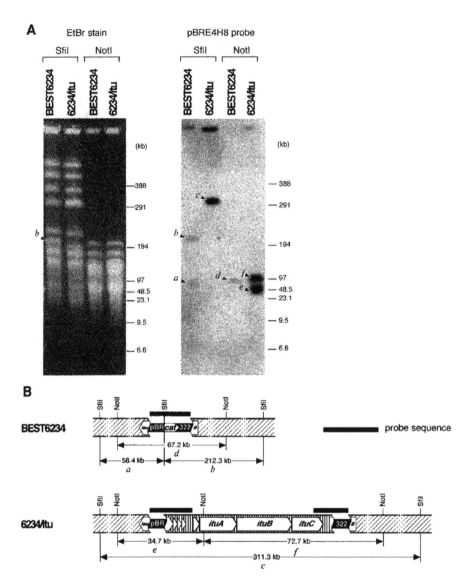

FIGURE 6.10 Confirmation of iturin A operon transfer. (A) Southern hybridization analysis. SfiI and NotI digestion of BEST6234 and 6234/itu genomic DNA prepared in an agarose gel block were fractionated by CHEF pulsed-field gel electrophoresis and subjected to Southern hybridization analysis using pBRE4H8 as the probe. Letters correspond to labeled portions of the map in panel B. EtBr, ethidium bromide. (B) Physical map around *leuB* of iturin A operon-transferred strain. Lettered fragments on the map were actually observed as bands on the photographs in panel A. This figure is drawn on the basis of the whole-genome sequence of strain 168 (DDBJ/EMBL/GenBank accession no. AL009126). Diagonal and vertical lines represent *B. subtilis* 168 DNA and *B. subtilis* RB14 DNA, respectively.

Since strain 168 lacks a functional 4'-phosphopantetheinyl transferase gene (*sfp*), a functional *sfp* gene is introduced by a Campbell-like insertion of the *sfp*-containing plasmid pMMN6 (51). When *sfp* was transferred to 6234/itu, the resulting strain, RM/iS2, produced iturin A at a concentration of 8 µg/ml in no. 3S medium at 30°C for 120 h, while the control strain, 6234/itu, and RM/Sp6 did not produce iturin A (Figure 6.11). However, the production of iturin A by RM/iS2 was 13-fold less than that by RB14 (105 µg/ml).

In the following Section 6.3, it was demonstrated that the introduction of the *degQ*$_{YB8}$ gene into the *sfp*+ strain 168 derivative causes plipastatin hyperproduction (11). Analogous with the previous study, the effect of *degQ*$_{YB8}$ on iturin A production by the *sfp*+ strain RM/iS2 was examined. By transformation with plasmid pUC19HP1NmrF, which carries *degQ*$_{YB8}$, 28 transformants (RM/iSd series) were obtained, and 4 of them, randomly selected (RM/iSd4, -12, -14, and -16), were subjected to further investigation. Southern hybridization analysis was performed to analyze the integration sites of pUC19HP1NmrF in these transformants. In each of them, the integration site of the plasmid was confirmed to be into one of the two ampicillin resistance genes that are in the genomic copies of pBR322 (RM/iSd4, -12, and -14)

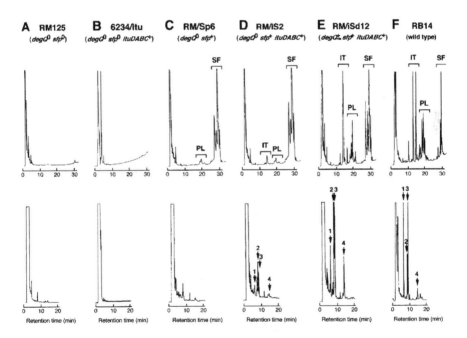

FIGURE 6.11 HPLC peak patterns of three lipopeptides (iturin A [IT], surfactin [SF], and plipastatin [PL]) (top) and those focused on iturin A (bottom) produced by (A) RM125, (B) 6234/itu, (C) RM/Sp6, (D) RM/iS2, (E) RM/iSd12 as represented by four RM/iSd strains, and (F) RB14 in no. 3S medium at 30°C for 120 h. Two distinct HPLC conditions were used to analyze one sample. The top chromatographs were obtained with a two-eluent gradient, while bottom chromatographs were obtained with one eluent. Peaks 1, 2, 3, and 4 correspond to iturin A whose β-amino acids are *n*-C$_{14}$-β-amino acid, *anteiso*-C$_{15}$-β-amino acid, *iso*-C$_{15}$-β-amino acid, and *n*-C$_{16}$-β-amino acid, respectively.

or pMMN6 (RM/iSdl6) and not into the *degQ* genomic allele (data not shown). This occurred probably because the length of *degQ*$_{YB8}$ in pUC19HP1NmrF is 0.5 kb, shorter than the 1 kb long ampicillin resistance gene present in the vector portion of pUC19HP1NmrF. In consideration of the two integration sites of pUC19HP1NmrF in RM/iS2, all of the four transformants of the RM/iSd series were assayed for iturin A production. The production levels by the four strains ranged from 51 to 64 µg/ml, which was six- to eightfold higher than that of RM/iS2 (not containing *degQ*$_{YB8}$) and one-half that of RB14. The time course of iturin A production, as well as those of surfactin and plipastatin production in the RM/iSd series strains, was similar to that of iturin A production in RB14 (data not shown). A slight difference was observed in the detailed peak compositions of the iturin A produced, which was caused by a structural difference of the β-amino acids in iturin A (Figure 6.11). Generally, the highest peak of iturin A produced by RB14 was peak 1, which corresponds to the *n*-C$_{14}$-β-amino acid, while that of the RM/iSd series strains was peak 2, which corresponds to the *anteiso*-C$_{15}$-β-amino acid. As well as the differences in iturin A production, the fatty acid compositions of surfactin and plipastatin produced by RB14 were different from those seen for the 168 derivatives.

6.2.3 DISCUSSION

Using a positive-selection system, a 42 kb region was transferred that contains a 38 kb complete iturin A operon that had no selectable markers. In the previous artificial transfer of the subtilin operon or the bacitracin operon into strain 168, transformants were obtained from a donor DNA that had a selection marker inserted into the relevant operon at the original host level for the direct selection of the transferred recombinant (3, 4). It is implied that the donor of this type of horizontal transfer is limited to the organisms that have established genetic transformation to introduce the selectable marker in the donor genome prior to horizontal transfer. The positive-selection system was performed, which employs the CI repressor and a Pr promoter-driven neomycin resistance gene (*Pr-neo*) to detect transfer. In this system, the CI repressor in the host strain usually represses *Pr-neo* expression, causing a neomycin-susceptible phenotype. However, once the transferred DNA is substituted for the cl gene, the host strain is no longer able to repress the *Pr-neo* expression and becomes neomycin resistant. Since this system does not require any selection marker in the donor DNA, it is suitable for the study of general horizontal transfer (43–45).

The utmost relevance in this study was to identify which genes are required for the conversion of a heterologous host into an antibiotic producer. Several genes are thought to be involved in antibiotic production: for example, regulator genes that control the conditional expression of antibiotic synthetase genes; antibiotic synthetase or structural genes; modification genes; and self-resistance, or efflux pump, genes. In general, these antibiotic production-related genes would form a cluster in the genome. In particular, synthetase genes are usually in one transcriptional unit that is expressed from one promoter in a polycistronic manner. This centered regulation appears to reflect the need for a strict expression control of synthetases. However, Guenzi et al. demonstrated that by inserting a secondary promoter into the surfactin operon, coordinate transcription of surfactin synthesis is not necessary for surfactin production (56). From the viewpoint of horizontal transfer, the cluster feature is reasonable, due to the very low frequency of transformation of multiple-donor regions. If these genes are separated into several parts, the simultaneous transfer of several genes necessary for conversion into a producer may be quite rare. When the cluster feature is a consequence of horizontal transfer, it is probable that the transfer of one cluster may be sufficient for conversion into a producer. Indeed, in a previous horizontal transfer of a bacitracin operon to strain 168, the transferred genes contained self-resistance genes and their regulator genes as well as synthetase genes, all of which are members of the cluster (54). In the transfer of the antibiotic subtilin, the transferred segment has not only the structural gene of the antibiotic but also the regulator genes of the structural gene, modification gene, immunity gene, and efflux gene, which are also of the same cluster (55, 57, 58). Thus, it was postulated that the iturin A operon, composed of one transcriptional unit encoding four synthetase genes, has the ability to completely convert a heterologous host into an iturin A producer upon transfer of this operon.

The 6234/itu transformant carrying the iturin A operon was not able to produce iturin A due to the presence of a null *sfp* allele in this strain. It was previously demonstrated that the *sfp* homologue gene *lpa-14* is essential for iturin A production in RB14 (3, 4). When *sfp* was introduced into the 6234/itu strain, the strain produced iturin A, although at a very low level (Figure 6.11). However, the integration of the additional pleiotropic regulator, *degQ*, which is also mutated in strain 168, resulted in about half of the iturin A production as that of RB14. Therefore, the conclusion is that the iturin A operon is the only cluster required for conversion of the non-iturin A producer *B. subtilis* into an iturin producer.

Even without the introduction of a self-resistance, or efflux pump, gene, the transferred strain produced iturin A in the heterologous host, indicating that it has an intrinsic resistance to or secretion mechanism for iturin A. The resistance, or efflux, mechanism in the transferred strain should be investigated.

6.3   GENETIC ANALYSIS FOR CONVERSION OF *B. SUBTILIS* 168 TO PLIPASTATIN PRODUCTION (11)

B. subtilis YB8 suppresses the growth of phytopathogenic fungi *in vitro* in Chapter 5 (7). The suppressive effect of strain YB8 is mainly due to production of the antifungal lipopeptide antibiotic plipastatin. Strain YB8 produces surfactin as well as plipastatin (59) and cloned and characterized the gene *lpa-8* (*sfp* [60]), which is required for the production of both plipastatin and surfactin in strain YB8 (7). The gene *lpa-8* encodes 4′-phosphopantetheinyl transferase, which converts inactive apoenzyme peptide synthetases to their active holoenzyme forms by posttranslational transfer of the 4′-phosphopantetheinyl moiety of coenzyme A to the synthetases (61). Strains 168 and MI113 have an inactive *sfp* allele (*sfp*⁰), which is why they cannot produce surfactin even though they have the intact surfactin operon. In this section, MI113 was converted into a coproducer of plipastatin and surfactin by transformation with YB8 chromosomal DNA.

The *B. subtilis* genome project determined the DNA sequence of strain 168 and revealed that there are two large operons that encode nonribosomal peptide synthetases (13). The surfactin operon is located between 32° and 35°. The other operon, located between 167° and 171° (*pps* operon), was thought to be the fengycin operon, because significant homology was observed between the fengycin synthetase gene of fengycin-producing *B. subtilis* F29-3 (14) and the operon from strain 168 (14, 29, 30).

Transposon mutagenesis was applied to the resultant transformant, strain 406, and it was determined that the *pps* operon and *lpa-8* are both essential for plipastatin production. To prove directly that the *pps* operon in strain 168 encodes plipastatin synthetases, the HPLC system was improved so as to enable the detection of a trace amount of plipastatin production from strain 168 supplied with *lpa-8*. This is the first report showing that although the *pps* operon in strain 168 is still active, this strain cannot produce plipastatin because of the sfp^0 mutation.

The *degQ* gene of YB8 (designated $degQ_{YB8}$) as an enhancer of extracellular protease production in strain 168 had been cloned. *degQ* is a pleiotropic regulatory gene that controls the production of degradative enzymes, an intracellular protease and several secreted enzymes (levansucrase, alkaline proteases and metalloproteases, α-amylase, β-glucanase, and xylanase) (62, 63). When the gene $degQ_{YB8}$ in strain YB8 was ablated, plipastatin production in YB8 was severely reduced but no significant change was observed in surfactin production, indicating that $degQ_{YB8}$ is necessary for plipastatin production in YB8. Therefore, the role of $degQ_{YB8}$ in the production of plipastatin was investigated in strain 168 expressing plasmid-borne *lpa-8*.

6.3.1 Materials and Methods

6.3.1.1 Strains and Media

The *B. subtilis* strains and plasmids used in this study are listed in Table 6.3. Plasmid pHV1249 was obtained from the Bacillus Genetic Stock Center of Ohio State University, and plasmids pNEXT24, pNEXT44, and pNEXT24A were obtained from M. Itaya (64). Low-salt Luria-Bertani (LB) medium contained (per liter) 10 g of Polypepton (Nippon Pharmaceutical Co. Ltd., Tokyo, Japan), 5 g of yeast extract, and 5 g of NaCl, and was adjusted to pH 7.2. ACS medium (65) containing (per liter) 100 g of sucrose, 11.7 g of citric acid, 4 g of Na_2SO_4, 5 g of yeast extract, 4.2 g of $(NH_4)_2HPO_4$, 0.76 g of KCl, 0.420 g of $MgCl_2$ $6H_2O$, 10.4 mg of $ZnCl_2$, 24.5 mg of $FeCl_3 \cdot 6H_2O$, and 18.1 mg of $MnCl_2$ $4H_2O$, was adjusted to pH 6.9 and was used in the production of plipastatin. When necessary, antibiotics were added at the following concentrations: ampicillin, 50 µg/ml; chloramphenicol, 5 µg/ml; erythromycin, 10 µg/ml; tetracycline, 20 µg/ml; and neomycin, 20 µg/ml.

Plipastatin production *in vitro* was detected by the formation of a clear inhibitory zone on LB agar medium containing a spore suspension of a phytopathogenic fungus, *Fusarium oxysporum* f. sp. *lycopersici* J1 SUF119, as previously described (7).

6.3.1.2 Purification of Plipastatins

The purification of plipastatin was carried out by using a modification of a method previously developed (7). *B. subtilis* 406 was cultivated in 200 ml of ACS medium in a 500 ml shake flask for 2 days at 30°C. The 60 liters of culture collected from

TABLE 6.3
Strains and Plasmids Used in This Study

Strain or Plasmid	Characteristic(s)[a]	Source or Reference
E. coli		
JM109	*relA1 supE44 endA1 hsdR17 gyrA96 mcrA mcrB*+ thi Δ(lac-proAB)/F' [*traD36 proAB*+ *lacI*q *lacZ* ΔM15]	
JM110	*dam dcm supE44 hsdR17 thi leu rpsL1 lacY galK galT ara tonAthr tsx Δ (lac-proAB)/F' [traD36 proAB+ lacIq lacZ ΔM15]*	
B. subtilis		
168	*trpC2*	
168(pC81AP)	168 with plasmid pC81AP; Cm^r	
MI113	*arg-15 trpC2 hsmM hsrM*	(66)
YB8	Wild-type plipastatin and surfactin coproducer	(7)
YB8*degQ*YB8::Nm^r	YB8 *degQ*YB8::Nm^r	This work
YB8*degQ*YB8::Nm^r(pCRV)	YB8 *degQ*YB8::Nm^r with plasmid pCRV; Nm^r Cm^r	This work
406	MI113::YB8 plipastatin and surfactin coproducer	This work
706	406 (*lpa-8*::mini-Tn*10*; Cm^r)	This work
703, 716, 724, 729	406 (plipastatin-deficient mini-Tn*10*-inserted mutant; Cm^r)	This work
751	406 (*ppsB*::Nm^r)	This work
752	406 (*ppsD*::Nm^r)	This work
753	406 (*ppsDE*::pNEXT44; Cm^r)	This work
790(pC81AP)	168(pC81AP) (*ppsB*:: Nm^r)	This work
791(pC81AP)	168(pC81AP) (*degQ*0::pUC19HP1Nm^rF; Nm^r)	This work
Plasmids		
pTB522	10.5 kb; Tc^r	(9)
pHV1249	Ap^r Em^r Cm^r; mini-Tn*10*	(8)
pNEXT24	Linker region between 12 kb and 23 kb *Not*I fragment from 168 chromosomal DNA clone in pBR322; Ap^r	(12)
pNEXT44	Linker region between 23 kb and 143 kb *Not*I fragment from 168 chromosome clone in pBR322; Ap^r Cm^r	(12)
pNEXT24A	Nm^r cassette inserted into *Not*I site of pNEXT24	(64)
pNEXT44B2	Nm^r cassette of pBEST502 inserted into *Not*I site of pNEXT44	This work
pBEST502	Nm^r Ap^r	(6)
pUC19	Ap^r	
pUCN1	Nm^r cassette from pBEST502 inserted into *Eco*RI site of pUC19	This work
pUC19HP1	pUC19 with 0.8 kb *Eco*RV fragment containing *degQ*YB8	This work
pUC19HP1Nm^rF	pUC19HP1 with Nm^r cassette from pUCN1 inserted into *Sse*83871 site	This work
pUC*degQ*YB8::Nm^r	pUC19HP1 with *degQ*YB8::Nm^r	This work
pE194	Em^r thermosensitive origin	
pE*degQ*YB8::Nm^r	pE194::pUC*degQ*YB8::Nm^r	This work

(*Continued*)

TABLE 6.3 (CONTINUED)
Strains and Plasmids Used in This Study

Strain or Plasmid	Characteristic(s)[a]	Source or Reference
pC194	Cm^r	
pCRV	pC194 with 0.6-kb *Eco*RI fragment containing $degQ_{YB8}$; Cm^r	This work
pC81AP	pC194 with 1.3-kb *Apa*I-*Pvu*II fragment containing *lpa-8* YB8; Cm^r	(7)

[a] Ap^r, ampicillin resistant; Cm^r, chloramphenicol resistant; Em^r, erythromycin resistant; Nm^r, neomycin resistant; Tc^r, tetracycline resistant.

300 flasks was adjusted to pH 2.0 with HCl, and the acid precipitate was collected and extracted with 3 liters of 95% ethanol. The extract was concentrated under reduced pressure and extracted with a solution of butanol and H_2O (1 liter:1 liter). The butanol phase was recovered, and 3 volumes (3.3 liters) of hexane were added. After separation of the suspension, the aqueous phase was recovered and 300 ml of butanol–hexane (1:1 [vol/vol]) was added. After centrifugation (8,000 × g for 10 min), the supernatant was concentrated to produce a brown crude powder, which was then dissolved in 12 ml of propanol. This solution was loaded on a propanol-filled column of Silica Gel 60 (Merck) (× 20 [wt/wt] powder) and was successively eluted with 1 column volume of propanol, 2 column volumes of 90% propanol, and 2.5 column volumes of 80% propanol. The 80% propanol eluate was further purified by reversed-phase preparative HPLC with a Prep-ODS column (GL Sciences, Tokyo, Japan; 2 cm diameter by 25 cm at a flow rate of 10 ml/min with detection at 205 nm) with a mixture of acetonitrile and 1 mM trifluoroacetic acid (1:1 [vol/vol]). The main peak was concentrated and subjected to the same HPLC system with a mixture of acetonitrile and 10 mM ammonium acetate (1:1 [vol/vol]). Two main fractions, called fractions A and B, were collected. Fraction A was further separated into peaks 1 and 2 by using the same HPLC system with a mixture of acetonitrile and acetate buffer (2% potassium acetate plus 6% acetic acid) (1:1 [vol/vol]). Fraction B was separated into peaks 3 and 4 by using the same HPLC system with a mixture of 40 mM ammonium acetate and acetonitrile (1:1 [vol/ vol]). All of these fractions were passed through Sephadex LH-20 columns with 80% methanol to remove the salts.

6.3.1.3 HPLC Analysis of Plipastatin

Plipastatin was extracted as follows. After 2 days' cultivation of a *B. subtilis* strain in 40 ml of ACS medium, the culture was acidified to pH 2.0 with 12 N HCl. Then the precipitate was collected by centrifugation and was extracted with ethanol. The extracted solution was filtered through a 0.2 μm pore size polytetrafluoroethylene membrane (Advantec, Tokyo, Japan).

Plipastatin was detected and quantified by reversed-phase HPLC as follows: The aforementioned filtrate was injected into an HPLC column (Inertsil ODS-2 [4.6 mm diameter by 250 mm]; GL Sciences), which was eluted at a flow rate of 1.0 ml/min with two solvent gradients of 0.05% trifluoroacetic acid (eluent A) and acetonitrile–isopropyl alcohol (3:7 [vol/vol]) plus 0.02% trifluoroacetic acid (eluent B). The gradient conditions were as follows: Starting at 60% eluent A and 40% eluent B, eluent A was linearly decreased to 0% by increasing the amount of eluent B over 30 min and was kept at 0% for over 5 min.

6.3.1.4 Mass Spectrometry and ¹H NMR

Collisionally activated dissociation (CAD) mass spectra of protonated molecular ions (M+ H]⁺ by fast atom bombardment (FAB) ionization were obtained by the *B/E* linked scanning method with a JMS-700 MStation (JEOL Ltd., Tokyo, Japan) under the following conditions: 6 keV of xenon for impacting particles, glycerol plus *m*-nitrobenzyl alcohol for the matrix, 0.1% trifluoroacetic acid for the solvent, and helium for the collision gas. ¹H nuclear magnetic resonance (NMR) spectra of purified substances were recorded on a JNM-GXSOO NMR spectrometer (JEOL Ltd.) at 500 MHz and 55°C in methanol-d_4.

6.3.1.5 Transformation and DNA Manipulation

The preparation of competent cells, transformation of *E. coli* and *B. subtilis*, and routine DNA manipulations were performed by the method of Anagnostopoulos and Spizizen (67) with slight modifications, as previously described (7).

6.3.1.6 Transposon Mutagenesis

Transposon mini-Tn*10* mutagenesis was carried out as described by Petit et al. (8). The transposon-carrying plasmid pHV1249 was introduced into *B. subtilis* 406 by the competent cell method (7), and the resultant chloramphenicol-resistant (Cmʳ) colonies were inoculated into LB medium containing chloramphenicol and incubated at 30°C. When the culture reached the logarithmic phase (optical density at 660 nm of 0.6 to 0.8), it was plated on LB agar medium containing chloramphenicol, followed by incubation at 51°C. Thermoresistant Cmʳ colonies were regarded as mini-Tn*10* insertional mutants.

6.3.1.7 Southern Blot Analysis

For *Not*I restriction analysis, DNA was purified in agarose plugs by a previously reported method (12). DNA fragments were separated by using a handmade contour-clamped homogeneous electric field pulsed-field gel electrophoresis (PFGE) apparatus (68) with a 1% agarose gel (Sigma; type II) with 0.5x TBE (lx TBE is 89 mM Tris-89 mM boric acid—2 mM EDTA) at 5 V/cm for 20 h at 13°C with a switching ramp of 6 to 15 s, which was designed to optimally separate DNA fragments in the range from 9 to 300 kb. Concatemers of λ phage (New England Biolabs, Inc.) were used as standard molecular weight markers. The digoxigenin DNA labeling and detection kit (Boehringer, Mannheim, Germany) was used for hybridization and detection according to the methods described by the manufacturer, and disodium 3-(4-methoxyspiro {1,2-dioxetane-3,2′-(5′-chloro)tricyclo[3.3.1.1³,⁷]decan}-4-yl) phenyl phosphate (CSPD; Tropix, Inc., Bedford, Massachusetts, USA) was used as a chemiluminescent substrate of alkaline phosphatase. Prehybridization was carried out for 1 h at 42°C in a solution containing 5x SSC (1x SSC is 0.15 M sodium chloride plus 0.015 M sodium citrate [pH 7.0]), 50% formamide, 50 mM sodium phosphate buffer (pH 7.0), 7% sodium dodecyl phosphate, 2% skim milk, 0.1% lauroylsarcosine, and 50 μg of fish sperm DNA per ml. Hybridization was performed overnight at 42°C in a prehybridization solution containing a denatured digoxigenin-labeled DNA probe. Filters were washed twice for 15 min at 68°C in 0.1x SSC containing 0.1% sodium dodecyl sulfate.

6.3.1.8   Construction of *degQ*-Related Mutants

The 0.8 kb *Eco*RV fragment containing *degQ* of YB8 (*degQ*$_{YB8}$) was cloned into the *Hinc*II site of pUC19, creating plasmid pUC19HP1. The *Bam*HI fragment of the neomycin resistance (Nm^r) gene cassette from pUCN1 was inserted into the *Fba*I site in the *degQ*$_{YB8}$ coding region of pUC19HP1, which was prepared from strain JM110, generating plasmid pUC *degQ*$_{YB8}$::Nm^r. Plasmid pUC *degQ*$_{YB8}$::Nm^r was digested with *Sse*8387I and ligated into the *Pst*I site of pE194, forming pE *degQ*$_{YB8}$::Nm^r (Figure 6.12).

The disruption of the *degQ*$_{YB8}$ coding region in YB8 was performed by using the thermosensitive replication origin of pE194 as follows (Figure 6.12). First, pE*degQ*$_{YB8}$::Nm^r was transformed into YB8 by electroporation, as previously described (7). The strain YB8(pE*degQ*$_{YB8}$::Nm^r) was plated on LB agar medium containing neomycin and erythromycin and then incubated at 48°C. The resultant strain, YB8::pE*degQ*$_{YB8}$::Nm^r, was cultivated in a liquid medium without selective pressure at 30°C for 10 generations. The culture was diluted and plated on LB agar medium to obtain single colonies. The neomycin and erythromycin resistance of the resultant colonies was assayed. Finally, the *degQ*$_{YB8}$–disrupted mutant, designated YB8 *degQ*$_{YB8}$::Nm^r, was isolated by the selection of neomycin-resistant and erythromycin-sensitive colonies. For complementation, plasmid pCRV was constructed by the insertion of a 0.8 kb *Eco*RV fragment containing *degQ*$_{YB8}$ into the *Pvu*II site of pC194 and transformed into YB8 *degQ*$_{YB8}$::Nm^r by electroporation.

Strain 791(pC81AP), which contains *degQ*$_{YB8}$, was constructed from strain 168(pC81AP) by the Campbell-type integration of plasmid pUC19HP1Nm^{r}F in which the Nm^r cassette was ligated into the *Pst*I site of pUC19HP1 (Table 6.3).

Strain 786, a *degQ*$_{YB8}$-disrupted mutant of strain 406, was constructed from strain 406 by double crossover recombination of plasmid pUC *degQ*$_{YB8}$::Nm^r.

6.3.1.9   DNA Sequencing of the *degQ* Region

The *degQ* region of each *B. subtilis* strain was cloned by PCR with primers 5′-CCTATTGAGATTTGCGGTGTCACGCAGGAC-3′ and 5′-CCCCCCTCCCAT TCCATTTT- ACTAAATGGGA-3′. PCR was performed with TaKaRa LA PCR kit, version 2 (Takara Shuzo, Kyoto, Japan). PCR cycling conditions were as follows: 4 min at 94°C, 30 cycles of 20 s at 98°C, and 1 min at 68°C in a model TP480 TaKaRa PCR thermal cycler.

The PCR products were blunted with the DNA blunting kit (Takara Shuzo) and were cloned into the *Hinc*II site of pUC19. The resultant plasmids were sequenced on a model 4000L DNA sequencer (Li-Cor) with the Thermo Seguenase cycle sequencing kit (Amersham) and IRD41-labeled primers.

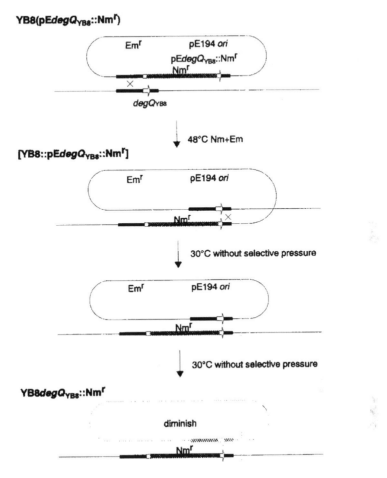

FIGURE 6.12 Disruption of $degQ_{YB8}$ in strain YB8 by insertional plasmid mutagenesis. Details are described in Section 6.3.1.9.

6.3.1.10 Nucleotide Sequence Accession Number

The DNA sequence data from the *degQ* region of YB8 have been deposited in the DDBJ, EMBL, and GenBank nucleotide sequence databases under accession no. AB010576.

6.3.2 RESULTS

6.3.2.1 Conversion of Strain MI113 into a Plipastatin Producer

In order to convert strain MI113 into a plipastatin producer, the transformation of MI113 with the chromosomal DNA of strain YB8 was necessary. However, due to the absence of proper markers, the detection of the transformants was not possible. Therefore, congression was done as follows. Cotransformation of MI113 with the chromosomal DNA of YB8 and a replicative plasmid DNA of pTB522 conferring tetracycline resistance (9) was performed and tetracycline-resistant transformants were selected. Colonies exhibiting growth suppression of the fungus *F. oxysporum* on a solid medium were selected as plipastatin producers. The plasmid was cured by overnight cultivation without selective pressure. A plipastatin-producing clone (designated strain 406) was selected and further characterized.

6.3.2.2   Transposon Mutagenesis of Strain 406

Strain 406 was transformed with the transposon-carrying plasmid pHV1249, and Tn*10* insertion mutants were isolated by selection for Cm^r. Among 6000 mini-Tn*10* insertion mutants assayed for the loss of growth suppression of *F. oxysporum in vitro*, 39 mutants showed defective production of plipastatin on the plate assay. All these mutants except one (mutant strain 706) exhibited hemolytic activity on blood-agar plates, which was evidence for surfactin production. The location of the mini-Tn*10* insertion in mutant strain 706, which produces neither plipastatin nor surfactin, was confirmed to be in *lpa-8* (*sfp*), which is required for the production of both plipastatin and surfactin (7), because the production of both lipopeptides was recovered by the introduction of an intact *lpa-8* gene carried on a plasmid. Nine of the 38 mutants with both Cm^r and Em^r were removed because their phenotypes might be due to integration of the entire pHV1249 into the chromosome as described earlier (8). The other 29 Cm^r and Em^s mutants were used in further experiments.

6.3.2.3   Isolation and Identification of the Antifungal
Antibiotic Produced by Strain 406

The HPLC peak pattern of the culture extract of strain 406 was compared with that of a mutant strain with a transposon insertion. Figure 6.13 shows HPLC peak

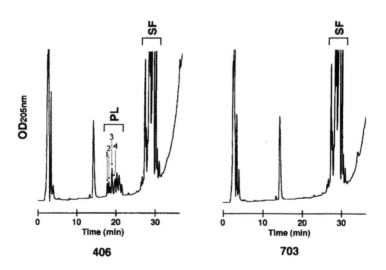

FIGURE 6.13   HPLC analysis of ethanol extracts of the acid precipitate from culture broths of *B. subtilis* 406 (left) and mutant strain 703 with a transposon inserted (right). Arrows indicate the peaks that were isolated and purified. SF and PL indicate the surfactin group and plipastatin group, respectively. OD, optical density.

patterns of strain 703, which is representative of 29 Cm^r and Em^s mutants, and of wild-type strain 406. There were several peaks from the extract of strain 406 that could not be detected from the extract of strain 703. Assuming that these peaks from strain 406 were antifungal substances, four substances corresponding to peaks 1 to 4 were isolated and purified as described in Section 6.3.1, and their structures were determined. Amino acid analysis of the hydrolysates of the purified substances, as well as chirality analysis, showed that the hydrolysate contained D-Ala(1) [D-Val(1) in peak 3], D-*allo*-Thr(1), L-Glx:(3), L-Pro(1), L-Ile(1), L-Tyr(1), D-Tyr(1), and D-Orn(1) in molar ratios.

The FAB mass spectrometry analysis resulted in protonated molecular ion peaks [M+H]$^+$ (*m/z* is 1463.8, 1463.8, 1491.8, and 1477.8 from compounds isolated from HPLC peaks 1 to 4, respectively). To determine the peptide sequences of these substances, FAB mass spectrometry with a *B/E* linked scanning method was performed. The representative collisionally activated dissociation (CAD) mass spectrum of the protonated molecular ion peak [M+H]$^+$ of HPLC peak 1 is shown in Figure 6.14B. In the case of proline-containing cyclic peptides, like iturin A (69, 70) or tyrocidine (71), the initial ring opening occurs preferentially at the N terminus of proline (72). As the plipastatin also contains proline in its cyclic portion, it was assumed that the ring opening occurred preferentially between alanine and proline, as shown in Figure 6.14A. In fact, the observed protein fragment ion peaks (*m/z* = 225.9, 389.0, 502.4, 1162.6, 1263.6, and 1392.7) fit the calculated mass of the fragmentation of a linear peptide with a ring opening located between alanine and praline (Figure 6.14A and B).

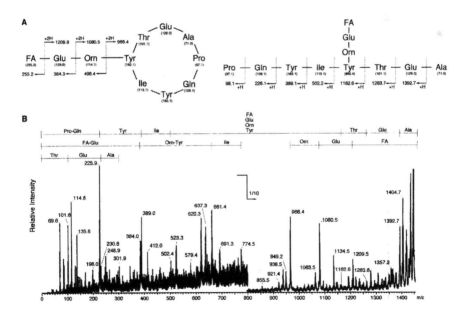

FIGURE 6.14 (A) Calculated mass from the fragment ion peaks of plipastatin A1 (left) and its ring form opened between Ala and Pro (right). (B) FAB-linked scanning CAD mass spectrum of the protonated molecular ion peak ([M+H]$^+$; *m/z*, 1463.8) of the substance purified from peak 1 as shown in Figure 6.13.

Furthermore, protein fragment ion peaks (*m/z* = 966.4, 1080.5, and 1209.5) coincided with the predicted mass of the fragmentation of the branched portion of plipastatin Al (Figure 6.14A and B).

These substances exhibited the same molecular weight and CAD mass spectra (data not shown), although peaks 1 and 2 were isolated separately in the HPLC preparation. It was thought that these compounds differed in the structure of their β-hydroxy fatty acids, so ^{1}H NMR spectra of these compounds were compared. They exhibited similar patterns except for the shape of the signals corresponding to the methyl substituent at high field. There was a methyl triplet signal at 0.88 ppm, which suggested the structure of 3-hydroxyhexadecanmc acid (*n*-C$_{16}$h^3) (73) in peak 2, whereas there were two methyl signals at 0.85 ppm (doublet) and at 0.86 ppm (triplet) in peak 1, which implied the structure of 13-methyl-3-hydroxypentadecanoic acid (a-C$_{15}$h^3) (73). Therefore, it was determined that peak 2 was plipastatin Al (74) and peak 1 was a new plipastatin A, which was an isomer of plipastatin Al. As the ^{1}H NMR spectra of peaks 3 and 4 were similar to those of peaks 2 and 1 at high field, respectively (data not shown), peak 3 was identified as plipastatin Bl (74) and peak 4 as plipastatin A2 (74). From these results, the antifungal substances produced by strain 406 were determined to be a group of plipastatins.

6.3.2.4   Identification of *Not*I Fragment with Mini-Tn*10* Inserted

Chromosomal DNAs from strain 406 and its plipastatin-defective mutants were digested with restriction enzyme *Not*I and separated by PFGE (Figure 6.15A). Following electrophoresis, the DNAs were transferred to a nylon membrane and were hybridized with a probe for mini-Tn*10* to determine the location of the insertion

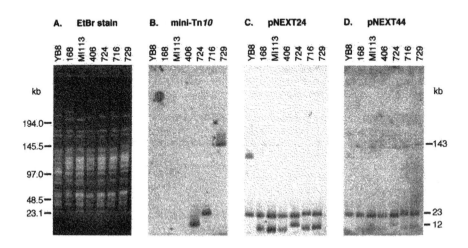

FIGURE 6.15   Southern blot analysis to identify the location of the gene cluster responsible for plipastatin production. Samples were digested with restriction enzyme *Not*Iprior to PFGE. (A) Ethidium bromide (EtBr) staining. (B) Detection of mini-Tn*10* insertion-containing *Not*I fragments by using the 1-kb *Eco*RV fragment of pHV1249 as a probe. (C) Identification of the pNEXT24 homologous fragment by using pNEXT24 as a probe. (D) Identification of the pNEXT44 homologous fragment by using pNEXT44 as a probe.

of mini-Tn*10* in the chromosome, as shown in Figure 6.15B. Three sizes of *Nat*I fragments, 13 kb (6 of 29 mutants), 24 kb (20 of 29 mutants), and 144 kb (3 of 29 mutants), were detected. The 144 kb fragment was judged to be equivalent to the *Nat*I fragment designated SN (12), because the value of 144 kb agreed well with the sum of 1 kb from the integrated mini-Tn*10* and 143 kb from the 5N *Not*I fragment (12) of strain 168. In addition, the existence of 12 kb (12,313 bp) and 23 kb (22,859 bp) *Not*I fragments between 5N and 10N (122 kb) in strain 168 is known (12, 48), and it was surmised that they correspond to the detected 13 and 24 kb fragments.

To identify these fragments precisely, rehybridization was performed by using pNEXT24 (47) (Figure 6.15C) and pNEXT44(47) (Figure 6.15D) as probes. Plasmid pNEXT24 contains a fragment of chromosomal DNA from strain 168, derived from the region containing the 12 and 23 kb *Not*I fragments that link fragments 5N and 10N. In contrast, plasmid pNEXT44 contains the 23 kb fragment linking region and the 5N *Not*I fragment, as shown in Figure 6.16. Since all three of the mini-Tn*10* fragments hybridized with pNEXT24 or pNEXT44, the gene cluster responsible for plipastatin production was located in the region between 5N and 10N (167° to 171°) in strain 168 (12).

6.3.2.5 Inactivation of Plipastatin Synthetase Gene with pNEXT24 and pNEXT44

The cloned fragments of strain 168 chromosomal DNA in pNEXT24 and pNEXT44 were sequenced. The sequences of the *Pst*I fragment of pNEXT24 and the *Eco*RI fragment of pNEXT44 matched that previously determined by Tognoni et al. (29). These results implied that the large open reading frames, believed to correspond to putative peptide synthetases in strain 168 (29, 30), encode plipastatin synthetases.

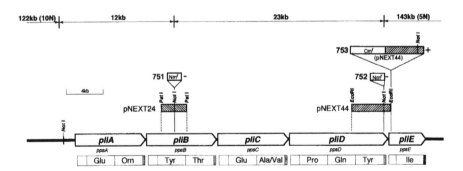

FIGURE 6.16 ORFs and *Not*I restriction map in the region between the two *Not*I fragments, 5N and 10N. Five ORFs which encode plipastatin synthetase are indicated. The putative amino acid–activating domain, racemase domain (dotted area), and thioesterase domain (black area) are also shown. Note that this amino acid–activating domain arrangement is different from that proposed by Tosato et al. (30). The cloned region in pNEXT24 and pNEXT44 and the insertions in the chromosome of strains 751, 752, and 753 are displayed above. + and − indicate plipastatin production positive and negative, respectively. (Data from Tognoni, A. et al., *Microbiology*, 141, 645–648 (1995); and Tosato, V., *Microbiology*, 143, 3443–3450 (1997).)

To confirm that the chromosomal fragments cloned in plasmids pNEXT24 and pNEXT44 were part of the plipastatin synthetase operon, pNEXT44 was integrated into strain 406 by Campbell-type insertion, but all the Cm^r transformants, which were produced by the insertion of pNEXT44, did not lose the ability to produce plipastatin. In fact, as pNEXT44 contains the end of *ppsD* and the beginning of *ppsE* (strain 753 in Figure 6.15), the duplication of this region by Campbell integration was thought to result in an intact *ppsE* gene, which was probably expressed by transcriptional read-through or by the activity of the Cm^r promoter of pNEXT44.

To disrupt *ppsD* completely, plasmid pNEXT44B2 was constructed by the insertion of an Nm^r cassette into the *Not*I site of pNEXT44 and was introduced into strain 406. All the Nm^r colonies that appeared on the selection plate were defective in the production of plipastatin. Similarly, plasmid pNEXT24A, which also had an Nm^r cassette inserted into the *Not*I site of pNEXT24, was introduced into strain 406. The resultant colonies exhibited no plipastatin production. Representative clones of these plipastatin-negative mutants are shown in Figure 6.16 as 752 and 751, respectively. From these results, we conclude that this region encodes plipastatin synthetase.

6.3.2.6 Production of Plipastatin by Strains Carrying *lpa-8*

To determine whether strain 168(pC81AP), which expresses *lpa-8* contained on a plasmid, can produce plipastatin, a *ppsB* disruption strain of strain 168(pC81AP) by a double crossover recombination of the linearized plasmid pNEXT24A was constructed. The newly constructed strain was named strain 790(pC81AP), and the HPLC peak pattern of 168(pC81AP) with that of 790(pC81AP) (Figure 6.17) was compared. They showed almost the same pattern except for the difference observed in the fractions taken at 18 to 22 min, when the plipastatin group is eluted (×16 area in Figure 6.17). Although the concentration of plipastatin Bl in 168(pC81AP) was very low (approximately 1 ppm), it was reproducible.

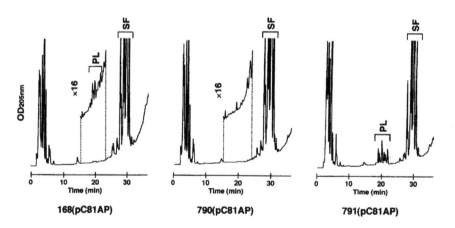

FIGURE 6.17   HPLC analysis of ethanol extracts of the acid precipitate from the culture broths of *B. subtilis* 168(pC81AP), 790(pC81AP), and 791(pC81AP). SF and PL indicate the surfactin group and plipastatin group, respectively. ×16, 16-fold magnification of the optical density (OD) at 205 nm.

6.3.2.7 Role of *degQ* of YB8 in Plipastatin Production

A 0.8 kb *Eco*RV fragment from YB8 was cloned by shotgun cloning, which had enhanced extracellular protease activity when expressed in strain 168 in a previous experiment. The sequence of this fragment completely matched that of *degQ* of strain 168, but the upstream sequence of *degQ* has several base substitutions compared with that of strain 168 (Figure 6.18). Especially, *degQ* of strain YB8 has a single base substitution of C for T at the promoter position –10 like *degQ36*, which has been known to lead to the hyperexpression of the *degQ* gene (75) (Figure 6.18). For convenience, we will refer to *degQ* of YB8 as $degQ_{YB8}$ and to *degQ* of 168 as $degQ^0$. To study the effect of $degQ_{YB8}$ on plipastatin production in YB8, a $degQ_{YB8}$ disruption mutant, YB8$degQ_{YB8}$::Nm^r was constructed. The plipastatin Bl productivity of YB8$degQ_{YB8}$::Nm^r was 5 ppm, which was about 10 times lower than that of YB8 (54 ppm). Introduction of a plasmid containing $degQ_{YB8}$ into YB8$degQ_{YB8}$::Nm^r restored plipastatin Bl production (48 ppm) to the level of YB8 (Figure 6.19). These results indicate that the $degQ_{YB8}$ gene was responsible for the hyperproduction of plipastatin in YB8. Furthermore, a plasmid containing $degQ_{YB8}$ was transformed into strain 168(pC81AP), and the obtained transformant 791(pC81AP), which possesses both *lpa-8* and $degQ_{YB8}$, produced plipastatin Bl at about 10 ppm (Fig. 6.17) and showed fungicidal activity *in vitro*. This productivity is 10-fold greater than that of 168(pC81AP) and the same as that of strain 406. From this result, it was determined that the introduction of $degQ_{YB8}$ into strain 168(pC81AP) on a plasmid enhances plipastatin production.

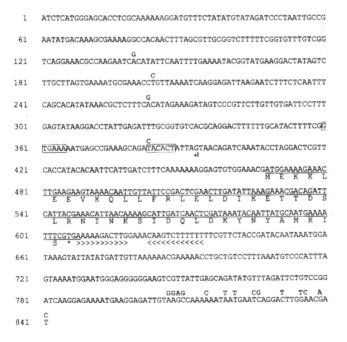

FIGURE 6.18 DNA sequence of the 0.8 kb *Eco*RV fragment containing $degQ_{YB8}$. The nucleotides of the strain 168 sequence that differ from that of strain YB8 are shown above. Boxes indicate –10 and –35 regions. *, termination codon; >> <<, inverted repeat sequence.

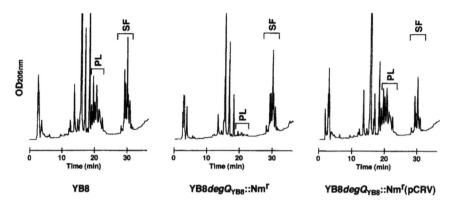

FIGURE 6.19   HPLC analysis of ethanol extracts of the acid precipitate from the culture broths of *B. subtilis* YB8, YB8 $degQ_{YB8}$, and YB8$degQ_{YB8}$(pCRV). SF and PL indicate the surfactin group and plipastatin group, respectively. OD, optical density.

6.3.3   Discussion

Tognoni et al. (29) and Tosato et al. (30) determined the DNA sequence of this region of the chromosome of strain 168 and suggested the existence of large proteins that contain structural motifs associated with the subunits of previously characterized peptide synthetases. From these reports, there are five ORFs (from *ppsA* to *ppsE*) organized into 10 amino acid–activating domains, and the second, fourth, sixth, and ninth domains have putative racemase activity (Figure 6.16). Furthermore, at some positions the first and fifth domains showed homology to the glutamic acid–activating domain of surfactin synthetase, and the second domain was remarkably similar to the ornithine-activating domain of gramicidin S synthetase. From these sequencing results, Tosato et al. proposed that this region is the fengycin operon (30). Fengycin is a lipopeptide antibiotic produced by *B. subtilis* F29-3, and it has the same amino acid composition as plipastatin except for one amino acid. One glutamic acid in fengycin is a glutamine in plipastatin, and it was concluded that fengycin contains three glutamic acid residues and no glutamine from elemental analysis (65). However, ambiguity still exists regarding the structure of fengycin. The fengycin operon had been cloned and sequenced from *B. subtilis* F29-3 (14, 76) and has significant homology with the putative peptide synthetase operon in strain 168 (30, 76). In this study, it was demonstrated that the antibiotic produced by strain 406 had the same sequence as plipastatin. Considering the biosynthesis of surfactin (82), we think that plipastatin is first synthesized in linear form, according to the amino acid–activating domain order on the *pps* operon, and then is cyclized by the formation of a lactone bond between the hydroxyl group of L-Tyr and the carboxyl group of L-Ile. The amino acid sequence of the linear form of the plipastatin precursor is thought to be as follows: β-hydroxy fatty acid-L-Glu- D-Orn L-Tyr-D-*allo*-Thr-L-Glu-D-Ala(Val)-L-Pro-L-Gln-D-Tyr-L-Ile. In the putative peptide synthetase as well, homology is observed among the first, fifth, and eighth amino acid–activating domains. The amino acid sequence of the first domain is 98% identical to that of the fifth. However, the amino acid sequence of the eighth domain is 49% identical to that of the first and fifth, indicating that the eighth domain activates glutamine but not glutamic acid. This peptide sequence and chiral pattern agree well with the domain arrangement of the putative peptide synthetase of strain 168. We propose the amino acid–activating domain assignment of the *pps* operon that is shown in Figure 6.16. This assignment is not the same as that for the fengycin operon proposed by Tosato et al. (30). Recently, *fenB*, which is a fengycin synthetase gene and a homolog of *ppsE*, was cloned and sequenced. FenB overexpression revealed that FenB is responsible for isoleucine activation (76). This result agrees with our prediction of activation of isoleucine by PpsE. Indeed, since the ^{13}C NMR spectrum of plipastatin resembles that of fengycin (data not shown), it is probable that fengycin has almost the same structure as plipastatin. From these results, conclusion is that the gene cluster located between 167° and 171° is the plipastatin synthetase operon and propose that *ppsA*, *ppsB*, *ppsC*, *ppsD*, and *ppsE* be renamed *pliA*, *pliB*, *pliC*, *pliD*, and *pliE*, respectively, as shown in Figure 6.16.

Strain 168 also has a peptide synthetase for surfactin biosynthesis (77) but could not produce surfactin because of a mutation in *sfp*, which encodes 4′-phosphopantetheinyl transferase (61). The *lpa-8* was cloned, which is identical to *sfp* from strain YB8, and was determined that this gene was essential for the production of both surfactin and plipastatin in YB8 (7). The introduction of a plasmid containing *lpa-8* into strain 168 resulted in both plipastatin and surfactin production, but plipastatin production was very poor (Figure 6.17), suggesting that there was some other gene responsible for plipastatin production.

The $degQ_{YB8}$ gene had been cloned, which enhanced the protease activity in strain 168 (data not shown), and it is supposed that $degQ_{YB8}$ is a hyperexpression type of *degQ*. When $degQ_{YB8}$ was introduced into strain 168(pC81AP), which has *lpa-8* on the plasmid, the resultant strain 168 derivative 791(pC81AP), which possessed both *lpa-8* and $degQ_{YB8}$, produced a detectable amount of plipastatin when analyzed by HPLC. This shows that $degQ_{YB8}$ is responsible for the enhancement of plipastatin production. This gene is not essential, because a $degQ_{YB8}$ disruption in YB8 reduced the plipastatin production by 1/10. A similar observation has been reported for degradative enzyme production (78, 79).

The *degQ* region of strains 406 and MI113 was sequenced and confirmed that plipastatin-producing 406 has the $degQ_{YB8}$ promoter while the host strain MI113 has a degQ0 promoter like strain 168. This result indicates that the degQ^0 region of strain 406 was exchanged with $degQ_{YB8}$ of YB8 by transformation. However, strain 786, which is a $degQ_{YB8}$-disrupted mutant derived from strain 406, did not exhibit a change in its plipastatin production (Table 6.4). To clarify whether $degQ_{YB8}$ of strain 406 is inactive or not, plasmid pCRV, which carries a copy of $degQ_{YB8}$, was introduced into strain 406, but no significant change in strain 406(pCRV) plipastatin production was observed. It would seem that there is another pathway that bypasses *degQ* in chimera strain 406.

degQ expression is known to be controlled by DegS–DegU and ComP–ComA modulator–effector pairs (63). The *srfA* operon, which comprises the surfactin synthetase gene and *comS*, is also regulated by ComP–ComA (80, 81). However, competence development requires nonphosphorylated DegU, whereas degradative enzyme

TABLE 6.4

Relevant Genotypes and Plipastatin Production

Strain		Relevant Allele		Level of Plipastatin Production[d]
	pps	*lpa-8* *(sfp)*	*degQ*	
YB8	*pps+*	*lpa-8+*	$degQ_{YB8}^{+}$	+ + +
YB8$degQ_{YB8}$::Nm^r	*pps+*	*lpa-8+*	$degQ_{YB8}^{-}$	+
YB8$degQ_{YB8}$::Nm^r (pCRV)	*pps+*	*lpa-8+*	$degQ_{YB8}^{+}$	+ + +
MI113	ND^a	*sfp*^{0b}	*degQ*^{0c}	—
406	*pps+*	*lpa-8+*	$degQ_{YB8}^{+}$	+ +
706	*pps+*	*lpa-8−*	$degQ_{YB8}^{+}$	—
751, 752	*pps−*	*lpa-8+*	$degQ_{YB8}^{+}$	—
786	*pps+*	*lpa-8+*	$degQ_{YB8}^{-}$	+ +
168	*pps+*	*sfp*^{0b}	*degQ*^{0c}	—
168 (pC81AP)	*pps+*	*lpa-8+*	*degQ*^{0c}	+
790 (pC81AP)	*pps−*	*lpa-8+*	*degQ*^{0c}	—
791 (pC81AP)	*pps+*	*lpa-8+*	$degQ_{YB8}^{+}$	+ +

ª ND, not determined.

ᵇ *sfp*^{0} is *sfp* of strain 168.

ᶜ *degQ*^{0} is *degQ* of strain 168.

ᵈ −, none detected; +, + +, and + + + indicate the relative levels of plipastatin production.

production needs phosphorylated DegU (82). In a previous study, plipastatin was produced by YB8 from late logarithmic phase to early stationary phase, whereas surfactin was produced from early to late logarithmic phase (59). This delay in plipastatin production may be due to the requirement for the phosphorylated form of DegU. This is the first report that relates *degQ* to the production of a metabolite other than degradative enzymes. The collation of genotypes and plipastatin-producing phenotypes is summarized in Table 6.4 and the pedigree tree of the key strains used in this study is shown in Figure 6.20.

Recently, researchers have developed targeted replacements of amino acid–activating domains within the *srfA* operon with a variety of other amino acid–activating domains cloned from bacteria and fungi by PCR (34, 83) and succeeded in the production of peptides with modified amino acid sequences (33, 34). On the other hand, de Ferra et al. replaced the integral thioesterase type I domain within surfactin synthetase and demonstrated that the synthesis truncated lipopeptides (53). Because of its high specificity and efficiency of genetic transformation (84), strain 168 is a good candidate as a recipient for recombinant peptide synthetase genes. Therefore, the identification of a new peptide synthetase is significant because it activates unique ammo acids in strain 168. In this respect, the presence of plipastatin synthetase will increase the production of a variety of modified peptides.

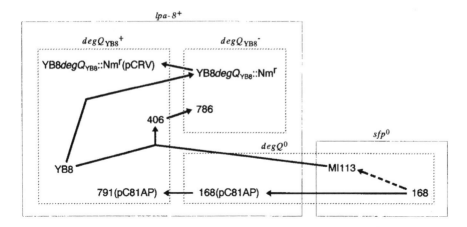

FIGURE 6.20   Pedigree tree of the key strains used in this study. *degQ* and *lpa-8* genotypes are also indicated in boxes. Arrows show the direction of evolution. The construction of strain MI113 from strain 168 (dotted arrow) was done by several steps (48, 66).

REFERENCES

1. Tsuge, T., Akiyama, T., and Shoda, M. Coning, sequencing, and characterization of the iturin A operon. *J. Bacteriol.*, 183, 6265–6273 (2001).

2. Duitman, E. H., Hamoen, L. W., Rembold, M., Venema, G., Seitz, H., Saenger, W., Bernhard, F., Reinhardt, R., Schmidt, M., Ullrich, C., Stein, T., Leenders, F., and Vater, J. The mycosubtilin synthetase of *Bacillus subtilis* ATCC6633: a multifunctional hybrid between a peptide synthetase, an amino transferase, and a fatty acid synthase. *Proc. Natl. Acad. Sci. U.S.A.*, 96, 13294–13299 (1999).

3. Hiraoka, H., Ano, T., and Shoda, M. Molecular cloning of a gene responsible for the biosynthesis of the lipopeptide antibiotics iturin and surfactin. *J. Ferment. Bioeng.*, 74, 323–326 (1992).

4. Huang, C. C., Ano, T., and Shoda, M. Nucleotide sequence and characteristics of the gene, *lpa-14*, responsible for biosynthesis of the lipopeptide antibiotics iturin A and surfactin from *Bacillus subtilis* RB14. *J. Ferment. Bioeng.*, 76, 445–450 (1993).

5. Hiraoka, H., Asaka, O., Ano, T., and Shoda, M. Characterization of *Bacillus subtilis* RB14, coproducer of peptide antibiotics iturin A and surfactin. *J. Gen. Appl. Microbiol.*, 38, 635–640 (1992).

6. Itaya, M., Kondo, K., and Tanaka, T. A neomycin resistance gene cassette selectable in a single copy state in the *Bacillus subtilis* chromosome. *Nucleic Acids Res.*, 17, 4410 (1989).

7. Tsuge, K., Ano, T., and Shoda, M. Isolation of a gene essential for biosynthesis of the lipopeptide antibiotics plipastatin B1 and surfactin in *Bacillus subtilis* YB8. *Arch. Microbiol.*, 165, 243–251 (1996).

8. Petit, M. A., Bruand, C., Janniere, L., and Ehrlich, S. D. Tn*10*-derived transposons active in *Bacillus subtilis*. *J. Bacteriol.*, 172, 6736–6740 (1990).

9. Imanaka, T., Himeno, T., and Aiba, S. Effect of *in vitro* DNA rearrangement in the NH$_2$-terminal region of penicillinase gene from *Bacillus licheniformis* on the mode of expression in *Bacillus subtilis*. *J. Gen. Microbiol.*, 131, 1753–1756 (1985).

10. Sambrook, J., Fritsch, E. F., and Maniatis, T. *Molecular Cloning: A Laboratory Manual*, 2nd edn. Cold Spring Harbor Laboratory, Cold Spring Harbor, New York (1989).

11. Tsuge, K., Ano, T., Hirai, H., Nakamura, Y., and Shoda, M. The genes *degQ, pps*, and *lpa-8 (sfp)* arc responsible for conversion of *Bacillus subtilis* 168 to plipastatin production. *Antimicrob. Agents Chemother.*, 43, 2183–2192 (1999).

12. Itaya, M., and Tanaka, T. Complete physical map of the *Bacillus subtilis* 168 chromosome constructed by a gene-directed mutagenesis method. *J. Mol. Biol.*, 220, 631–648 (1991).

13. Kunst, F. et al. The complete genome sequence of the gram-positive bacterium *Bacillus subtilis*. *Nature*, 390, 249–256 (1997).

14. Chen, C. L., Chang, L. K., Chang, Y. S., Liu, S. T., and Tschen, J. S. M. Transposon mutagenesis and cloning of the genes encoding the enzyme of fengycin biosynthesis in *Bacillus subtilis*. *Mol. Gen. Genet.*, 248, 121–125 (1995).

15. Venvoert, I. I. G. S., Verbree, E. C., van der Linden, K. H., Nijkamp, H. J. J., and Stuitje, A. R. Cloning, nucleotide sequence, and expression of the *Escherichia coli fabD* gene, encoding malonyl coenzyme A-acyl carrier protein transacylase. *J. Bacteriol.*, 174, 2851–2857 (1992).

16. Morbidoni, H. R., de Mendoza, D., and Cronan, Jr., J. E. *Bacillus subtilis* acyl carrier protein is encoded in a cluster of lipid biosynthesis genes. *J. Bacteriol.*, 178, 4794–4800 (1996).

17. Rose, M., and Entian, K. D. New genes in the 170 region of the *Bacillus subtilis* genome encode DNA gyrase subunits, a thioredoxin, a xylanase and an amino acid transporter. *Microbiology*, 142, 3097–3101 (1996).

18. Isogai, I., Takayama, S., Murakoshi, S., and Suzuki, A. Structures of β-amino acids in antibiotics iturin A. *Tetrahedron Lett.*, 23, 3065–3068 (1982).

19. McKenzie, T., Hoshino, T., Tanaka, T., and Sueoka, N. The nucleotide sequencing of pUB110: some salient features in relation to replication and its regulation. *Plasmid*, 15, 93–103 (1986).

20. Muller, A. K., Rojo, F., and Alonso, J. C. The level of the pUB110 replication initiator protein is autoregulated, which provides an additional control for plasmid copy number. *Nucleic Acids Res.*, 23, 184–1900 (1995).

21. Revill, W. P., Bibbe, M. J., and Hopwood, D. A. Purification of a malonyl transferase from *Streptomyces coelicolor* A3(2) and analysis of its genetic determinant. *J. Bacteriol.*, 177, 3946–3952 (1995).

22. Summers, R. G., Ali, A., Shen, B., Wessel, W. A., and Hutchinson, C. R. Malonyl-coenzyme A:acyl carrier protein acyltransferase of *Streptomyces glaucescens*: a possible link between fatty acid and polyketide biosynthesis. *Biochemistry*, 34, 9389–9402 (1995).

23. Hourdou, M. L., Besson, F., and Michel, G. Specific inhibition of iturin biosynthesis by cerulenin. *Can. J. Microbiol.*, 36, 164–168 (1989).

24. Marahiel, M. A., Stachelhaus, T., and Mootz, H. D. Modular peptide synthetases involved in nonribosomal peptide synthesis. *Chem. Rev.*, 97, 2651–2673 (1997).

25. Mootz, H. D., and. Marahiel, M. A. The tyrocidine biosynthesis operon of *Bacillus brevis*: complete nucleotide sequence and biochemical characterization of functional internal adenylation domains. *J. Bacteriol.*, 179, 6843–6850 (1997).

26. Cosmina, P., Rodriguez, F., de Ferra, F., Grandi, G., Perego, M., Venema, G., and van Sinderen, D. Sequence and analysis of the genetic locus responsible for surfactin synthesis in *Bacillus subtilis*. *Mol. Microbiol.*, 8, 821–831 (1993).

27. Konz, D., Doekel, S., and Marahiel, M. A. Molecular and biochemical characterization of the protein template controlling biosynthesis of the lipopeptide lichenysin. *J. Bacteriol.*, 181, 133–140 (1999).

28. Yakimov, M. M., Kroger, M., Slepak, T. N., Giuliano, I., Timmis, K. N., and Golyshin, P. N. A putative lichenysin A synthetase operon in *Bacillus licheniformis*: initial characterization. *Biochim. Biophys. Acta*, 1399, 141–153 (1998).

29. Tognoni, A., Franchi, E., Magistrelli, C., Colombo, E., Cosmina, P., and Grandi, G. A putative new peptide synthase operon in *Bacillus subtilis*: partial characterization. *Microbiology*, 141, 645–648 (1995).

30. Tosato, V., Albertini, A. M., Zotti, M., Sonda, S., and Bruschi, C. V. Sequence completion, identification and definition of the fengycin operon in *Bacillus subtilis* 168. *Microbiology*, 143, 3443–3450 (1997).

31. Lin, T. S., Chen, C. L., Chang, L. K., Tschen, J. S., and Liu, S. T. Functional and transcriptional analyses of a fengycin synthetase gene, *fenC*, from *Bacillus subtilis*. *J. Bacteriol.*, 181, 5060–5067 (1999).

32. Mootz, H. D., and Marahiel, M. A. Design and application of multi-modular peptide synthetases. *Curr. Opin. Biotechnol.*, 10, 341–348 (1999).

33. Schneider, A., Stachelhaus, T., and Marahiel, M. A. Targeted alteration of the substrate specificity of peptide synthetases by rational module swapping. *Mol. Gen. Genet.*, 257, 308–318 (1998).

34. Stachelhaus, T., Schneider, A., and Marahiel, M. A. Rational design of peptide antibiotics by targeted replacement of bacterial and fungal domains. *Science*, 269, 69–72 (1995).

35. Stachelhaus, T., Schneider, A., and Marahiel, M. A. Engineered biosynthesis of peptide antibiotics. *Biochem. Pharmacol.*, 52, 177–186 (1996).

36. Tsuge, K., Inoue, S., Ano, T., Itaya, M., and Shoda, M. Horizontal transfer of iturin A operon, *itu*, to *Bacillus subtilis* 168 and conversion into an iturin A producer. *Antimicrob. Agents Chemother.*, 49, 4641–4648 (2005).

37. Peypoux, F., Bonmatin, J. M., and Wallach, J. Recent trends in the biochemistry of surfactin. *Appl. Microbiol. Biotechnol.*, 51, 553–563 (1999).

38. Umezawa, H., Aoyagi, T., Nishikiori, T., Okuyama, A., Yamagishi, Y., Hamada, M., and Takeuchi, T. Plipastatins: new inhibitors of phospholipase A_2, produced by *Bacillus cereus* BMG202-fF67. I. Taxonomy, production, isolation and preliminary characterization. *J. Antibiotics*, 39, 737–744 (1986).

39. Maget-Daoa, R., and Peypoux, F. Iturins, a special class of pore-forming lipopeptides: biological and physicochemical properties. *Toxicology*, 87, 151–174 (1994).

40. Peypoux, F., Guinand, M., Michel, G., Delcambe, L., Das, B. C., and Lederer, E. Structure of iturine A, a peptidolipid antibiotic from *Bacillus subtilis*. *Biochemistry*, 17, 3992–3996 (1978).

41. Koumoutsi, A., Chen, X.-H., Henne, A., Licsegang, H., Hitzeroth, G., Franke, P., Valer, J., and Borriss, R. Structural and functional characterization of gene clusters directing nonribosomal synthesis of bioactive cyclic lipopeptides in *Bacillus amyloliquefaciens* strain FZB42. *J. Bacteriol.*, 186, 1084–1096 (2004).

42. Lambalot, R. H., Gehring, A. M., Flugel, R. S., Zuber, P., LaCelle, M., Marahiel, M. A., Reid, R., Khosla, C., and Walsh, C. T. A new enzyme superfamily—the phosphopantetheinyl transferases. *Chem. Biol.*, 3, 923–936 (1996).

43. Itaya, M. Effective cloning of unmarked DNA fragments in the *Bacillus subtilis* genome. *Biosci. Biotechnol. Biochem.*, 63, 602–604 (1999).

44. Itaya, M., Nagata, T., Shiroishi, T. T., Fujita, K., and Tsuge, K. Efficient cloning and engineering of giant DNAs in a novel *Bacillus subtilis* genome vector. *J. Biochem.* 128, 869–875 (2000).

45. Itaya, M., Fujita, K., Ikeuchi, M., Koizumi, M., and Tsuge, K. Stable positional cloning of long continuous DNA in the *Bacillus subtilis* genome vector. *J. Biochem.* 134, 513–519 (2003).

46. Anagnostopoulos, C., and Spizizen, J. Requirements for transformation in *Bacillus subtilis*. *J. Bacteriol.*, 81, 741–746 (1961).

47. Nakano, M. M., and Zuber, P. Mutational analysis of the regulatory region of the *srfA* operon in *Bacillus subtilis*. *J. Bacteriol.*, 175, 3188–3191 (1993).

48. Itaya, M., and Tanaka, T. Predicted and unsuspected alterations of the genome structure of genetically defined *Bacillus subtilis* 168 strains. *Biosci. Biotechnol. Biochem.*, 61, 56–64 (1997).
49. Nagahari, K., and Sakguchi, K. Cloning of *Bacillus subtilis* leucine A, B, and C genes with *Escherichia coli* plasmids and expression of the *leuC* gene in *E. coli*. *Mol. Gen. Genet.*, 158, 263–270 (1978).
50. Itaya, M. Integration of repeated sequences (pBR322) in the *Bacillus subtilis* 168 chromosome without affecting the genome structure. *Mol. Gen. Genet.*, 241, 287–297 (1993).
51. Nakano, M. M., Corbell, N., Besson, J., and Zuber, P. Isolation and characterization of *sfp*: a gene that functions in the production of the lipopeptide biosurfactant, surfactin, in *Bacillus subtilis*. *Mol. Gen. Genet.*, 232, 313–321 (1992).
52. Hiraoka, H., Ano, T., and Shoda, M. Molecular cloning of a gene responsible for the biosynthesis of the lipopeptide antibiotics iturin and surfactin. *J. Ferment. Bioeng.*, 74, 323–326 (1992).
53. de Ferra, F., Rodriguez, F., Tortora, O., Tosi, C., and Grandi, G. Engineering of peptide synthetases: key role of the thioesterase-like domain for efficient production of recombinant peptides. *J. Biol. Chem.*, 272, 25304–25309 (1997).
54. Eppelmann, K., Doekel, S., and Marahiel, M. A. Engineered biosynthesis of the peptide antibiotic bacitracin in the surrogate host *Bacillus subtilis*. *J. Biol. Chem.*, 276, 34824–34831 (2001).
55. Liu, W., and Hansen, J. N. Conversion of *Bacillus subtilis* 168 to a subtilin producer by competence transformation. *J. Bacteriol.*, 173, 7387–7390 (1991).
56. Guenzi, E., Galli, G., Grgurina, I., Pace, E., Ferranti, P., and Grandi, G. Coordinate transcription and physical linkage of domains in surfactin synthetase are not essential for proper assembly and activity of the multienzyme complex. *J. Biol. Chem.*, 273, 14403–14410 (1998).
57. Izaguirre, G., and Hansen, J. N. Use of alkaline phosphatase as a reporter polypeptide to study the role of the subtilin leader segment and the SpaT transporter in the posttranslational modifications and secretion of subtilin in *Bacillus subtilis* 168. *Appl. Environ. Microbiol.*, 63, 3965–3971 (1997).
58. Liu, W., and Hansen, J. N. Enhancement of the chemical and antimicrobial properties of subtilin by site-directed mutagenesis. *J. Biol. Chem.*, 267, 25078–25085 (1992).
59. Tsuge, K., Ano, T., and Shoda, M. Characterization of *Bacillus subtilis* YB8, coproducer of lipopeptides surfactin and plipastatin. *J. Gen. Appl. Microbiol.*, 41, 541–545 (1995).
60. Nakano, M. M., Corbell, N., Besson, J., and Zuber, P. Isolation and characterization of *sfp*: a gene that functions in the production of the lipopeptide biosurfactant, surfactin, in *Bacillus subtilis*. *Mol. Gen. Genet.*, 232, 313–321 (1992).
61. Lambalot, R. H., Gehring, A. M., Flugel, R. S., Zuber, P., LaCelle, M., Marablel, M. A., Reid, R., Khosal, C., and Walsh, C. T. A new enzyme superfamily—the phosphopantetheinyl transferase. *Chem. Biol.*, 3, 923–936 (1996).
62. Kunst, F., Msadek, T., and Rapoport, G. Signal transduction network controlling degradative enzyme synthesis and competence in *Bacillus subtilis*, pp. 1–20. In Piggot, P., Moran, Jr., C. P., and Youngman, P. (eds.), *Regulation of Bacterial Differentiation*. American Society for Microbiology, Washington, D.C. (1994).
63. Msadek, T., Kunst, F., Klier, A., and Rapoport, G. DegS-DegU and ComP-ComA modulator-effector pairs control expression of the *Bacillus subtilis* pleiotropic regulatory gene *degQ*. *J. Bacteriol.*, 173, 2366–2377 (1991).
64. Itaya, M. Personal communication (1999).
65. Vanittanakom, N., Loeffter, W., Koch, U., and Jung, G. Fengycin: a novel antifungal lipopeptide antibiotic produced by *Bacillus subtilis* F-29-3. *J. Antibiotics*, 39, 888–901 (1986).

66. Tanaka, T. Personal communication (1999).
67. Anagnostopoulos, C., and Spizizen, J. Requirements for transformation in *Bacillus subtilis*. *J. Bacteriol.*, 81, 741–746 (1961).
68. Chu, G., Vollrath, D., and Davis, R. W. Separation of large DNA molecules by contour-clamped homogeneous electric fields. *Science*, 234, 1582–1585 (1986).
69. Ishikawa, K., Niwa, Y., Hatakeda, K., and Gotoh, T. Computer-aided peptide sequencing of an unknown cyclic peptide from *Bacillus subtilis*. *Org. Mass Spectrom.*, 23, 290–291 (1988).
70. Niwa, Y., and Ishikawa, K. Mass spectrometry of proteins: peptide sequencing and precise molecular weight measurement. *J. Natl. Chem. Lab. India*, 86, 95–102 (1991).
71. Tang, X. J., Thibault, P., and Boyd, R. K. Characterization of the tyrocidine and gramicidin fraction of the tyrothricin complex from *Bacillus brevis* using liquid chromatography and mass spectrometry. *Int. J. Mass Spectrom. Ion Process.*, 122, 143–151 (1992).
72. Eckart, K. Mass spectrometry of cyclic peptides. *Mass Spectrom. Rev.*, 13, 23–55 (1994).
73. Nishikori, T., Naganawa, H., Muraoka, Y., Aoyagi, T., and Umezawa, H. Plipastatins: new inhibitors of phospholipase A_2, produced by *Bacillus cereus* BMG302-fF67. II. Structure of fatty acid residue and amino acid sequence. *J. Antibiotics*, 39, 745–754 (1986).
74. Nishikori, T, Naganawa, H., Muraoka, Y., Aoyagi, T., and Umezawa, H. Plipastatins: new inhibitors of phospholipase A2, produced by Bacillus cereus BMG302-fF67. III. Structural elucidation of plipastatins. *J. Antibiotics*, 39, 755–761 (1986).
75. Yang, M., Ferrari, E., Chen, E., and Benner, D. J. Identification of the pleiotropic *sacQ* gene of *Bacillus subtilis*. *J. Bacteriol.*, 16, 113–119 (1986).
76. Lin, G. H., Chen, C. L., Tschen, J. S. M., Tsay, S. S., Chang, Y. S., and Liu, S. T. Molecular cloning and characterization of fengycin synthetase gene *fenB* from *Bacillus subtilis*. *J. Bacteriol.*, 180, 1338–1341 (1998).
77. Cosmina, P., Rodriguez, F., de Ferra, F., Grandi, G., Pergo, M., Venema, G., and van Sinderen, D. Sequence and analysis of the genetic locus responsible for surfactin synthesis in *Bacillus subtilis*. *Mol. Microbiol.*, 8, 821–831 (1993).
78. Kunst, F., and Rapoport, G. Salt stress is an environmental signal affecting degradative enzyme synthesis in *Bacillus subtilis*. *J. Bacteriol.*, 177, 2403–2407 (1995).
79. Ogura, M., and Tanaka, T. Expression of alkaline protease gene in *Bacillus subtilis* mutants that lack positive regulatory genes *degR*, *degQ*, *senS*, *tenA*, and *proB*. *Biosci. Biotechnol. Biochem.*, 61, 372–374 (1997).
80. Nakano, M. M., and Zuber, P. Mutational analysis of the regulatory region of the *srfA* operon in *Bacillus subtilis*. *J. Bacteriol.*, 175, 3188–3191 (1993).
81. Roggiani, M., and Dubnau, D. ComA, a phosphorylated response regulator protein of *Bacillus subtilis*, binds to the promoter region of *srfA*. *J. Bacteriol.*, 175, 3182–3187 (1993).
82. Dahl, M. K., Msadek, T., Kunst, F., and Rapoport, G. The phosphorylation state of the DegU response regulator acts as a molecular switch allowing either degradative enzyme synthesis or expression of genetic competence in *Bacillus subtilis*. *J. Biol. Chem.*, 267, 14509–14514 (1992).
83. Turgay, K., Krause, M., and Marahiel, M. A. Four homologous domains in the primary structure of GrsB are related to domains in a super-family of adenylate-forming enzymes. *Mol. Microbiol.*, 6, 529–546 (1992).
84. Dubnau, D. A. Genetic transformation in *Bacillus subtilis*, pp. 175–220. In Dubnau, D. A. (ed.), *The Molecular Biology of Bacilli*, vol. 1. Academic Press, New York (1982).

7 Optimization Study of Production of Antifungal Substances and Spores in Submerged Fermentation (SmF) or in Solid-State Fermentation (SSF)

7.1 PRODUCTION OF ITURIN BY FOAM SEPARATION IN SUBMERGED FERMENTATION (1)

The newly isolated *Bacillus subtilis* NB22 produces an antifungal antibiotic, iturin.

When *B. subtilis* NB22 was cultivated in a complex medium, it was found that iturin was condensed only in the foam formed during cultivation. This indicates that continuous separation and condensation of the product is possible only by collecting the foam under properly controlled foaming conditions. Thus, foaming control factors such as the aeration rate, temperature, and nutrient compositions were investigated to assess the optimal production of iturin in association with the distribution pattern of each component of iturin condensed in the foam.

7.1.1 MATERIALS AND METHODS

7.1.1.1 Strain and Medium Used

The details of the characteristics of *B. subtilis* NB22 were described in Chapter 2. NB22 was grown in the no. 3 medium, which was shown in Chapters 2 and 3.

7.1.1.2 Operation of Jar Fermentor

A computer-controlled 7-liter jar fermentor (working volume 5 liters) with a standard flat blade turbine impeller was used for the main cultivation of the bacterium. Precultivation was carried out at 30°C by a reciprocal shaker (120 strokes per min) in a shaken flask (working volume 100 ml) in no. 3 medium for 1 day, and 100 ml of the culture broth was inoculated into 5 liters of fresh medium in a jar fermentor. When the aeration rate and temperature were changed, no. 3 medium was used. When the concentration of Polypepton or glucose in no. 3 medium was varied, the temperature and aeration rate were fixed at 30°C and 0.1 vvm (volume of air/volume of the reactor/min), respectively. The foam produced during fermentation was freely expelled from the fermentor through a sterilized silicone tube (i.d. 0.4 mm) to a foam-trapping unit consisting of a conical flask set up outside the fermentor. This trap contained 10 ml of 1% antifoam agent (Toshiba Silicone Co. Ltd., Tokyo) put in beforehand. When the foam reached the unit together with air, it was instantly liquefied. The volume of the liquefied foam was measured periodically and taken as the volume of foam produced during fermentation.

7.1.1.3   Assay of Iturin

Entrapped foam in the foam trapping unit was centrifuged at 10,000 × g for 10 min to remove the entrained cells and the supernatant was acidified to pH 2 with 12 N HCl. Then, the solution was centrifuged and the precipitate was extracted with 10 ml methanol for 3 h. After the extracted solution was centrifuged at 10,000 × g for 10 min, the supernatant was injected into a high-performance liquid chromatography (HPLC) (Column: Gasukurokogyo Unisil NG Cl8, 10.7 Φ × 250 mm). The mobile phase was acetonitrile–ammonium acetate (10 mM) 2:3 (v/v), which was pumped at a flow rate of 1 ml/min. The elution pattern monitored at 280 nm is shown in Figure 7.1. The components are denoted as peaks 1 to 6 according to the elution time order. The area of each peak was measured and the concentration of each component of iturin was determined by a calibration curve that was prepared by each component. When peak 6 is extremely small, peak 6 is omitted.

7.1.1.4   Cell and Sugar Concentrations

The cell concentration of the culture broth in the fermentor was measured by turbidity (OD_{660}), and the glucose concentration in the broth was determined by enzymatic assay (B-test, Wako Co. Ltd., Tokyo) after periodical sampling.

7.1.2   RESULTS

7.1.2.1   Effect of Aeration Rate

The aeration rate was varied from 0.1 vvm to 0.3 vvm, while the temperature and agitation speed were maintained at 30°C and 400 rpm, respectively. Figure 7.2 shows the time course of iturin production. The ordinate represents the concentration of iturin in the foam. No iturin was detected in the liquid broth in the fermentor. It is obvious that a lower aeration rate was more favorable for a higher production of iturin. Figure 7.3 represents the changes of cell concentration (OD_{660}) and pH at 0.1 and 0.3 vvm. During each cultivation, the pH and OD_{600} increased with the increase of the aeration rate. This indicates that the supply of oxygen limited the growth of the bacterium, and thus the increase of oxygen supply accelerated the growth of the cells. The glucose consumption rate was largest at 0.3 vvm, as shown in Figure 7.4. This corresponds to the larger growth rate at 0.3 vvm shown in Figure 7.3. The relationship between the iturin obtained and the foam produced is shown in Figure 7.5. The units of the ordinate and the abscissa are expressed as per liter of culture broth. Less foam was produced at 0.1 vvm and more iturin was accumulated. The reason why the increase of oxygen gave no stimulation to the increase of iturin production is partly because iturin production is not associated with the growth of the cells.

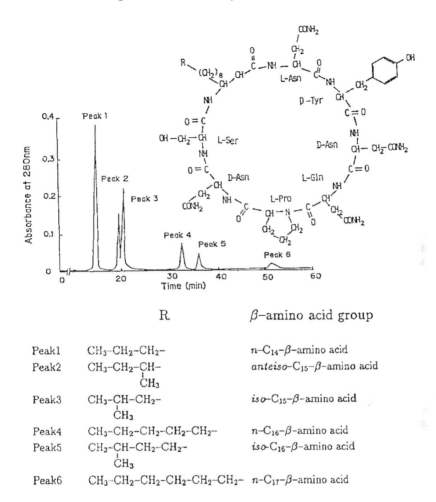

	R	β–amino acid group
Peak1	CH_3–CH_2–CH_2–	n–C_{14}–β–amino acid
Peak2	CH_3–CH_2–$\underset{\underset{CH_3}{\mid}}{CH}$–	$anteiso$–C_{15}–β–amino acid
Peak3	CH_3–$\underset{\underset{CH_3}{\mid}}{CH}$–$CH_2$–	iso–C_{15}–β–amino acid
Peak4	CH_3–CH_2–CH_2–CH_2–CH_2–	n–C_{16}–β–amino acid
Peak5	CH_3–$\underset{\underset{CH_3}{\mid}}{CH}$–$CH_2$–$CH_2$–	iso–C_{16}–β–amino acid
Peak6	CH_3–CH_2–CH_2–CH_2–CH_2–CH_2–	n–C_{17}–β–amino acid

FIGURE 7.1   HPLC elution pattern and iturin structure.

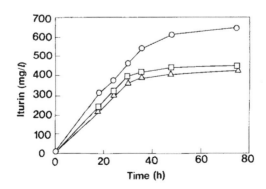

FIGURE 7.2   Time course of iturin production at different aeration rates. Symbols: o, 0.1 vvm; □, 0.2 vvm; Δ, 0.3 vvm.

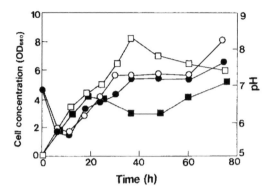

FIGURE 7.3 Changes of cell concentration (OD_{660}) and pH at aeration rates of 0.1 and 0.3 vvm. Cell concentration: ■, 0.1 vvm; □, 0.3 vvm. pH: ●, 0.1 vvm; ○, 0.3 vvm.

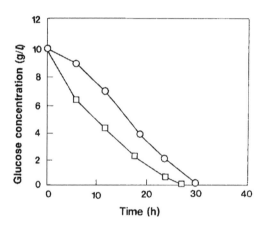

FIGURE 7.4 Glucose consumption rate at aeration rates of 0.1 and 0.3 vvm. Symbols: ○, 0.1 vvm; □, 0.3 vvm.

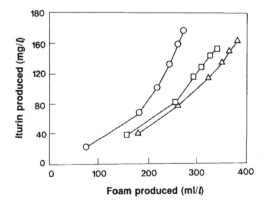

FIGURE 7.5 Relationship between iturin and foam produced at different aeration rates. Symbols: ○, 0.1 vvm; □, 0.2 vvm; Δ, 0.3 vvm.

Glucose was used preferentially for cell synthesis in the earlier stage of cultivation, as shown in Figure 7.4. This is reflected in the decline of pH seen in Figure 7.3. Then, Polypepton was consumed as the main nutrient of iturin synthesis in the later stage of the cultivation and this led to an increase of pH, as shown in Figure 7.3. When the aeration rate exceeded 0.3 vvm, too rapid foaming during cultivation resulted in a loss of fresh medium entrained with the foam. In this sense, a lower aeration rate was favorable.

7.1.2.2   Effect of Temperature

The effect of temperature on foaming and iturin condensation was investigated at 20°C, 30°C, and 40°C. Figure 7.6 represents the relationship between the foam produced and the iturin concentration. Although glucose consumption and growth of the cells was higher at 40°C than at 30°C (data not shown), 30°C was optimal in the production of iturin. An example of the minimum inhibitory concentration (MIC) of each component of iturin to some phytopathogenic fungi and bacteria is shown in Table 2.8 in Chapter 2. The longer the side aliphatic chain R in Figure 7.1, the lower is the MIC to the fungi. As peak 6 was extremely small in amount and it was hard to purify enough for the MIC test, the MIC value was not determined. It is preferable to obtain components with a larger molecular weight of iturin. The distribution of each component at different temperatures is shown in Figure 7.7. The total amount of iturin produced is shown above each bar. At 30°C–40°C, the percentage of peak 3 is significantly larger than at 20°C, reflecting some metabolic changes at each temperature. However, in the case of peaks 4 and 5, no such relatively larger proportions were not observed at 30°C–40°C.

7.1.2.3   Effect of Glucose and Polypepton Concentrations

Concentrations of glucose and Polypepton were varied, and the ratio of glucose to Polypepton and the productivity of iturin against the unit of Polypepton were correlated as shown in Figure 7.8. The iturin concentration in this figure was determined

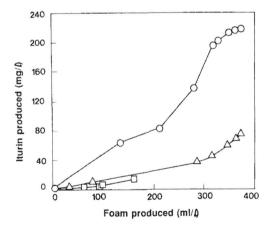

FIGURE 7.6   Relationship between iturin and foam produced at different temperatures. Symbols: Δ, 20°C; o, 30°C; □, 40°C.

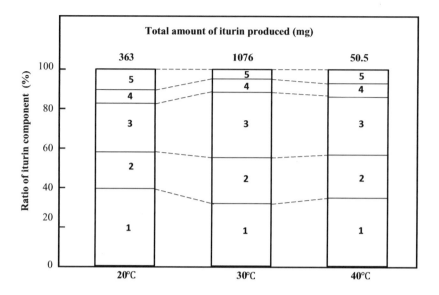

FIGURE 7.7   Distribution of iturin components at different temperatures. Peaks 1–5 correspond to those in Figure 7.1.

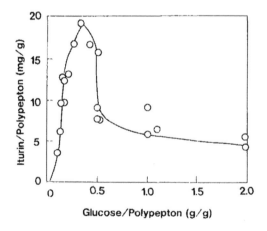

FIGURE 7.8   Relationship between iturin produced per g of Polypepton and ratio of glucose to Polypepton.

in the final stage of cultivation at each experimental condition. It is obvious that the optimal ratio of glucose to Polypepton was approximately 4. The distribution of each component of iturin in different Polypepton concentrations is shown in Figure 7.9. The total amounts of iturin produced are also shown above each bar. In lower concentrations of Polypepton, a larger ratio of peaks 4 and 5 was observed, in spite of the smaller amount of iturin produced. This suggests that the ratio of glucose to

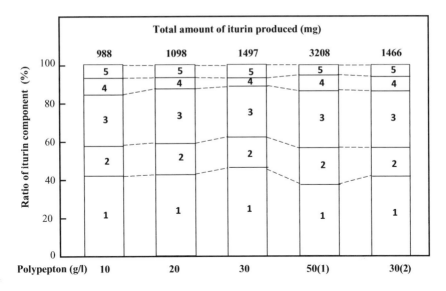

FIGURE 7.9 Distribution of iturin components at different Polypepton concentrations. 50(1) and 30(2) correspond to the first and second cultivations in Figure 7.11, respectively.

Polypepton, which approximately corresponds to the ratio of carbon to nitrogen, is a clue to larger production of peaks 4 and 5. The total amount of iturin produced increased in accordance with the increase in the initial concentration of Polypepton in no. 3 medium. However, the increase of Polypepton to more than 50 g/l caused a significant increase of loss of medium in the fermenter due to vigorous foaming, and it was practically impossible to carry out the fermentation.

7.1.2.4   Introduction of a New Basket-Type Agitation Unit

Thus, a new basket-type agitation unit was introduced for controlling the foam production in the high concentration of Polypepton. The structure of the unit is shown in Figure 7.10 and the details of its performance are described by Hayashi et al. (2). The iturin obtained and foam produced at 50 g/l of Polypepton is shown in Figure 7.11A. The concentration of iturin reached up to 3100 mg, which corresponds to 620 mg per liter of broth and 815 mg per ml of foam produced. This value was the largest in these experiments. However, as is clear from Figure 7.11A, nearly 4 liters of foam was produced, and the loss of volume in the medium was still significant. Thus, a repeat cultivation was carried out by supplementing fresh medium to the culture broth at the final stage in the first cultivation. The additional volume of the fresh medium in the second cultivation was nearly equal to that of the foam lost in the first cultivation. The initial concentration of Polypepton at the second fermentation was fixed at 30 g/l. The result is shown in Figure 7.11B. The iturin produced was about 1500 mg, which is 500 mg per liter of the broth. This productivity per unit of Polypepton was almost the same level of iturin concentration as that in the first cultivation. However, a third cultivation was not successful because iturin production was much less than in the first and second ones. The reason for the lower production of iturin is not

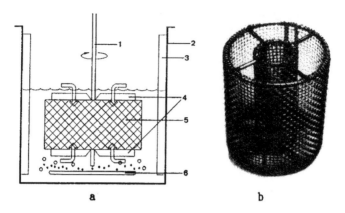

FIGURE 7.10 Configuration of jar fermenter equipped with a basket-shaped unit for agitation. (a) 1, rotation shaft; 2, fermenter vessel; 3, baffle plates; 4, flat blade turbine type impellers; 5, basket-shape unit; 6, sparger. (b) Photograph of the unit.

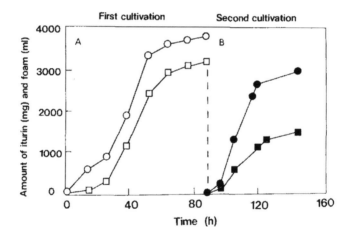

FIGURE 7.11 Foam and iturin production in (A) first cultivation and (B) second cultivation. Symbols: o and ●, foam produced; □ and ■, iturin produced.

clear at present but is presumably because the concentrated products or nutrient had an adverse effect on the activity of the bacterium. Thus, the fed-batch cultivation of Polypepton is one possibility. It is interesting that the distribution of each component of iturin at the first cultivation was almost the same as that at 10–20 g/l of Polypepton as shown in Figure 7.9. This result demonstrates that both the total amounts of iturin produced and the ratio of peaks 3, 4, and 5 were maximum at 50 g/l of Polypepton.

With the advances in bioreactor-based biotechnological processes, it has become important to develop methods for the isolation and purification of products from cultures or cells that are suitable for large-scale application. Some sophisticated separation methods, such as membrane separation or chromatographic procedures, are successful as laboratory methods but have limitations in large-scale processes. Foam separation is presented here as a possible simpler separation method. Foam separation has been reported for the removal of several microorganisms from aqueous suspension (3–5). Also, foam separation of extracellular proteins produced by *Saccharomyces cerevisiae* was investigated using a bioreactor of a cylinder type to clarify the separation ratio between the top and bottom of the reactor (6). However, no foam separation and concentration of peptidolipidic substances in a jar fermentor have been reported. Foaming in a jar fermentor is generally a nuisance and the use of an antifoam agent is in most cases unavoidable. The feature of the method introduced here is closely related to the chemical nature of the product, which contains both hydrophobic and hydrophilic parts in its structure. The advantage of this method also lies in the fact that the use of an antifoam agent, which sometimes causes physiological and economical disadvantage in fermentation, will be minimized. Scale-up seems possible if the proper cultivation methods and conditions are established.

7.2 PRODUCTION OF ITURIN IN SOLID-STATE FERMENTATION (7)

For the production of iturin by NB22 in submerged fermentation (SmF), we developed a productive liquid medium referred to as "no. 3 medium" and a new foam separation system that is able to produce approximately 1 g of total iturin per 5 liters of the culture broth in a 7-liter scale SmF (1).

In the present study, the production of iturin by NB22 during the solid-state fermentation (SSF) of okara, a soybean curd residue, which is a by-product of tofu manufacturing, was investigated. Approximately 700,000 tons of okara was produced in Japan in 1986, and now most okara is incinerated as an industrial waste. Several ideas have been presented for reutilizing okara. However, no positive results have been achieved for using okara as a substrate. Fungi have been widely utilized in SSF, not only in the practical production of enzymes or chemicals, but also in laboratory-scale experiments (8, 9). A few attempts to apply bacteria to SSF have been made (10). However, specific bacteria for the production of useful products have not been used in SSF. *B. subtilis*, which is an obligate aerobe, has not been one of them.

7.2.1 Materials and Methods

7.2.1.1 Preservation of *B. subtilis* NB22

NB22 was cultivated in 10 ml of no. 3 medium in a 50 ml flask by shaking at 120 rpm at 30°C for 24 h. Then 500 µl of the culture broth were mixed with 500 mg of sterile glycerol in 1.5 ml Eppendorf tubes and the tubes were stored at –40°C. At least one day prior to use, one tube was transferred to a freezer (–40°C), then used as inocula.

7.2.1.2 Preparation of Liquid Inoculum of *B. subtilis* NB22 for SSF at 15 g Scale

No. 3 medium containing 1% Polypepton S (Nippon Pharmaceuticals Co.) was used as the liquid source for the NB22 cells throughout the SSF experiments. Polypepton

S was used instead of Polypepton because the growth of NB22 was found to be superior to that for Polypepton. Also, 100 μl of glycerol-preserved NB22 described earlier were inoculated to 100 ml medium in 500 ml shaken flasks, and incubated at 30°C at 120 rpm for 24 h.

7.2.1.3 Preparation of Okara as a Solid Substrate

Okara was kindly provided by a tofu manufacturing company. Okara collected in 10–20 kg lots was divided into 100–120 g pieces, wrapped in a sheet of commercial wrapping film, and frozen at –20°C to –40°C. Then, 24 h prior to autoclaving, each block of okara was thawed out in a 4°C refrigerator. Next, 150 g of okara was placed in a glass petri dish (25 cm Φ, 6 cm deep), autoclaved for 20 min at 120°C, and then cooled to room temperature for 8 to 12 h. It was then autoclaved again for 20 min at 120°C, cooled to room temperature for 3 to 5 h, and used as a substrate. Next, 15 g of sterile okara was placed in 100 ml sterile conical flasks, and 75 μl of 1 M KH_2PO_4, 150 μl of 1 M $MgSO_4$, 833 μl of 0.45 g/ml glucose, and 367 μl of sterile distilled water were added to each flask to maintain initial water content at 82%. The addition of these nutrients has previously been found to be effective for the production of iturin. Each petri dish was stirred with a sterile stainless steel spatula to achieve uniformity.

7.2.1.4 SSF at 15 g Scale

First, 3 ml of the culture broth of NB22 (OD_{660} =6 to 12) was added to each solid substrate prepared in the aforementioned manners and homogenized with a stainless steel spatula in a safety cabinet. All flasks were incubated in a water bath at a fixed temperature for the specified time.

7.2.1.5 Sampling for the Measurement of pH, Cell Population, and Water Content

At certain times of cultivation (usually a 12 h interval), 1 g of the solid was sampled for the measurement of pH and cell population. pH was measured after suspending the solid in 2 ml of distilled water. After suspending 1 g of the solid in 9 ml of distilled water followed by a serial dilution, cell population was determined by plating each diluted cell suspension onto no. 3 medium agar plates. At the end of cultivation, 1.0 g of the solid was sampled and dried in an 85°C drying oven for at least 12 h, and its weight was measured for the calculation of water content.

7.2.1.6 Extraction and Measurement of Iturin

At the end of cultivation, methanol was added at the ratio of 3 ml methanol to 1 g okara. It was shaken for 1 h at room temperature on a reciprocal shaker at 92 strokes/min. More than 90% of iturin was previously confirmed to be extracted under these extraction conditions. The suspension was then centrifuged at 4°C at 11,000 × g for 10 min and the supernatant was filtered with a 0.20 μm polytetrafluoroethylene (PTFE) filter (JP020, Advantec Toyo Ltd., Tokyo). The filtrate was injected into an HPLC to determine the concentration of each component of iturin at 205 nm. A calibration curve was prepared using purified components of iturin. Conditions for quantification were as follows: mobile phase, acetonitrile/10 mM ammonium acetate = 3/4 (v/v); column, 4.6 mm Φ × 250 mm ODS-2 (GL Sciences Inc., Tokyo); column temperature, 30°C, flow rate, 1.0 ml/ min.

7.2.1.7 Suspension Fermentation (SF) of Okara

Nonhomogeneous suspension fermentation (SF) of okara was carried out in order to compare with the SSF. Then, 15 g each of okara was suspended in 45 ml of distilled water in a 200 ml conical flask with a cotton plug. Next, 600 μl of the 24 h culture of NB22 in no. 3 medium (Polypepton S 1%) was inoculated into the suspension of okara and incubated at temperatures ranging from 25°C to 37°C by shaking at 120 rpm for 48 h. At the end of cultivation, the liquid and solid portions were separated by filtration with conventional cotton gauze and then centrifuged at 11,000 × g. The centrifugate was treated with concentrated HCl to adjust the pH to 2.0 for peptide precipitation and then extracted with 4.5 ml methanol. The precipitate was extracted with 45 ml of methanol with shaking on a reciprocal shaker in the same manner as the extraction of iturin from the SSF culture. Iturin was quantified as described earlier.

7.2.1.8 SSF at 300–500 g Scale

For the inoculation of NB22, 100 and 120 ml of NB22 culture broth were prepared in 500 ml shaking flasks, shaken at 120 strokes/min for 18–24 h. Then, 300 and 500 g of okara were each placed in glass jars (8.5 cm inner diameter, 2 mm thick, 210 mm high, 1350 ml nominal volume). They were twice autoclaved at 120°C for 1 h with an interval of 10–12 h, and allowed to stand for 3 to 5 h in a safety cabinet for cooling before preparation. Glucose, KH$_2$PO4, MgSO$_4$, and distilled water were added proportionally as previously described. SSF was initially carried out statically in a water incubator. Later, a conventional air pump whose air-flow rate was fixed at 2 l/min was employed for the enhanced production of iturin (see Figure 7.15). As shown in Figure 7.15, air was supplied through a silicon tube, with a pinchcock at the end, and air at the outlet came out of five pairs of pinholes serially arranged at distances of 2 cm between each pair. This air-flow rate corresponded approximately to 2 vvm for the 500 g scale.

7.2.1.9 Sampling Position

For the 300 g okara experiments, sampling was done according to the illustration in Figure 7.15 as follows: First, the uppermost 19.5 g of the SSF culture (which corresponded to 15 g of okara before the addition of nutrients and inoculum) was sampled into a 100 ml conical flask and was named T (top). Likewise, the middle 19.5 g, named M (middle) and the bottom 19.5 g, named B (bottom), were sampled.

For the 500 g SSF, sampling was carried out likewise except that the top-down partition was split into five sections, namely, U (uppermost), T (top), M (middle), B (bottom), and L (lowermost).

7.2.1.10 SSF at 3.0 kg Scale

First, 600 ml of NB22 culture was prepared in 2 liter shaking flasks at 30°C at 100 strokes/min for 26 h in a water incubator. The culture was inoculated into 3 kg of okara placed in 7 liter glass jars (17 cm inner diameter, 5 mm thick, 400 mm high), which were autoclaved twice at 120°C for 1 h with an interval of 12 h. Six hours were allowed for cooling before preparation and inoculation. Preparation of the solid medium was carried out in the following manner: The sterile substrate was transferred into a UV-sterilized polyethylene bag (950 mm × 750 mm). The addition of nutrients and inoculation of the cells were done in the bag and the mixture was homogenized with vigorous shaking and mixing by hand from the exterior of the bag, following which the substrate was transferred into the glass jars. They were placed in a water bath where temperature was controlled at 25°C.

For aeration of the 7 liter jars, a compressor was connected with three fermentor control units, which varied the air-flow rate from 6.5 l/min (0.93 vvm) to 14 l/min (2 vvm).

7.2.1.11   Sampling

For the 3.0 kg okara scale, the whole solid culture was first divided into five layers for sampling (Figure 7.16). Each layer was then divided coaxially into two portions with an equal cross-sectional area. Three 19.5 g samples were taken from each portion and stored at –20°C. They were thawed at room temperature just before the extraction of iturin. All measurements were done in triplicate, with the average of the two closer values being adopted.

7.2.1.12   Measurement of Temperature

For the 3.0 kg scale with aeration at 2 vvm, temperature was monitored by inserting a thermometer into each section at 12 h intervals.

7.2.1.13   Measurement of pH, Cell Concentration, Iturin, and Water Content

For the 300–500 g scale, the measurement of pH, cell concentration, iturin, and water content was done in the same way as described for the 15 g scale experiment. In the 3.0 kg scale experiment, periodical sampling for the measurement of these quantities was performed using a sterile 19 mm Φ stainless steel pipe by poking the substrate.

7.2.2   Results

7.2.2.1   15 g Scale SSF

Figure 7.12 shows the effect of temperature on iturin production in the 15 g scale SSF of okara where the temperature was controlled at fixed values within a ±0.1°C deviation. It is clear that 25°C was optimum. In order to determine whether this optimum temperature depends on either the substrate, i.e., okara, or the type of cultivation system, SSF, okara was suspended in water and SF was carried out. The SF is different from SSF in that the dissolution rate of suspended okara into water is higher and the free bacteria cells have easy access to the substrate or dissolved nutrients of okara. The results of Figure 7.13 indicate that iturin production in SF of okara was better at 25°C than at 30°C. Since the optimal temperature in SMF using no. 3 medium was 30°C (1), the shift of optimal temperature may reflect differences in the quality of the two substrates.

Figure 7.14 shows the change in the number of viable cells and iturin production at 25°C. The number of viable cells increased by the order of 1 to 2 during

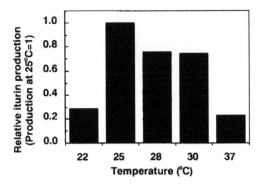

FIGURE 7.12   Effect of temperature on iturin production in solid-state fermentation (SSF) of soybean curd residue (okara).

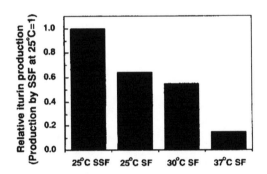

FIGURE 7.13   Effect of temperature on iturin production in the suspension fermentation (SF) of okara in comparison with 25°C solid-state fermentation (SSF).

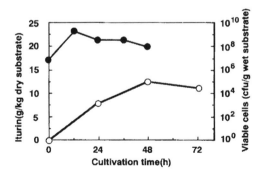

FIGURE 7.14   Time courses of iturin production and viable cell number of NB22 in SSF at 25°C. Symbols: o, iturin production; •, viable cell number.

the first 12 h, but then decreased, in a manner similar to that observed in SmF (1), with further incubation. Iturin production attained a maximum at 48 h. Actually, the growth of the bacterium was found to have leveled off in 5–6 h in another experiment, suggesting that the majority of the iturin was produced by the cells in a stationary phase even in SSF. Total iturin yield for the 15 g scale was within 10–16 g/kg dry substrate, with an average of 12 g/kg dry substrate for several experiments.

7.2.2.2 Static 300–500 g SSF

Figure 7.15 presents the results of a 300 g scale okara experiment using a glass jar as a reactor at 25°C under static conditions. Sampling was carried out as shown in Figure 7.15 from three top-down sections. Only 35%–60% of the iturin produced at the 15 g scale was obtained without aeration in the T and M sections, and 10%–15% in the B section is shown by the slashed bar.

7.2.2.3 Aerated 300–500 g SSF

The results obtained by aeration using the air pump from the bottom section of the glass jar are also presented in Figure 7.15. Improvements in iturin production, especially in the top section of the reactor, were observed and the amount of iturin produced was similar to the level for the 15 g okara scale (Figure 7.14), indicating that

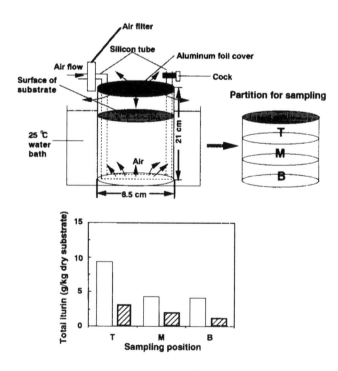

FIGURE 7.15 Method for 300 g scale SSF and the production of iturin at different sampling positions: T (top), M (middle), and B (bottom). White filled and slashed bars indicate total iturin production with and without aeration, respectively.

aeration is a crucial factor for improving iturin production in the 20-fold scale-up from 15 g scale. Iturin production in the 500 g scale was virtually comparable to the 300 g scale (data not shown).

7.2.2.4   3 kg Scale SSF

Figure 7.16 shows the productivity of iturin for the 3 kg scale experiment. Air was supplied from outside in a way similar to that shown in Figure 7.15. The sampling position is shown along the side. "C" and "R" were defined as the central region and its circumambient area that bisects the cross-sectional area. A maximum of 8 g and an average of 6 g iturin/kg dry substrate was obtained in the 3 kg scale SSF with aeration at 2 vvm. However, productivity varied according to the sampling positions. Iturin production was larger in the circumambient region in the reactor than in the central positions. Figure 7.17 presents the counts of viable cells at 0, 24, and 48 h at sampling positions UC, MC, and LC. There were no significant differences in cell numbers in the UR–LR positions (data not shown), suggesting that the number of viable cells did not depend upon sampling positions. The deviation in iturin production was not considered to be due to changes in cell numbers.

The changes in temperature at several sampling positions are shown in Figure 7.18. The temperature rose to 37°C–45°C in 24 h due to heat evolution during SSF, and those temperatures are not appropriate for iturin production. MR, BR, and LR, which are closer to the wall of the jar, exhibited lower temperature rises, mainly because the cooling effect of the water bath was more efficient compared with MC, BC, and LC, respectively. This resulted in a relatively higher production of iturin in the circumambient areas as shown in Figure 7.16. This verifies that temperature control and homogeneous temperature distribution are crucial in SSF.

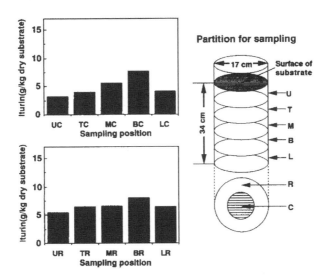

FIGURE 7.16   Iturin production in 3 kg scale SSF with aeration at 2 vvm and partition for sampling. See text for the labels of sampling positions. For example, UC and UR denote uppermost center and uppermost round positions, respectively.

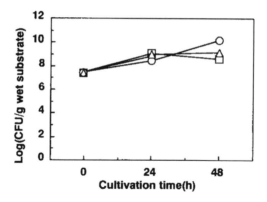

FIGURE 7.17   Changes in viable cell numbers in some sampling positions: UC (o), MC (□), and LC (Δ) in Figure 7.16.

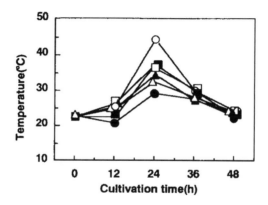

FIGURE 7.18   Temperature changes in each sampling position in Figure 7.16: MC (o), MR (□), BC (■), BR (▲), LC (Δ), and LR (●).

However, LC in Figure 7.16, where the temperature rise was significantly lower than other sampling positions, showed a lower production of iturin. Figure 7.19 shows the changes in water content in the central areas of the sampling positions. Water content did not vary except for the LC position, which was close to the outlet of air. Thus, a decline in iturin production in LC may be mostly due to a drastic decrease in water content and partly due to nonuniform aeration because the water content of the LR was about 80%.

7.2.2.5   Comparison of Iturin Production in SmF and SSF

Since the solid-state fermentation of okara is thought to be a biochemical transformation of organic matter present in okara by *B. subtilis* NB22, whose metabolism occurs in the water-soluble phase. The water-soluble portion of okara was prepared by homogenizing the material with a homogenizer at $11,000 \times g$ for 10 min, and filtrating the homogenate with a paper filter. The filtrate was dried at 105°C for 12–16 h and the carbon and nitrogen contents were measured by elemental analysis.

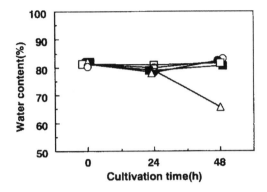

FIGURE 7.19   Changes in water content in some sampling positions. Positions other than LC did not exhibit much change: UC (o), TC (●), MC (o), BC (■), and LC (Δ).

TABLE 7.1

Carbon and Nitrogen Contents of Polypepton and Water-Soluble Fraction of Okara Determined by Elementary Analysis

	C (%)	N (%)
Polypepton	45.5	13.4
Water-soluble fraction of okara	48.3	4.96

Table 7.1 shows the data together with those of Polypepton used for liquid culture; 19.2% of dried okara was a water-soluble fraction. Referring to the data in Table 7.1, the contents of carbon and nitrogen of the water-soluble fraction of okara were calculated to be 21.3 g-C/l okara and 2.3 g-N/l okara, where the water content of the original okara (77%) was used.

In no. 3 medium in SmF, which contains 3% Polypepton and 1% glucose, the contents of carbon and nitrogen were calculated as 17.7 g-C/liter medium and 4 g-N/liter medium, respectively.

Since 1 liter of okara (almost equal to 1 kg in weight) and 1 liter of no. 3 medium are similar with respect to soluble carbon and nitrogen contents, a comparison of the two methods may be plausible. Iturin productivity in 5-liter scale SmF, where no. 3 medium with 3% Polypepton was used in Section 7.1 (1), and in 3 kg scale SSF were compared (Figure 7.20). The comparison was based on equivalent volumes for both cultivation methods. It is obvious that total iturin production was preferable in SSF.

In the comparison of the ratio of each of the five homologues of iturin obtained in SmF and in 3 kg scale SSF, the production of peak 4 was observed to be significantly larger in SSF (Figure 7.20). The ratios of peaks 2 to 5 to total iturin are 50% and 70% in SmF and SSF, respectively. We have already reported that the longer the side chain of the homologue of iturin, the stronger is the antibiotic activity (1), indicating that SSF has an advantage over SmF in this point.

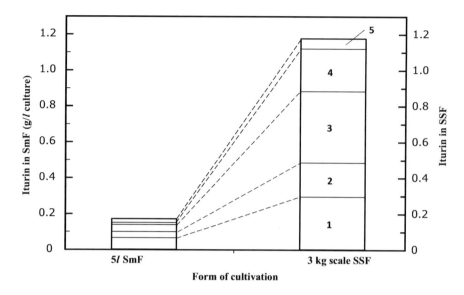

FIGURE 7.20 Comparison of iturin productivity in 5-liter SmF in Section 7.1, and 3 kg scale SSF of okara. Numbers in the graph represent the five iturin homologues corresponding to peaks 1–5 in Figure 7.1.

TABLE 7.2

Comparison between Iturin Production in Solid-State Fermentation (SSF) and Submerged Fermentation (SmF) of No. 3 Medium (Polypepton 3%)

	SSF			SmF
Scale	**15(g)**	**300–500(g)**	**3(kg)**	**5(*l*)**
Cultivation time (days)		2		5
Iturin productivity (mg/g wet culture/day)	1.0–1.65	0.5–1.0	0.55–0.8	0.032–0.044
Total iturin (mg) per container	47	550–910	5500	900
Culture volume (*l*) per 1 g iturin	1.35	1.4	1.5	7.2

Table 7.2 is a summary of a comparison of the production and purification of iturin between SSF and SmF. Cultivation time for the SSF takes only 2 days for the maximum production of iturin, while that for SmF takes 5 days (1). When the production of iturin is evaluated per unit gram of wet substrate, SSF is found to be six to eight times more efficient than SmF.

As mentioned earlier, some controlling factors such as temperature, aeration, and moisture content have to be optimized and controlled for further study in the scale-up of the production of iturin from okara. Based on these data, a new improved solid-state fermentor, which reduces radical temperature rise and the drying of solid culture by modifying the air supply method for the antibiotic production, is under investigation.

Since okara is one of the safest wastes produced by the food industry, the okara treated by *B. subtilis* NB22 itself can be utilized as an organic material that functions as both an organic fertilizer and a microbial pesticide with which to suppress plant diseases.

The extraction procedure of the solid material to extract each component of iturin is much simpler in SSF than in SmF, and less solvent is necessary in SSF. This is mainly because the water content of okara is much lower than the liquid medium, and that the accumulated iturin concentration in SSF is higher than that in SmF. In this respect, the solid fermentation is promising if proper operational conditions are assessed.

7.3 SOLID-STATE FERMENTATION USING DEHYDRATED MATERIAL (11)

Soybean curd residue, known as okara, is produced from the tofu industry in Japan. A typical sample of okara is rich in water-insoluble ingredients as shown in Table 7.3 (protein, 4.8%; fat, 3.6%; starch and sugar, 6.4%; fiber, 3.3%; ash, 0.8%; water, 81%) making it a potentially useful substrate for microbial fermentation. The main disadvantage of okara is natural spoilage when storage is not under refrigeration. Then, dehydration of the soybean curd residue has been attempted to improve utilization by SSF.

7.3.1 Materials and Methods

7.3.1.1 Microorganism Strain

B. subtilis NB22 was used. This strain was characterized in Chapter 2 (11, 13, 14).

7.3.1.2 Preparation of Seeding Culture

B. subtilis NB22 grown in no. 3S medium (10 g Polypepton S [Nippon Pharmaceuticals Co., Tokyo], 10 g glucose, 1 g KH_2PO_4, and 0.5 g $MgSO_4$ in 1 liter of distilled water) was mixed with glycerol at 1:1 (w/w) and stored at $-40°C$. Then 0.1 ml of the glycerol suspension was inoculated into a 500 ml shake flask containing 100 ml of no. 3 medium (0.1% inoculation) and cultivated at 30°C and 120 strokes/min as preculture.

TABLE 7.3
Properties of Soybean Curd Residue, Okara

	Wet Okara(%)		Lyophilized Okara (%)
Protein	4.8	H	6.99
Fat	3.6	C	46.3
Starch and sugar	6.4	N	3.99
Fiber	3.3	S	0.25
Ash	0.8	Metals	3.59
Water	81	O	38.9

7.3.1.3   Preparation of Okara as a Solid Substrate

Soybean curd residue consists mostly of the water-insoluble components of soybean, since the water-soluble portion is separated in an aqueous suspension known as soy milk, which is used in the tofu manufacturing process. The elemental composition of okara varies depending on the characteristics of the original crop of soybean. However, typical analysis of lyophilized okara is shown in Table 7.3.

When intact okara was obtained from a local manufacturer, 100 g lots were covered with commercial wrapping film and stored at –40°C before autoclaving to prevent spoilage. In small-scale cultivation, 15 g of okara was placed in a 100 ml conical flask and autoclaved twice at 120°C, for 20 min at 8–12 h intervals. On a 3 kg scale cultivation, 3 kg of okara was placed in an 8 liter glass jar and autoclaved twice at 120°C for 60 min at 8–12 h intervals. In both cases, interval sterilization was adopted to ensure the killing of spore-forming, relatively heat-resistant microorganisms in the material.

After autoclaving, sterile solutions were added to every 15 g of okara as follows: 75 µl of 1 M KH_2PO_4, 150 µl of 1 M $MgSO_4$, 833 µl of 0.45 g/ml glucose, and 367 µl of deionized distilled water, for the supplementation of nutrients and adjustment of moisture content to 82.8%.

7.3.1.4   Partially Dehydrated Okara

Partially dehydrated okara supplied from Fukuyama Technical Exchange Center (Fukuyama, Hiroshima) was prepared using a newly developed drying oven IE-308 (Ishihara Electronics Co. Ltd., Fukuyama, Hiroshima) as shown in Figure 7.21.

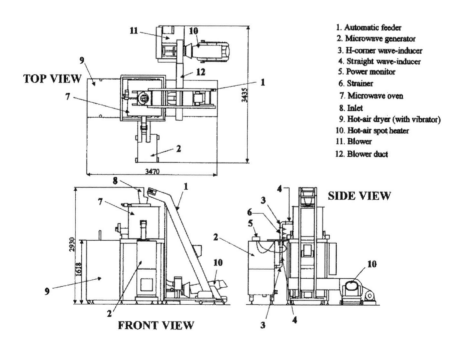

1. Automatic feeder
2. Microwave generator
3. H-corner wave-inducer
4. Straight wave-inducer
5. Power monitor
6. Strainer
7. Microwave oven
8. Inlet
9. Hot-air dryer (with vibrator)
10. Hot-air spot heater
11. Blower
12. Blower duct

FIGURE 7.21   Schematic plan of the okara drying oven IE-308 (units: mm).

Intact okara is fed by the automatic feeder (1) to the inlet of the microwave oven (7), exposed to microwaves, and the vapor is removed in the hot-air dryer with a vibrator (9) by the hot air, which is heated by the hot-air spot heater (10) and enters through the blower duct (15). Water contents from 12.5% to 80% of the dehydrated sample were achievable, and in this experiment a sample of 38.2% water content was used. The elemental analysis did not differ significantly from the intact okara described earlier. The pH of this sample was 6.3 compared with 6.6 for fresh okara. Dehydrated okara was used after adjustment of the water content by the addition of deionized distilled water and incubation for 12 h at 4°C. Then, 100–150 g of reconstituted okara was placed in a 300 ml glass beaker, with the addition of deionized distilled water to adjust the initial moisture content to 77.6%, and was allowed to soak for 6–12 h at 4°C prior to autoclaving twice at 120°C for 30 min with an interval of 8–12 h. After autoclaving, 15 g aliquots of the soaked okara were placed into 100 ml sterile conical flasks with cotton plugs. Next, 833 μl of 0.45 g/ml glucose, 75 μl of 1M KH_2PO_4, and 225 μl of 1M $MgSO_4$ were added for the fortification of nutrients. The preculture for seeding was mixed thoroughly with a sterile stainless spatula at a ratio of 3 ml to every 15 g of okara with a moisture content of 77.6%. The initial pH of the prepared okara was measured after adding 2 ml of deionized distilled water to 1 g of fermented okara.

In the case of the 3 kg scale cultivation, an 8-liter glass jar was used for the container. This was soaked in a water bath and air was supplied through silicon-rubber tubing connected to a compressor. The temperature profile was obtained as described in Section 7.2.

7.3.1.5   Measurement of Viable Cells

Viable cell number was measured by colony counting at 12 h intervals up to 48 h as follows: 1 g of each of the solid cultures was sampled into a sterile 18 mm diameter test tube, mixed thoroughly with 9 ml of sterile distilled water with a vortex, diluted serially up to 10^{-6}, and spread onto no. 3S agar plates. After 18 h of incubation at 30°C, the number of colonies was counted.

7.3.1.6   Temperature Dependency of Iturin A Production

The temperature dependency of iturin A production was investigated on a 15 g scale. Incubation was carried out at 23°C, 25°C, 30°C, 37°C, 42°C, and 48°C.

7.3.1.7   Extraction and Measurement of Iturin A

A crude extract of iturin A from the 15 g scale SSF culture was prepared by adding 45 ml of methanol to the whole fermented product, i.e., 3.36 g of initial dry weight of the intact okara, and the mixture was shaken at 92 strokes/min for 60 min with a reciprocal shaker (Eyela-shaker SS-8; Tokyo Rikakikai Co., Tokyo). On a 3 kg scale, 15 g of the solid culture was sampled from 10 parts of the jar. Samples were treated as described earlier, and the average was taken as the result of quantification of iturin A.

The methanol extract of iturin A obtained was filtered after centrifugation at 15,500 × g for 10 min at room temperature, through a 0.20 μm PTFE membrane filter unit. Then, 20 μl of the filtrate was injected into HPLC (880 PU; Japan Spectroscopic Co.,

Tokyo) using an ultraviolet (UV) detector (870 UV; Japan Spectroscopic Co., Tokyo) to determine the concentration of iturin A at 205 nm. A calibration curve was prepared using independently purified homologues of iturin A. For HPLC analysis, the mobile phase was acetonitrile/10 mM ammonium acetate (3/4 [v/v]) used on a reverse-phase, hydrophobic column, (4.6 mm diameter × 250 mm ODS-2, GL Sciences Inc., Tokyo). The column temperature was 30°C and the flow rate was 1.0 ml/min.

7.3.2   RESULTS AND DISCUSSION

7.3.2.1   Measurement of Viable Cells

Figure 7.22 shows the viable cell number of *B. subtilis* NB22 in the SSF of intact and dehydrated okara. Cell populations were identical, showing that dehydration did not influence the growth of the cells.

7.3.2.2   Optimal Temperature for Iturin A Production

The optimal temperature for iturin A production in this SSF was 25°C (Figure 7.23). This suggests that the reactor for iturin A production should be operated in a relatively cool environment.

7.3.2.3   Temperature Distribution in the 3 kg Scale Reactor

Figure 7.24 shows the profiles of temperature distribution within the 3 kg scale reactor near the center of gravity of the whole reactor. Accumulation of heat is one of the factors that should be considered in the scaling-up of the SSF.

7.3.2.4   Iturin A Production in Intact Okara

Figure 7.25 shows the amount of iturin A produced in the SSF of intact okara in the 15 g and 3 kg scale fermentations. The total amount of iturin A produced on a 15 g

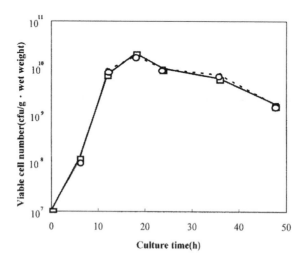

FIGURE 7.22   Viable cell number of *B. subtilis* NB22 in the solid-state fermentation (SSF) of intact and dehydrated okara at 80% moisture content. o, intact okara; □, dehydrated okara.

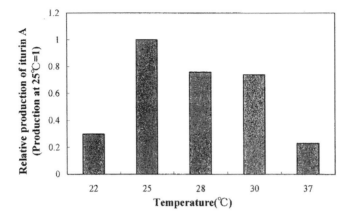

FIGURE 7.23 Effect of temperature on iturin A production in the solid-state fermentation (SSF) of okara.

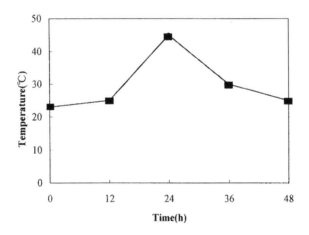

FIGURE 7.24 Time course of temperature in the 3 kg scale solid-state fermentation (SSF) of okara at the center of gravity of the 8-liter glass jar.

scale was 1.65 g/kg/wet weight and on a 3 kg scale was about 60% of the 15 g scale. The decreased productivity on a 3 kg scale compared to the 15 g scale could be due to the higher temperature, which rose as high as 45°C near the center of the reactor, in the 3 kg scale cultivation. This can be solved by using a higher aeration rate.

7.3.2.5 Effect of Moisture Content on Iturin A Production by *B. subtilis* NB22

Figure 7.26 shows the effect of initial moisture content on iturin A production with partially dehydrated okara at 25°C. A moisture content of 82% was optimal. Little change in moisture content was observed during the 48 h period (data not shown). Iturin A was produced at a similar level with intact okara also at a water

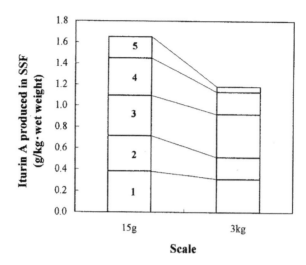

FIGURE 7.25   Iturin A production at 48 h in the solid-state fermentation (SSF) of intact okara in 15 g and 3 kg scale fermentations. Numbers in the graph represent the five major homologues produced corresponding to peaks 1–5 in Figure 7.1.

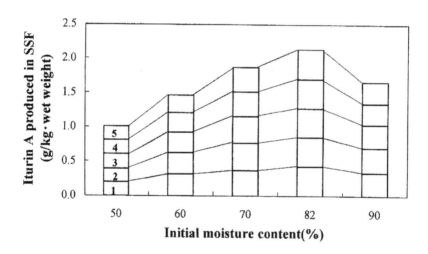

FIGURE 7.26   Effect of initial moisture content of dehydrated okara on iturin A production. Numbers in the graph represent the five major homologues produced corresponding to peaks 1–5 in Figure 7.1.

content of 82%, suggesting that dehydrated okara can be used as a solid-state substrate.

Solid-state fermentation for iturin A production is simple to operate, as well as requires a shorter time (48 h) for the completion of production, compared with 120 h in liquid fermentation (1). The concentration of total iturin A produced is six to eight times higher than that in liquid fermentation, implying that a smaller amount of solvent would be necessary for the purification of iturin A from the fermented product. Moreover, the solid culture of *B. subtilis* NB22 could be used directly as a functional fertilizer in agriculture with the ability to control plant diseases caused by various fungi and bacteria (12, 13, 14). Thus, extensive utilization of okara as a solid substrate for antibiotic production will give an additional value to the industrial waste.

7.4 DUAL PRODUCTION OF ITURIN AND SURFACTIN IN SOLID-STATE FERMENTATION (15)

Although coproduction of iturin and surfactin was first reported by Sandrin et al. (16), optimal conditions for production of surfactin (17–20) and iturin A (1) have been investigated separately only in submerged fermentation. It was shown that SSF is effective for the production of iturin A (7). There have been no reports on the production of the two compounds in the SSF by dual producers. *B. subtilis* RB14 produces cyclic lipopeptide antibiotics, iturin A and surfactin, simultaneously. Temperature dependency of the production of iturin A and surfactin by a dual producer, *B. subtilis* RB14, in the solid-state fermentation of okara (soybean curd residue) was investigated.

7.4.1 Materials and Methods

7.4.1.1 Strain Used

B. subtilis RB14 was explained in detail in Chapter 3.

7.4.1.2 Solid-State Fermentation of Okara

Basically the experimental procedure is the same as described in Section 7.2.1. Fifteen grams of okara was placed in a 100 ml Erlenmeyer flask, the mouth of which was stoppered with a cotton plug, and autoclaved twice at 120°C for 30 min at an interval of 8–12 h to kill the spore-forming microorganisms inhabiting the material. After autoclaving 833 μl of 0.45 g/ml glucose, 75 μl of 1 M KH_2PO_4, 225 μl of 1 M $MgSO_4$, and 367 μl of deionized distilled water were aseptically added for the fortification of nutrients. Then 3 ml of a preculture of RB14 grown in the no. 3S medium was inoculated. The flasks were immersed in a water incubator for 48 h without shaking. Cultivation temperature was varied from 23°C, 25°C, 30°C, 37°C, 42°C, to 48°C in accordance with the purpose.

7.4.1.3 Measurement of Iturin and Surfactin Concentrations

Iturin A and surfactin were extracted from the cultivated okara with methanol. The extract was then centrifuged, and the supernatant was filtered prior to the injection into HPLC (either 880 PU or TRI ROTAR-V, Japan Spectroscopic Co., Tokyo) using a UV detector set at 205 nm (870 UV or 875 UV, Japan Spectroscopic Co., Tokyo) to determine the concentrations of iturin A and surfactin. Calibration curves were prepared by using individually purified five peaks of iturin A (11, 12) and the commercial product of surfactin (Wako Pure Chemicals Industry, Osaka). Conditions for quantification by HPLC are as follows. For iturin A: mobile phase, acetonitrile/10 mM

ammonium acetate = 4/3 (v/v); column, 4.6 mm $\Phi \times$ 250 mm ODS-2; column temperature, 30°C; flow rate, 1.0 ml/min. For surfactin: mobile phase, acetonitrile/3.8 mM
trifluoroacetic acid = 4/1 (v/v), the same column at the same temperature with that of
iturin A was used; flow rate, 1.5 ml/min.

7.4.2   RESULTS

7.4.2.1   Effect of Temperature on Production of Iturin and Surfactin

Figure 7.27 shows the effect of temperature on the production of iturin A and surfactin by RB14 and on the growth of RB14. Figure 7.27a shows the actual amounts of

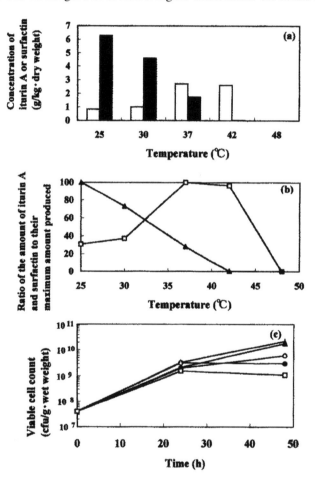

FIGURE 7.27   Effect of temperature on the production of iturin A and surfactin and on the
growth of *B. subtilis* RB14 in solid-state fermentation of okara. (a) Concentrations of iturin
A and surfactin after 48 h at different temperatures. Symbols: ■, iturin A; □, surfactin. (b)
Ratio of the amount of iturin A and surfactin to their maximum amount produced at different
temperatures. Symbols: ▲, iturin A; □, surfactin. (c) Count of viable cells; o, 25°C; •, 30°C;
Δ, 37°C; ▲, 42°C; and □, 48°C.

iturin A and surfactin produced. Figure 7.27b shows relative productivity when the maximum value in Figure 7.27 is taken to be 100%. Production of iturin A decreased as the temperature was increased. Surfactin production, in contrast, increased up to 37°C, and then decreased beyond that temperature. Between 25°C and 37°C, the production profiles of the two compounds were very different. Figure 7.27c indicates that cell population varied within one order of magnitude among the temperatures tested, and the final viable cell counts of RB14 at each temperature were proportional to the concentration of surfactin. This may correspond to the fact that surfactin is produced by RB14 in a growth-associated pattern.

Although the precise metabolic pathways for the syntheses of iturin A and surfactin are not elucidated, their quite different temperature dependencies were clearly shown in our results (Figure 7.27a and b). Although the same gene, *lpa-14*, coregulates the production of iturin A and surfactin (21, 22), the results suggest that the common pathways involving Lpa-14 protein are not significantly affected by the temperature, whereas the pathways that are specific for each compound are affected by the temperature.

From the data showing the effect of temperature on the production of iturin A and surfactin, it is possible to produce iturin A and surfactin at different ratios in the fermented okara. As the optimum ratio of iturin A to surfactin observed to inhibit the growth of a phytopathogenic fungus, *F. oxysporum* f. sp. *lycopersici* race J1 SUFI19, is 1 to 4 when the concentration of surfactin is fixed at 100 mg/l (23), and the antiphytopathogenic activity may vary with the ratio of the concentrations of the two substances depending on plant pathogens; different activities as a biological pesticide against other phytopathogens can be expected from the solid culture fermented at different temperatures.

7.4.2.2 Change in the Ratio of the Iturin Homologues at Different Temperatures

There are five homologues of iturin A produced that can be distinguished and quantified using the HPLC system described previously (11). In the course of the investigation of temperature effects, changes among the relative ratios of the five major homologues of iturin A were found as shown in Figure 7.28. The effect of temperature on the ratio was marked for peaks 1 and 4, which possess normal aliphatic chains. The ratio of peak 1 decreased as the temperature was increased, whereas a completely opposite tendency was observed for peak 4. The effect of temperature was not so marked on 2, 3, and 5, which possess branched aliphatic chains such as *iso* or *anteiso* chains, and the peak ratios of these peaks were almost constant at different temperatures. No information is available concerning the biosynthesis of iturin A, but the fatty acids of side chains can be postulated to be derived from the synthetic pathways of fatty acids of cellular membranes. The composition of cellular membrane fatty acids is altered by varying the growth temperature to maintain the proper membrane rigidity at a given temperature. It has been reported that the relative amount of long-chain fatty acids increases with increasing temperature for five *Bacillus* strains (24) as well as Clostridia (25), mainly because the longer the chain of a fatty acid, the higher the melting temperature of the fatty acid becomes. In this experiment, the decrease in relative ratio of 1 with increasing temperature is similar

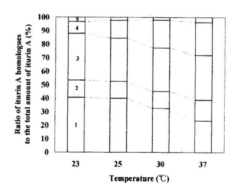

FIGURE 7.28.   The ratios of the amount of iturin A homologue to the total amount of iturin A produced by *B. subtilis* RB14 in the solid-state fermentation of okara. Numbers 1–5 in the figure represent homologues of iturin A with n-C_{14}-, *anteiso*-C_{15}-, *iso*-C_{15}-, n-C_{16}-, and *iso*-C_{16}- β-amino acid residues, respectively.

to the result mentioned earlier. However, it is interesting that the ratio of 1 consisting of n-C_{14}-β-amino acid residue decreased with increasing temperature, while that of 4 consisting of n-C_{16}-amino acid residue increased. This may indicate that there is some specific metabolic change that occurs in the RB14 intracellularly.

Although the total amount of iturin A produced decreased as temperature was increased, the efficacy of the activity of the fermented okara as a biological pesticide to inhibit plant pathogens, and thus to prevent plant diseases, may not vary so significantly, mainly because the longer the side chains of an iturin A homologue, the lower the minimum inhibitory concentration for various plant pathogens becomes, as has been previously reported (11).

Not only was the effect of temperature on iturin A and surfactin production revealed but the effect of temperature on the relative ratios among the five iturin A homologues was also shown. Precise mechanisms of the temperature dependencies described earlier should be explained by biochemical and/or genetic studies.

7.5   ENHANCED ITURIN PRODUCTION IN SUBMERGED FERMENTATION AND PLANT TEST (26)

In previous Chapters it was demonstrated that *B. subtilis* RB14 or RB14-C, a streptomycin-resistant mutant derived from the parental strain *B. subtilis* RB14, suppressed the damping-off of tomato plants in soil (27) by producing iturin A and surfactin (23). Although surfactin has a relatively weak antibiotic activity, the antibiotic activity of iturin A is very strong.

B. subtilis RB14 and RB14-C in submerged culture produce iturin A at the level of a few hundred milligrams per liter of culture (17, 23, 27). To enhance iturin A production and its suppressive effect on plant pathogens, a spontaneous surfactin-null mutant of RB14-C (*B. subtilis* RB14-CS) was isolated, and by developing a liquid medium capable of enhancing iturin A production the improved suppression of damping-off in tomato pot tests by RB14-CS was shown.

7.5.1 MATERIALS AND METHODS

7.5.1.1 Microorganisms

The *B. subtilis* RB14-CS (RB14-CS) used was a spontaneous mutant of *B. subtilis* RB14-C. The iturin A productions of RB14 and RB14-C were confirmed to be similar to the previously reported levels (27). RB14-CS cells were preserved at –80°C in 30% glycerin solution until use. The plant pathogenic fungus *R. solani* K-1 was isolated from cockscomb as a severe damping-off pathogen of many plants at the Kanagawa Horticultural Experiment Station, Kanagawa, Japan (28).

7.5.1.2 Media

The L medium contained 10 g/l of Polypepton (Nihon Pharmaceutical, Tokyo, Japan), 5 g/l of yeast extract, and 5 g/l of NaCl, adjusted to pH 7.0 and solidified with 2.0% agar when necessary. The no. 3S medium used for iturin A production contained 30 g/l of Polypepton S (Nihon Pharmaceutical), 10 g/l of glucose, 1 g/l of KH_2PO_4, and 0.5 g/l of $MgSO_4$ $7H_2O$, adjusted to pH 6.8. To investigate the effect of various concentrations of nitrogen and carbon sources on iturin A production by RB14-CS, modified no. 3S media were prepared by changing the concentrations of Polypepton S and glucose to 40, 80, and 120 g/l, or those of glucose to 33, 67, and 100 g/l. The different media containing a combination of 80 g/l soybean meal (Honen, Tokyo, Japan), 80 g/l of Polypepton S, and 67 g/l of maltose or glucose were examined, and the medium containing 80 g/l soybean meal and 67 g/l maltose designated as SM medium was mainly used. Potato dextrose agar (PDA) medium containing 200 g/l of potato infusion, 20 g/l of glucose, and 15 g/l of agar, adjusted to pH 5.6, was used for the stock culture of *R. solani* K-1. Potato dextrose Polypepton (PDP) medium, containing 200 g/l of potato infusion, 20 g/l of glucose, and 10 g/l of Polypepton (pH 5.6), was used for the cultivation of *R. solani* K-1 (27).

7.5.1.3 Cultivation of RB14-CS

Ten microliters of RB14-CS cell suspension preserved at –80°C were inoculated to 5 ml of L medium in a test tube and cultivated at 30°C at 120 strokes per minute (spm) for 16 h. This preculture was then inoculated in 40 ml of no. 3S, modified no. 3S, SM, and other media mentioned earlier at 1% inoculation in conical flasks, and the flasks were shaken at 30°C and 120 spm. On days 2, 4, 6, and 8, iturin A concentrations were determined from individual flasks. On day 8, each culture was serially diluted in 0.85% NaCl solution and plated to L agar plates containing streptomycin 20 mg/l. The plates were incubated at 37°C for 16 h, the numbers of colonies were counted, and log colony-forming units (cfu) per milliliter of culture was expressed as the cell number.

7.5.1.4 Quantitative Analysis of Iturin A in Liquid Medium

A 100 μl sample of *B. subtilis* B14-CS culture was suspended in a microtube containing 900 μl of a mixture of acetonitrile and 10 mM ammonium acetate 35:65 (vol/vol) and shaken for 20 min. The solution was them centrifuged at 18,000 × g for 10 min, and the supernatant was filtered through a 0.2 μm pore-size hydrophilic PTFE-type disposable syringe filter unit (DISMIC 13HP020AN; Toyo Roshi,

Tokyo, Japan). The amount of iturin A in filtrate was quantified using a reverse-phase HPLC column (Chromolith Performance RP-18e 100-4.6; Merck KGaA, Darmstadt, Germany) on an LC-800 HPLC system (JASCO, Tokyo, Japan) operated at a flow rate of 2.0 ml/min. A mixture of acetonitrile and 10 mM ammonium acetate (35:65, vol/vol) was used as the eluent, and elution was monitored at 205 nm. The five peaks corresponding to the major iturin A homologues (27) were regarded as the total iturin A produced because the other peaks fell below the levels of detection in this system, 0.3 μg/ml. The concentration of iturin A was determined by a standard curve prepared by using standard iturin purchased from Sigma (St. Louis, Missouri, USA).

7.5.1.5   Elementary Analysis of the Water-Soluble Fraction

The suspension containing 80 g/l of soybean meal was homogenized with a homogenizer (ACE homogenizer, Nihonseiki, Tokyo, Japan) at 10,000 rpm for 10 min, autoclaved, and filtered through a paper filter. The filtrate was dried at 105°C for 24 h, and the carbon and nitrogen contents were determined by elemental analysis. At the same time, the solution containing 80 g/l of Polypepton S was analyzed in the same manner.

7.5.1.6   Inoculation of *R. solani* into Soil

The soil used in this study was a low humic andosol collected from a field at the Kanagawa Horticultural Experiment Station (14). The main characteristics of the prepared soil were as follows: texture, low humic andosol; moisture content, 12.7%; maximum water holding capacity, 137/100 g dry soil; pH 5.9; and bulk density, 0.522 g/cm. These properties were measured as previously described (29). The soil was sieved through an 8-mesh screen (approximately 2 mm pore size) and air-dried. Soil and vermiculite were mixed at a ratio of 4:1 (wt/wt) and amended with chemical fertilizer so that the final concentrations of N, P_2O_5, and K_2O were 70, 240, and 70 mg per 100 g of dry soil, respectively. The soil was put into a sterilizable polypropylene bag and autoclaved for 60 min at 121°C four times at 12 h intervals before use. Sterilized soil (150 g) was disposed to plastic pots with a diameter of about 90 mm and a height 80 mm.

R. solani K-1 was grown on PDA plates, and 5 mm plugs were inoculated to sterile PDP medium (50 ml in 200 ml Erlenmeyer flasks). The flasks were incubated without shaking in the dark at 30°C for 6 days, and the formed mycelial mats were suspended in sterile water and homogenized at 4000 rpm for 2 min using an ACE homogenizer (Nihonseiki). The homogenate was inoculated to the soil in pots at 4.1 g of wet mat per pot. After inoculation, the pots were incubated at 28°C and the moisture was maintained at 60% of the maximum water-holding capacity. Two days later, the soil was mixed with RB14-CS.

7.5.1.7   Effects of RB14-CS on the Plant Growth in Soil

B. subtilis RB14-CS was incubated for 16 h in L medium at 37°C at 120 spm in a shaker, and then 1 ml of this preculture was inoculated to 100 ml of no. 3S or SM medium in an Erlenmeyer flask (nominal volume, 500 ml). The flask was incubated for 5 days at 30°C at 120 spm, and then 3, 10, 20, and 30 ml volumes of each RB14-CS culture were mixed with 150 g of soil in a pot. For the inoculations of cell suspension of RB14-CS culture, 30 ml of 5-day-old cultures prepared with SM medium was centrifuged at 12,000 × g for 15 min at 4°C. The pellet was then washed three times with sterile distilled water and resuspended with 20 ml of sterile distilled water and used for inoculation as described earlier. In each condition, the water content was maintained at 60% of the maximum water-holding capacity by daily addition of sterilized water. Three days after RB14-CS inoculation, germinated tomato seeds were planted in each pot. Three pots were prepared for each treatment and each experiment was repeated at least three times.

Tomato seeds (Ogata Fukuju; Takii Seed, Kyoto, Japan) were surface-disinfected for 1 min with 70% ethanol, rinsed five times with sterile distilled water, and then surface disinfected again for 5 min with 0.5% sodium hypochlorite. After at least 10 rinses with sterile distilled water, the seeds were germinated on a 2% agar plate at 30°C for 2 days. For the pot test, each pot was sown with nine germinated seeds and placed in a growth chamber at 28°C with 90% relative humidity under 16 h of light (about 8000 lux). At least three pots were prepared for each treatment. Soil moisture was kept at 60% of the maximum soil water-holding capacity. After 17 days, the percentage of diseased seedlings per pot was determined, and the lengths and dry weights of the shoots were measured by clipping them at the soil surface.

7.5.1.8   Cell Recovery from Soil and Determination of Cell Number

The soils were sampled before sowing and at 0, 3, and 17 days after planting. In the 17-day sampling, the roots and surrounding soil were collected together. For each sample, 3 g of soil was suspended in 8 ml of 0.85% NaCl solution (pH 7.0) in a 50 ml Erlenmeyer flask and then shaken for 15 min at 140 spm at room temperature. The suspension was serially diluted in 0.85% NaCl solution and plated onto L agar plates containing 20 mg/l of streptomycin. After 12 h of inoculation at 37°C, the number of colonies was counted and expressed as log cfu per gram of dry soil. At the same time, 1 ml of each suspension was heated at 80°C for 15 min to count the number of spores.

7.5.1.9   Quantitative Analysis of Iturin A Recovered from Soil

Three grams of soil was suspended in 21 ml of a mixture of acetonitrile and 3.8 mM trifluoroacetic acid (4:1, vol/vol) in a 50-ml Erlenmeyer flask, and then shaken for 1 h at 140 spm at room temperature. The soil was removed by filtration using filter paper (Toyo Roshi) and the filtrate was evaporated. The resulting precipitate was extracted with 2 ml of pure methanol for 2 h. The extracted solution was centrifuged at 10,000 × g for 2 min, and then the supernatant was filtered through a 0.2 μm pore-size PTFE membrane (JP020; Advantec, Tokyo, Japan) and applied for HPLC as described earlier.

7.5.1.10   Statistical Analysis

Each plant test was repeated at least three times, and the mean of data was analyzed by Fisher's protected least significant difference analysis.

7.5.2 Results

7.5.2.1 Enhancement of Iturin A Production by Changing and Increasing Nitrogen and Carbon Sources in Media

Control experiments confirmed that no surfactin was produced by RB14-CS cells in culture using the previously reported quantitative analytical method (23). Then, the production of iturin A by RB14-CS in no. 3S medium, which was used in the previous reports using *B. subtilis* RB14 or RB 14-C (21, 22, 27) was examined. The production level of iturin A, 280 mg/l by RB14-CS in no. 3S medium, as is shown in Figure 7.29a, was higher than those previously reported for RB14 and RB14-C, 100–140 mg/l (22, 27). To enhance iturin A production in no. 3S medium, concentrations of glucose and Polypepton S were increased. As a result, the highest iturin A production in cultures with modified no. 3S medium containing 80 g/l of Polypepton S and 67 g/l of glucose was obtained (Figure 7.29). Then, the use of SM medium containing 67 g/l of maltose and 80 g/l of soybean meal, which is an inexpensive and effective nitrogen source for *B. subtilis*, was examined. Figure 7.30 shows the chronological change of iturin A concentration in RB-14CS cultures grown in SM medium. The maximum iturin A concentration obtained in SM medium (4400 mg/l) was about 16 times higher than that obtained in standard no. 3S medium (280 mg/l) and 50% higher than that obtained in the modified no. 3S medium as shown in Figure 7.29b. Fructose, sucrose, galactose, and lactose as the sole carbon sources in the SM medium were used, but iturin A production was the highest with maltose (data not shown). When iturin A production in media containing a combination of 80 g/l soybean meal, 80 g/l of Polypepton S, and 67 g/l of maltose or glucose was measured, iturin A production in the SM medium was the highest and was 2.2-fold higher than that in the medium containing 80 g/l of Polypepton S and 67 g/l of glucose and 20% higher than that in the medium containing 80 g/l of Polypepton S and 67 g/l of maltose as shown in Figure 7.30. The cell number in the medium with soybean meal and maltose after 8 days of cultivation was 2.83×10^9 cfu/ml, and those

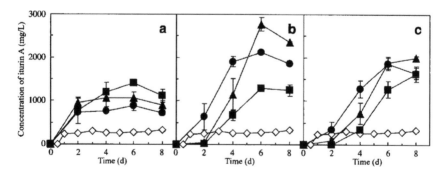

FIGURE 7.29 Iturin A production by *B. subtilis* RB14-CS in no. 3S medium (open diamonds) and the modified no. 3S media containing various concentrations of Polypepton S and glucose. (a) 40 g/l of Polypepton S, (b) 80 g/l of Polypepton S, and (c) 120 g/l of Polypepton S. Closed circles, 33 g/l of glucose; closed triangles, 67 g/l of glucose; closed squares, 100 g/l of glucose. Two flasks were prepared for each condition and each experiment was repeated twice.

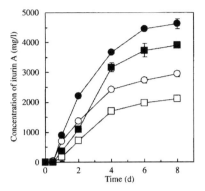

FIGURE 7.30 Iturin A production by *B. subtilis* RBl4-CS in the medium containing 80 g/l soybean meal (Soy-M) or Polypepton S (Poly-S) as nitrogen sources and 67 g/l maltose (Mal) or 67 g/l glucose (Glu) as carbon sources. Open squares, Poly-S and Glu; filled squares, Poly-S and Mal; open circles, Soy-M and Glu; filled circles, Soy-M and Mal. Two flasks were prepared for each condition and each experiment was repeated twice.

in the medium with other combinations and no. 3S medium were 2.57–8.75 × 10^9 and 3.09 × 10^9 cfu/ml, respectively. These results indicate that the components in each medium were associated with the production of iturin A rather than with differences in cell number.

7.5.2.2 Carbon and Nitrogen Contents in the Water-Soluble Portions of the Media

The carbon and nitrogen contents in the water-soluble fraction of soybean meal were 35.9% and 4.41%, respectively. Therefore, in SM medium containing 80 g/l soybean meal and 67 g/l maltose, the total carbon and nitrogen contents of soluble fraction were calculated to be 37.4 and 1.30 g/l, respectively. Similarly, the carbon and nitrogen contents in Polypepton S were determined as 32.7% and 7.52%, respectively. In modified no. 3S medium containing 80 g/l Polypepton S and 67 g/l glucose, the calculated carbon and nitrogen contents were 53 and 6 g/l, respectively. These results indicate that the modified no. 3S medium contained 1.42-fold more carbon and 4.61-fold more nitrogen than SM medium.

7.5.2.3 Suppressive Effects of Cultures Grown in SM and no. 3S Media in Plant Tests

Table 7.4 shows the occurrence of damping-off of tomato seedlings when 30 ml of culture of RB14-CS grown in no. 3S medium or 3–30 ml of culture grown in SM medium were applied to the soil infested with *R. solani*. When only *R. solani* was supplied to the soil, 85.2% of the plants suffered from damping-off. The 30 ml culture grown in no. 3S medium showed a slight decrease in the disease occurrence (77.8%). In contrast, the 20–30 ml of culture of RB14-CS grown in SM medium showed significantly reduced disease occurrences (37% and 22%, respectively). The disease occurrences were similar in 3 ml of SM-based culture and in 30 ml of no. 3S-based culture (Table 7.4), but the disease occurrences were reduced by the

TABLE 7.4

**Suppressive Effects of No. 3S and SM Media Culture of *B. subtilis*
RB14-CS on the Occurrence of Tomato Damping-Off Caused by
R. solani K-1 17 Days after the Application**

R. solani	Volume of RB14-CS Cultured in No. 3S Medium (ml/pot)	Volume of RB14-CS Cultured in SM Medium (ml/pot)	Shoot Length (mm)	Dry Weight of Shoots (mg)	Disease Occurrence (%)
+	–	–	16.2[a]	4.4[a]	85.2[a]
+	30	–	21.4[a]	5.0[ab]	77.8[ab]
+	–	30	64.4[c]	14.9[c]	22.2[c]
+	–	20	54.9[bc]	12.3[bc]	37.0[c]
+	–	10	32.2[ab]	6.1[ab]	63.0[b]
+	–	3	23.9[a]	5.2[ab]	75.9[ab]
–	–	–	119.6[d]	36.7[d]	0.0[d]
LSD $p = 0.05$			22.9	7.7	20.9

Notes: For each treatment, each datum is the average of the results obtained from six replications. The different letters represent significant differences ($p = 0.05$) according to Fisher's protected least significant difference analysis. LSD, least significant difference.

increase in the dose of SM-based culture. The shoot lengths and dry shoot weights were not significantly different between pots receiving 20 and 30 ml of SM-based culture, but the disease occurrence was smaller, and the shoot lengths and dry weights of shoots were larger in these two treatments than those in the other two treatments. Table 7.5 shows the disease occurrence when culture (approximately 30 ml) using SM medium or cell suspension of the culture of 30 ml collected by centrifugation was applied to the soil. When the cell suspension was applied to the soil, disease occurrence was not significantly different from that of the control pots with *R. solani*, while initial cell number of RB14-CS in soil was approximately equal to that with the culture. These results strongly suggest that the suppressive effects of culture were due to the presence of iturin A in the culture media.

The changes in the concentration of iturin A extracted from the soil at each sampling day are shown in Table 7.6. Control experiments showed that no iturin A peak was observed in untreated soil. The recovery error for iturin A from the soil using this method was almost ±20%, which was consistent with a previous report (30). After 3 days, nearly 60% of the initial iturin A remained in the soil, but almost no iturin A was detected on day 17. Similar results were obtained in pots without *R. solani*, indicating that the decrease in iturin A concentration in the soil was not influenced by the presence of the pathogen.

Table 7.7 shows the RB14-CS cell number in the soil. The cell number decreased to one-third that of the initial value by day 17. However, a significantly high cell number was maintained in the soil, with more than 95% of the population existing as spores (data not shown).

TABLE 7.5

Suppressive Effects of Culture Using SM Medium and Cell Suspension of the Culture of *B. subtilis* RB14-CS on the Occurrence of Tomato Damping-Off Caused by *R. solani* K-1 17 Days after the Application

R. solani	RB14-CS	Estimated iturin A Concentration in Soil on Day 0 (μg/g-dry soil)*	Initial Number of RB14-CS in Soil (×10⁸) cfu/g dry soil)	Shoot Length (mm)	Dry Weight of Shoots (mg)	Disease Occurrence (%)
+	–	0	–	11.3ᵃ	2.4ᵃ	87.0ᵃ
+	Culture	959	14.3	40.8ᵇ	9.9ᵇ	38.9ᵇ
+	Cell suspension	21.7	17.9	18.1ᵃ	4.5ᵃ	77.8ᵃ
–	–	0	–	99.7ᶜ	25.0ᶜ	0ᶜ
LSD *p* = 0.05				12.4	4.1	19.4

Notes: Thirty milliliters of culture and the cell suspension obtained from 30 ml of culture were used. For each treatment, each datum is an average of the results obtained from six replications. The different letters represent significant differences (*p* = 0.05) according to Fisher's protected least significant difference analysis. LSD, least significant difference.

* These values were estimated from the concentrations of iturin A in SM medium and cell suspension obtained from SM medium prepared as described in Section 7.5.1.9 measured by HPLC.

TABLE 7.6

Concentration of Iturin A Recovered from Soil Sampled on days 0, 3, and 17 after Addition of 30 ml of *B. subtilis* RB14-CS Culture Prepared in SM Medium

	Iturin A Concentration (μg/g dry soil)		
Treatment	0 days	3 days	17 days
R. solani + RB14-CS (SM medium)	522 (88)	329 (85)	0 (0)
RB14-CS (SM medium)	458 (109)	326 (41)	4 (4)

Notes: Each value is an average of results obtained from six replications, and the values in brackets indicate the standard deviations.

7.5.3 DISCUSSION

Although iturin A is known to be a strong growth inhibitor of various plant pathogens *in vitro* (27, 31), it is not clear whether higher iturin A concentrations are more effective in increasing the efficacy of iturin A in soil and persistence of iturin A in soil. Thus, media that produce higher levels of iturin A were sought, and SM medium was found to be the most effective among the test media for iturin A production, yielding a maximum production level of 4.4 g/l, which is higher than any previously

TABLE 7.7

Bacillus subtilis RB14-CS Cell Counts in Soil on days 0, 3, and 17 after Addition of 30 ml of B. subtilis RB14-CS Culture Prepared in No. 3S Medium or SM Medium

Treatment	Cell Number ($\times 10^8$ cfu/g dry soil)		
	0 days	3 days	17 days
R. solani + RB14-CS (no. 3S medium)	18.3	6.4	2.3
R. solani + RB14-CS (SM medium)	27.8	28.0	6.6

Note: Each value is an average of results obtained from six replications.

reported iturin A production level. Furthermore, this production level was correlated with biological control activity because application of the iturin A–containing culture suppressed dose-dependently damping-off caused by *R. solani* in pot tests (Table 7.4). In contrast, the application of isolated cells had no such effect (Table 7.5), indicating that secreted iturin A in the culture medium was likely to account for the biocontrol. Soil levels of iturin A are important for their effect against the plant pathogen because germinating young seedlings are killed by the pathogen *R. solani* before or soon after they emerge from the soil (32).

The quantity and quality of lipopeptides produced by *B. subtilis* have been shown to be dependent on the nutritional conditions (33). As the production of iturin A in SM medium was significantly higher than that in modified no. 3S medium, the soluble carbon and nitrogen contents between the media were compared. Because soybean meal is solid and partly solubilized during cultivation, it was assumed that the bacterium consumed the solubilized portion of soybean meal. Elementary analyses of the water-soluble portions of soybean meal and Polypepton S showed that the modified no. 3S medium contained 1.42-fold more carbon and 4.61-fold more nitrogen than SM medium, suggesting that the enhanced production of iturin A in SM medium was due to factors other than the carbon and nitrogen contents.

The iturin synthesis genes were cloned (34). Thus, as another method to enhance iturin A production, it may be possible in the future to obtain higher iturin A production and biocontrol activity through gene amplification, as previously shown for mycosubtilin (35).

The co-application of *B. subtilis* with organic matter (13) or solid-state fermentation is an alternative to enhance the activity of *B. subtilis*. In a previous study, solid-state fermentation using okara, a by-product of tofu manufacturing, as a substrate showed 10-fold higher iturin A production than that in submerged fermentation (7). However, the industrial use of solid-state fermentation has been limited by practical problems such as difficulties in controlling temperature, oxygen, and moisture, as well as issues with up-scaling the reactor. In contrast, submerged fermentation offers the advantages of high homogeneity of heat and mass transfer and easy control of oxygen and temperature. Therefore, enhanced production of iturin A by *B. subtilis* in submerged fermentation may provide a more effective system for industrial production of iturin A as a biocontrol agent.

Our monitoring of iturin A in soil revealed that, despite the relatively high input iturin A concentration, iturin A was not detectable in soil on day 17. The disappearance of iturin A from soil was likely due to chemical degradation, irreversible adsorption to soil, and/or potentially biological degradation via airborne contaminants introduced into the presterilized soil. This soil degradability and brief persistence of iturin A may be an advantage for biological control agents over chemical pesticides that are persistent in soil for longer periods.

The persistence of *B. subtilis* in soil is also important in maintaining suppressiveness to plant pathogens. Although vegetative cells of *B. subtilis* are prone to dying in soil, spores are stably maintained for more than 2 months (30). Therefore, when spores are applied to soil and germinate to iturin A–producing vegetative cells, biological control of *B. subtilis* can be expected in older or mature plants.

Furthermore, *B. subtilis* can be co-applied with chemical pesticides (36–40). In Chapter 9, it is demonstrated that *B. subtilis* is resistant to flutolanil and that the use of this bacterium with flutolanil reduced the amount of flutolanil to one-fifth of the single use of the fungicide (41). Herein, the medium for enhanced production of iturin A was shown, which reflected the higher suppressive effect on damping-off. This finding suggests that the use of *B. subtilis* may be useful for further reducing the use of agricultural chemicals.

7.6 ENHANCED ITURIN PRODUCTION IN SOLID-STATE FERMENTATION (42)

As okara consists of insoluble fiber and approximately 70% water, it can be used as a substrate in solid-state fermentation. Thus far, most microbial cultivation has been conducted in submerged fermentation. However, solid-state fermentation has several advantages over submerged fermentation in that (i) it requires less energy for cultivation because it does not need any agitation unit, (ii) it can be used for treatment of solid waste, (iii) it has a long history of production of traditional foods by using mainly fungi, especially in Japan, and (iv) it uses a smaller amount of solvent for product extraction. Therefore, solid-state fermentation can be used to produce secondary metabolites (43, 44). In this section, one derivative of *B. subtilis* RB14, *B. subtilis* RB14-CS, was cultivated in solid-state fermentations using okara as a main solid substrate, and a product that has a dual function of microbial pesticide and organic fertilizer was produced, and the efficacy of the product to plant testing was confirmed.

7.6.1 MATERIALS AND METHODS

7.6.1.1 Strains

B. subtilis RB14-CS, which is a spontaneous mutant derived from RB14-C and was selected by morphological change of RB14-C in colony formation, is a single iturin A producer and a non-surfactin producer. *B. subtilis* RB14-C is a streptomycin-resistant mutant from a parent strain RB14 and is a coproducer of the antibiotics iturin A and surfactin (27). *Rhizoctonia solani* K-1, isolated at Kanagawa Horticultural Experiment Station (Kanagawa, Japan) (27), was used as a fungal plant pathogen that causes a severe damping-off in many plants. Characters of damping-off of tomato have been previously described in detail (45, 46).

7.6.1.2   Media Used

The L medium used for the growth of the bacterium contains 10 g of Polypepton (Nippon Pharmaceutical, Tokyo), 5 g of yeast extract, and 5 g of NaCl (per liter). One milliliter of L medium culture broth, after 24 h cultivation at 30°C, was inoculated into 100 ml of no. 3S medium consisting of 30 g of Polypepton S (Nippon Pharmaceutical), 10 g of glucose, 1 g of KH_2PO_4, 0.5 g of $MgSO_4 \cdot 7H_2O$ (per liter) (pH 6.8), and incubated at 120 spm at 30°C for 24 h in a shaking flask and used as a seed for solid-state fermentation.

7.6.1.3   Solid-State Fermentation

Okara, was divided into 150 to 200 g pieces, wrapped in a sheet of commercial wrapping film and frozen at –15°C. Fifteen grams of thawed okara was placed in a 100 ml conical flask and autoclaved twice at 120°C for 20 min at an interval of 8–12 h to kill spore-forming microorganisms inhabiting the material. After cooling to room temperature, the following solutions were added as nutrient supplements for every 15 g of okara, and moisture content was adjusted to 79%: 833 μl of 0.45 g glucose/ml, 75 μl of 1 M KH_2PO_4, 150 μl of 1 M $MgSO_4 \cdot 7H_2O$, and 367 μl of deionized distilled water. Then, 3 ml of an RB14-CS culture grown in no. 3S medium was added to 15 g of okara and mixed with a stainless steel spatula. All flasks were incubated statically in a water incubator at 25°C, and at a specified time, one flask was taken and the whole okara in a flask was used as a sample for analysis.

7.6.1.4   Determination of Viable Cell Number and pH

For the determination of viable cell number and pH during solid-state fermentation, 1 g of okara was placed in a sterile 18 mm diameter test tube. It was mixed thoroughly, with the use of a vortex, with 9 ml of sterile distilled water and shaken at 150 spm for 5 min at room temperature. Then, the mixture was serially diluted and spread onto L agar plates consisting of L medium and 15 g/l agar. After 18 h of incubation at 37°C, the number of colonies was determined and expressed as colony-forming units (cfu). The suspension was also heated at 80°C for 15 min and spread onto L agar plates, and the number of colonies including the spores only was expressed as cfu. The cell suspension obtained for cfu measurement was used for pH measurement. The viable cell number and pH in submerged fermentation were determined after sampling the culture broth in the similar method described earlier.

For the determination of viable cell number in soil, 3 g of soil containing RB14-CS was suspended in 8 ml of 0.85% NaCl solution (pH 7.0) in a 50 ml conical flask and shaken at 150 spm for 15 min. The suspension was used to determine the viable cell number and spore number in soil by using L agar plates containing 20 μg/ml streptomycin in a similar manner to that described earlier.

7.6.1.5 Extraction and Quantitation of Iturin A

For the extraction of iturin A, 45 ml of methanol was added to the whole fermented product in a flask, and the mixture was shaken at 150 spm for 1 h. The methanol extract obtained was centrifuged at $18,000 \times g$ at 4°C for 10 min, and filtered through a 0.2 μm pore-size polytetrafluoroethylene membrane (JP020, Advantec, Tokyo), and the filtrate was injected into a HPLC with a column (Chromolith Performance RP-18eb, 4.6 mm diameter × 100 mm height, Merck, Germany) for measuring iturin A concentrations. The details of the measurement conditions were described (27).

7.6.1.6 Soil Used

The soil used in this study was a low humic andosol taken from a field at the Kanagawa Horticultural Experiment Station. The soil was sieved through an 8-mesh (about 2 mm pore size) screen and air-dried. The soil and vermiculite were mixed in the ratio 4:1 (wt/wt) and nutrient-amended so that the final concentrations of N, P_2O_5, and K_2O were 70, 240, and 70 mg per 100 g of dry soil, respectively. The prepared soil was kept in plastic bags at room temperature. The soil was placed in a sterilizable polypropylene bag and autoclaved for 60 min at 121°C four times at 12 h intervals. The main characteristics of the soil thus prepared were as follows: texture, low humic andosol; moisture content, 12.7%; maximum water-holding capacity, 137 g/100 g dry soil; pH 5.9; and bulk density, 0.522 g/cm^3. The measurements of these properties followed previously described methods (29).

7.6.1.7 Plant Test

The sterilized soil (150 g), prepared by the method described earlier, was placed in a plastic pot with a diameter of approximately 90 mm and a height of 80 mm, and its moisture content was kept at 60% of the maximum water-holding capacity by the daily addition of sterilized water.

R. solani was incubated statically in the dark at 28°C for 7 days in PDP medium consisting of 200 g of potato infusion, 20 g of glucose, and 10 g of Polypepton (per liter) (pH 5.6). The mycelial mats that formed on the surface of the medium were homogenized by a homogenizer (ACE homogenizer, Nihonseiki, Tokyo, Japan) at 4000 rpm for 2 min in sterile water and inoculated into the soil at a ratio of 3.6 g-mat to one pot 6 days before planting the germinated tomato seeds.

The plant test was performed according to the following procedure: Tomato seeds were disinfected with 70% ethanol and then with 0.5% sodium hypochlorite. After rinsing with sterile water, the seeds were germinated on a 2% agar plate at 30°C for 2 days in the dark. Each pot was sown with nine germinated seeds 3 days after the okara culture of RB14-CS was introduced into soil and placed in a growth chamber at 30°C with 90% relative humidity under 16 h of light (about 8000 lux). After 2 weeks, the percentage of diseased seedlings per pot was determined.

7.6.1.8 Fertilizer Effect of Cultured Okara

The effects of raw okara and cultured okara on the growth of tomato seedlings were investigated in the soil without the inoculation of a plant pathogen. The experimental procedure was the same as that described earlier. The shoots were clipped off at the soil surface level, and their length and dry weight were measured after 2 weeks.

7.6.1.9 Quantitative Analysis of Iturin A Recovered from Soil

Three grams of soil were suspended in 21 ml of a mixture of acetonitrile:3.8 mM trifluoroacetic acid (4:1 vol/vol) in a 50 ml conical flask and shaken at 150 spm for 1 h. The soil in the suspension was then removed using a filter paper (Toyo Roshi, Tokyo) and the filtrate was evaporated. The precipitate was subjected to extraction using 2 ml of methanol for 2 h, and the extract was applied to HPLC as described earlier. The validity of this method was described in a previous paper (27).

7.6.1.10 Fertilizer Analysis

The fertilizer analysis of raw okara and okara cultured with *B. subtilis* RB14-CS was performed at the Kanagawa Prefectural Agricultural Research Institute.

7.6.1.11 Statistical Analysis

Each plant test was repeated at least three times, and each mean of data was analyzed by Fisher's analysis of variance.

7.6.2 RESULTS

7.6.2.1 Iturin A Production by *B. subtilis* RB 14-CS in Solid-State Fermentation

Figure 7.31A shows the growth of RB14-CS and iturin A concentration produced by RB14-CS in solid-state fermentation using okara. The data are means of three independent experiments. After the initial rapid growth, the viable cell number declined due to decrease in pH. The pH declined first was thought to be mainly due to acid production by consumption of carbon substrates during the growth of vegetative RB14-CS cells. Then, pH increased to 8.7 mainly by ammonification of degradation of nitrogenous compounds in okara (47). The viable cell number leveled off after 5 days at 2×10^{10} CFU/g wet substrate. The number of spores increased from the start of the experiment, and on day 4, almost all the viable cells were spores, resulting in termination of iturin A production. The iturin A concentration reached approximately

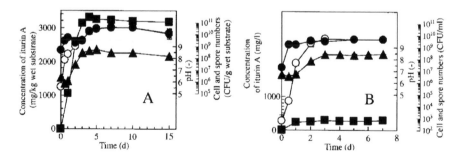

FIGURE 7.31 Chronological changes of the growth of *B. subtilis* RB14-CS, concentration of iturin A and pH in solid-state fermentation using (A) soybean curd residue (okara) as the substrate and (B) submerged fermentation using no. 3S medium. Filled squares (■) indicate iturin A concentration, filled circles (●) indicate viable cell number, open circles (○) indicate spore number, and filled triangles (▲) denote pH.

3300 mg/kg wet substrate (14 g/kg dry substrate) in 4 days. Figure 7.31B shows the growth and the iturin A concentration in submerged fermentation using no. 3S liquid medium. The growth leveled off in 1 day, and the number of spores increased, and on day 2, the number of spores was almost equal to that of viable cells. The increase in iturin A concentration leveled off in 1 day. As is the case with solid-state fermentation, the pH was decreased during initial 12 h by quick consumption of carbon source and then the increase in pH continued until 3 days by ammonification of degradation of Polypepton S. The maximum iturin A concentration was only 300 mg/l. It is obvious that the iturin A concentration in solid-state fermentation is remarkably high compared with that in submerged fermentation.

7.6.2.2   Suppressive Effects of Okara Cultivated with RB14-CS on Damping-Off of Tomato

Table 7.8 shows the results of the plant test using 5- and 10-day okara cultures in solid-state fermentation. In the soil without the plant pathogen, *R. solani*, all seeds grew and no damping-off occurred. In the soil infested with *R. solani*, 88% of the plants showed damping-off. When 1 g of 5-day-cultured okara was introduced into the soil, no significant suppressive effect was observed, but the introduction of 2 g of 5-day-cultured okara and 1 g of 10-day-cultured okara reduced the disease occurrence to 76% and 72%, respectively. Two grams of 10-day-cultured okara significantly reduced the disease occurrence to 41%. An experiment, in which raw okara was used to treat soil infested with *R. solani*, confirmed that disease occurrence was almost 100%, indicating that the raw okara has no suppressive effect of the growth of *R. solani*.

TABLE 7.8

Suppressive Effects of Okara Cultured with *B. subtilis* RB14-CS on the Occurrence of Damping-Off of Tomato Caused by *R. solani* K-1 (*N* = 6)

| | | Okara Cultured with RB14-CS | | |
| | | Cultivation Time (days) | Amount (g/pot) | Disease Occurrence (%) |
Run	*R. solani*			
1	−	−	−	0.0[a]
2	+	−	−	88[d]
3	+	5	1	91[d]
4	+	5	2	76[c]
5	+	10	1	72[c]
6	+	10	2	41[b]
LSD *P* = 0.05				11

Notes: For each treatment, each datum is an average of the results from experiments repeated six times. Means in any column with different letters are significantly different (*P* = 0.05) according to Fisher's protected least significant difference (LSD) analysis.

7.6.2.3 Viable Cell Number and Iturin A Concentration in Soil

The viable cell number and iturin A concentration in the soil are shown in Table 7.9. The viable cell number among four samples was decreased during incubation, but at least half remained, indicating that RB14-CS cells stably survive in soil mainly as spores. Although the soil was sterilized initially, the plant test was conducted in an open-air growth chamber, and thus the soil was contaminated with airborne contaminants. The number of contaminants reached 10^9 CFU/soil after 3 days, but the number of viable RB14-CS cells was not affected.

When 1–2 g of the okara solid culture were introduced into the soil, after 3 days, 30%–50% of iturin A introduced disappeared and almost no iturin A was detected after 17 days, indicating the weak persistence of iturin A in soil.

7.6.2.4 Fertilizer Effects of Okara Cultured with RB14-CS on the Growth of Tomato

Table 7.10 presents the relative shoot lengths and dry weights of the seedlings 14 days after planting when only okara cultured with RB14-CS was introduced into the soil. Run 1 shows the result when no okara was introduced. When raw okara was introduced to soil, the shoot length and dry weight of each seedling was lower than Run 1 and the shoot length was reduced by 15% and the dry weight by 15%–25%. On the other hand, the introduction of 5- and 10-day-cultured okara increased the shoot length by more than 10% and the dry weight of the seedling by more than 25%.

7.6.2.5 Comparison of Iturin A Production in Solid-State Fermentation and Submerged Fermentation

Comparison between solid-state fermentation and submerged fermentation was conducted using different bases. Results are summarized in Table 7.11. Iturin A productions in solid-state fermentation per weight and apparent volume of medium were 11 and 4 times higher than those in submerged fermentation, respectively. For the calculation of apparent volume of okara, bulk density of okara 0.367 kg/l was used.

TABLE 7.9

Populations of *B. subtilis* RB14-CS and Concentrations of Iturin A in Soil 0, 3, and 17 Days after Inoculation

Okara Culture of RB14-CS		Populations of *B. subtilis* (CFU/g dry soil)			Concentrations of Iturin A (µg/g dry soil)		
Cultivation Time (days)	Amounts (g/pot)	0 Days	3 Days	17 Days	0 Days	3 Days	17 Days
5	1	1.78×10^9	1.29×10^9	9.77×10^8	46	18	0
5	2	2.75×10^9	2.34×10^9	1.51×10^9	84	51	1
10	1	1.32×10^9	1.07×10^9	9.33×10^8	67	24	0
10	2	2.04×10^9	2.04×10^9	1.58×10^9	99	54	0

Note: Average values of six experiments.

TABLE 7.10

Effects of Okara Cultured with *B. subtilis* RB14-CS on the Growth of Tomato Seedlings 14 Days after Planting (*N* = 6)

	Okara Treatment		Relative	Relative
Run	Conditions	Amount (g/pot)	Shoot Length	Dry Weight
1	—	—	100[c]	100[c]
2	Raw	1	87[d]	85[d]
3	Raw	2	85[d]	75[d]
4	Cultured for 5 days	1	108[b]	126[b]
5	Cultured for 5 days	2	110[b]	128[b]
6	Cultured for 10 days	1	110[b]	130[b]
7	Cultured for 10 days	2	119[a]	155[a]
LSD *P* = 0.05			6	14

Notes: Each datum is a relative value when the result of Run 1 was defined as 100. For each treatment, each datum is an average of the results from experiments repeated six times. Means in any column with different letters are significantly different (*p* = 0.05) according to Fisher's protected least significant difference (LSD) analysis.

TABLE 7.11

Comparison of Iturin A Production between Solid-State Fermentation and Submerged Fermentation Using Different Bases

	Unit	Solid-State Fermentation	Submerged Fermentation
Per weight of medium	mg/kg	3300	298
Per apparent volume of medium	mg/l	1210	298
Per maximum viable cell number	$\mu g/(10^8 \times CFU)$	9.09	6.27
Per unit of total carbon	mg/g total C	23.1	16.9
Per unit of soluble carbon	mg/g soluble C	82.5	16.9

In addition to these bases, the calculations of iturin A production was performed using the basis of maximum viable cell number of RB14-CS and initial total carbon. Iturin A production per maximum viable cell number and that per initial total carbon were 45% and 37% higher in solid-state fermentation than those in submerged fermentation, respectively.

In the solid-state fermentation of okara, iturin A production is thought to occur mainly in the water-soluble phase. Therefore, calculation in the light of productivity per soluble carbon provided was also performed. For this analysis, the water-soluble portion of okara was prepared by homogenizing okara with a homogenizer at 10,000 rpm for 10 min and filtering the homogenate with a paper filter. The filtrate was dried at 105°C for 12 h, and the carbon content was determined by elemental analysis. One kilogram of okara contained 62.7 g of water-soluble fraction, and the carbon content of the water-soluble fraction of okara was determined as 48.3% (w/w) by elemental analysis. Referring to the aforementioned data, the carbon content of the water-soluble fraction of okara medium containing glucose used in solid-state fermentation was calculated to be 40.1 g-C/kg wet substrate. The carbon content of Polypepton S in no. 3S medium, measured by elemental analysis, was 45.5% (w/w). As no. 3S medium contains 3% Polypepton S and 1 % glucose, the calculated carbon content was 17.7 g-C/l medium. Using these data, iturin A production per g-carbon in solid-state fermentation was approximately fivefold higher than that in submerged fermentation as shown in Table 7.11. These analyses show the superiority of solid-state fermentation to submerged fermentation in iturin A production by RB14-CS.

7.6.3 Discussion

The high iturin A production was obtained in solid-state fermentation using the by-product of tofu manufacture. When the dual producers *B. subtilis* NB22 or RB14, which produce iturin A and surfactin was cultivated, it was found that the total amounts of iturin A and surfactin produced are almost constant despite the use of different culture conditions, and the maximum concentration of iturin A was approximately 300 mg/l (23, 27, 47). Several reasons why productivity of iturin A in solid-state fermentation is enhanced can be speculated.

Secondary metabolites, such as iturin A, are generally produced after a logarithmic growth phase in which nutrients become scarce. In submerged fermentation, the nutrients and oxygen are homogeneously distributed abundantly in the liquid medium. This induces the RB14-CS growth rather than the production of iturin A as a second metabolite, and the growth leveled off in 1 day, as shown in Figure 7.31B. On the other hand, RB14-CS in solid-state fermentation consumes only a portion of nutrients that gradually dissolve in the water phase from the solid substrate, and the dissolution of nutrients may be a limiting step. Therefore, cell division occurs on the surface of a solid substrate under the limitation of nutrient supply, which accelerates the synthesis of secondary metabolite iturin A, which lasts much longer than submerged fermentation. This reflected the differences in iturin A production in the two fermentation processes (Figure 7.31). In solid-state fermentation, it is speculated that biofilm of RB14-CS was formed on the okara, and this biofilm formation may have contributed to enhanced iturin A production, such as in the case of other antimicrobial compounds production (48, 49).

Cells grown in solid culture are generally less susceptible to product inhibition than those grown in submerged fermentation (50). Thus, the high accumulation concentration of iturin A in solid-state fermentation was speculated to be not harmful for the synthetic activity of iturin A. This is also one of the reasons RB14-CS enhances iturin A production. There are a few reports dealing with biocontrol using the product of solid-state fermentation (43, 51). The effectiveness of okara cultivated with RB14-CS of suppression of damping-off was proven in results shown in Table 7.8. The details of the changes in the properties of these okara samples are not clear. When the fertilizer components of raw okara and 10-day-cultured okara were analyzed, the ash contents were 4.6% and 7.0%, respectively, and the ratios of total carbon to total nitrogen (C/N ratio) were 10.4 and 9.4, respectively. This small change in the C/N ratio suggests the degradation of high-molecular weight organic matter into low-molecular weight and stabilization of organic matter. Therefore, the chemical or physical change of okara, which is beneficial for disease suppression, further proceeded in 10 days. This may have contributed to the sound growth of the plant, resulting in the increase in the suppressive activity of RB14-CS (29, 52).

The iturin A introduced into the soil was degraded in 17 days. We have already confirmed that the purified iturin A introduced into soil was degraded completely within 2 weeks (27). Therefore, a short persistence of iturin A in soil is preferable as a biocontrol agent in sharp contrast to the long persistence of a chemical pesticide.

The concentration of the pathogen in soil has not been determined. However, the increase in iturin A concentration in soil reflected the stronger suppressiveness against the occurrence of the disease. Therefore, it is not clear whether iturin A decreased the concentration or virulence of the plant pathogen (53); it is speculated that the concentration of iturin A is crucial for biocontrol.

The use of B. subtilis seems to be a safe and ecological means of controlling plant diseases, but poor survival when vegetative B. subtilis cells were introduced into soils was reported (44, 54). In the previous report, when only the spores of B. subtilis were introduced into soil, they showed persistence at a high level and showed no loss in their viability until the end of the experiment at 50 days in both sterile soil and non-sterile soil (30). This result suggests that spores are tolerant, not only to abiotic stresses, but also to biotic stresses, such as predation, parasitism, and lytic enzyme reactions by the indigenous microbial community in soil. As shown in Figure 7.31, 100% of the RB14-CS cells were spores on day 5 and these cells were introduced into soil for the plant tests. Therefore, this guaranteed a high and stable survival of the bacterium for a long period, and this will be advantageous to B. subtilis over other non-spore-forming bacteria in control of plant diseases.

The shoot length and the dry weight of each seedling increased significantly when 2 g of 10-day-cultured okara was introduced to the soil without R. solani (Table 7.10). As the introduction of raw okara into the soil decreased the growth of tomato seedlings, the cultivation of okara with RB14-CS may have accelerated the degradation of inhibitory components for plant growth in okara and stabilized organic matter.

Solid organic waste from the food industry is produced at approximately 30 million tons in Japan. If this waste is treated with RB14-CS, and the scale-up of solid-state fermentation by RB14-CS using this waste is established, the by-product, which has a dual function as an organic fertilizer and a biological control agent, will contribute to the recycling of organic matter between the agricultural and industrial sectors and to the materialization of environmentally friendly agriculture.

7.7   OPTIMIZATION OF ITURIN PRODUCTION
IN SOLID-STATE FERMENTATION (55)

In Section 7.6, B. subtilis RB14-CS produced a higher concentration of iturin A in SSF containing okara, soybean curd residue, compared to the production by SmF. In addition, the product from SSF by RB14-CS increased the suppressive effect on a plant disease in soil. As a higher iturin A concentration was more effective for suppressing plant diseases, enhancement of the SSF-based production of iturin A under optimized conditions could increase the usefulness of this potential biocontrol agent.

Traditional methods of optimization often require a considerable amount of work and time, and frequently fail to yield optimized conditions because they do not consider possible interactions among factors. Recently, statistical experimental strategies including factorial design and response surface methodology (RSM) have been successfully employed for optimization of SSF (56–59). These strategies are particularly suited for optimization of SSF because they are generally subjective to less accidental errors, which are often introduced by the heterogeneous nature of the solid components in SSF. In this study, statistical strategies to optimize the SSF medium conditions for maximum production of iturin A by RB14-CS were used and the products generated under the optimized conditions for their suppressive effects on plant disease control was examined.

7.7.1   MATERIALS AND METHODS

7.7.1.1   Strains

B. subtilis RB14-CS, a streptomycin-resistant mutant of the original RB14, was used as a producer of antifungal iturin, and a plant pathogen, *R. solani* K-1, was used (27).

7.7.1.2   Preparation of Seeding Culture

The procedure of preparation of seeding culture was the same as that in the media described in Section 7.6.1.

7.7.1.3   Solid-State Fermentation

Raw okara was obtained from Marusho Sho-ten (Machida, Tokyo) and frozen below −15°C until use. For SSF under the basal medium condition, 15 g of okara was placed in a 100 ml conical flask and autoclaved twice at 120°C for 20 min each at an interval of 8–12 h to completely kill the spore-forming microorganisms inhabiting the material. The flask was cooled to room temperature, and the following sterilized solutions were added to supplement the nutrient and adjust the moisture content to approximately 79%; 833 μl of 0.45 g/ml glucose, 75 μl of 1 M KH_2PO_4, 150 μl of 1 M $MgSO_4 \cdot 7H_2O$, and 367 μl of deionized distilled water. Then, 3 ml of the RB14-CS seeding culture described earlier was added, and the contents of the flask were mixed with a stainless steel spatula. The flask was immersed in a water incubator and incubated at 25°C with a relative humidity of 75%. To determine the significant factors affecting the production of iturin A, initial pH, initial water content, temperature, relative humidity, and volume of inoculum were changed. Approximately 1 g of solid substrate was suspended homogeneously in nine folds weight of distilled water, and the initial pH was measured by a pH meter. Initial pH was changed by adding the appropriate amount of 1 N HCl to solid medium. For the relative humidity, an air incubator equipped with a humidity regulator (AE-215, Advantec, Tokyo) was used. Two flasks were prepared for each condition, and each experiment was repeated twice.

7.7.1.4   Extraction and Measurement of Iturin A

This procedure was the same as the procedure described in Section 7.6.1.

7.7.1.5   Selection of Carbon and Nitrogen Sources

Several potential carbon sources were tested, including glucose and fructose as monosaccharides, maltose and sucrose as disaccharides, and dextrin (FZ-100, Honen, Tokyo) and soluble starch as polysaccharides. After adding 0.375 g of each carbon source to flasks containing 15 g of wet okara, SSF was performed as described earlier. Iturin A production was measured, and the production was compared among the different carbon sources.

Similarly, several nitrogen sources were tested, including a hydrolyzed plant protein solution from soybean and wheat having N content of 3.03% (w/v) and C content of 11.2% (w/v; Mieki R, Ajnomoto, Tokyo), Polypepton S having N content of 8.59% (w/w) and C content of 36.1% (w/w; Nihon Pharmaceutical, Tokyo), soybean meal having N content of 8.13% (w/w) and C content of 42.0% (w/w; Honen), urea, $(NH_4)_2HPO_4$, NH_4Cl, NH_4OCOCH_3, and NH_4NO_3. Each nitrogen source was added to a flask containing 15 g of wet okara and 0.375 g of glucose to adjust the C/N ratio of 11, a value shown to be the most appropriate for the maximum iturin A production in preliminary experiments (data not shown). Two flasks were prepared for each condition, and each experiment was repeated twice.

7.7.1.6   Measurement of Reducing Sugar Concentration

Of the SSF product, 0.5 g was collected from flasks, placed into a test tube, and mixed with 4.5 ml of distilled water. Solid particles were homogenized with a small spatula, and the mixtures were vortexed thoroughly and centrifuged at $18,000 \times g$ for 10 min at room temperature. The concentration of reducing sugar in the supernatant was determined with the Somogyi–Nelson method with glucose as a standard.

7.7.1.7   Experimental Design

A central composite design with five levels coded in duplicate was used to determine the optimum conditions for production of iturin A. The quadratic model for predicting the optimal point was expressed according to Equation 7.1:

$$Y = b_0 + \sum b_i x_i + \sum b_{ii} x_i^2 + \sum b_{ij} x_i y_j \qquad (7.1)$$

where y is the response variable, b is the regression coefficients, and x is the coded levels of the independent variable. Design Expert 6.0.10 (Stat-Ease) was used for the regression analysis of the experimental data obtained. The ranges and the levels of variables investigated in this study are given in Tables 7.12 and 7.13. The concentration of iturin A was taken as the dependent response variable. The statistical significance of the second-order model equation (Equation 7.1) was determined by the F value, and the proportion of variance explained by the model was given by the multiple coefficient of determination, R^2.

TABLE 7.12

Range of Variables at Different Levels for the Central Composite Design

	Variables	Levels				
		−1.414	−1	0	+1	+1.414
X_1	Glucose (g/15 g wet okara)	0.093	0.3	0.8	1.3	1.507
X_2	Soybean meal (g/15 g wet okara)	0.086	0.5	1.5	2.5	2.914

TABLE 7.13

Experimental Design and Results of the Central Composite Design

Runs	Glucose (g/15 g wet okara)	Soybean Meal (g/15 g wet okara)	Iturin A Conc. (µg/g IWO[a])	
			Sample 1	Sample 2
1	1.3 (1)[b]	2.5 (1)[b]	4,921	4,848
2	1.3 (1)	0.5 (−1)	4,322	4,199
3	0.3 (−1)	0.5 (−1)	3,158	3,059
4	0.3 (−1)	2.5 (1)	3,666	3,724
5	0.8 (0)	2.914 (1.414)	4,400	5,139
6	0.093 (−1.414)	1.5 (0)	2,518	2,312
7	0.8 (0)	0.086 (−1.414)	3,238	3,305
8	1.507 (1.414)	1.5 (0)	4,756	4,543
9	0.8 (0)	1.5 (0)	5,336	5,468
10	0.8 (0)	1.5 (0)	5,368	5,393
11	0.8 (0)	1.5 (0)	5,346	5,375
12	0.8 (0)	1.5 (0)	5,432	5,375

[a] *IWO*, initial wet okara.

[b] Values in the brackets indicate coded levels.

7.7.1.8   Determination of Viable Cell Number

For the determination of viable cell number during solid-state fermentation, 1 g of solid culture was placed in a sterile 18 mm diameter test tube. It was mixed thoroughly with 9 ml of sterile distilled water and shaken at 150 spm for 5 min at room temperature. Then, the mixture was serially diluted and spread onto L agar plates consisting of L medium and 15 g/l agar. After 18 h of incubation at 37°C, the number of colonies was determined and expressed as colony-forming units (cfu).

7.7.1.9   Disease Suppression in Tomato Seedlings

Disease suppression experiments of tomato seedlings were performed as previously reported (27, 42). Volcanic soil and vermiculite were mixed with a ratio of 4:1 (w/w) and supplemented with chemical fertilizers, including (per kilogram dry soil), 8 ml of Hyponex® (liquid fertilizer , N 6%, P 10%, and K 5%; Hyponex Japan, Osaka, Japan), 2 g of Kudosekkai® (CaO 55%, MgO 16%, Hinokuni, Saitama, Japan), 3 g of $MgSO_4 \cdot 7H_2O$, 0.5 g of K_2HPO_4, and 80 mg of $FeSO_4 \cdot 7H_2O$. The soil (150 g) was dispensed to plastic pots (90 mm diameter), and the moisture was kept at 60% of the maximum water-holding capacity of the soil by daily addition of sterilized water.

A pathogen, *R. solani*, was incubated statically in the dark at 28°C for 7 days in PDP medium containing (per liter) potato infusion 200 g, glucose 20 g, and Polypepton 10 g (pH 5.6). The mycelial mats formed on the surface of the medium were homogenized at 4000 rpm for 2 min in sterile water and inoculated into the soil at a ratio of 3.6 g mat per pot 6 days before planting pregerminated tomato seeds, which had been disinfected with 70% ethanol for 1 min, followed by 0.5% sodium hypochlorite, rinsed with sterile water, and germinated on a 2% agar plate at 30°C for 2 days. Each pot was sown with nine pre-germinated seeds 3 days after SSF culture of RB14-CS was introduced into soil and placed in a growth chamber at 30°C with 90% relative humidity under 16 h of light (about 8,000 lux). After 2 weeks, the percentage of diseased seedlings per pot was determined as the disease occurrence.

7.7.1.10   Statistical Analysis

Each plant test was repeated nine times, and each mean of data was analyzed using Fisher's analysis of variance.

7.7.2   RESULTS

7.7.2.1   Factors Affecting the Production of Iturin A in the SSF

Before optimizing SSF conditions using statistical experimental design, significant factors affecting the production of iturin A were determined experimentally. At first, effects of culture conditions, including initial pH, initial water content, temperature, relative humidity, and volume of inoculum on iturin A production in SSF were investigated. In these experiments, the basal culture conditions described in Section 7.7.1.3 were used, and one factor was altered per experiment. As shown in Figure 7.32, changes in initial water content, relative humidity, and volume of inoculum had no significant effect on iturin A production in SSF after 96 h. In contrast, iturin A production was affected by changes in temperature and initial pH, with maximum production at 25°C and pH 5.9–6.3, respectively.

Next, the effects of different carbon and nitrogen sources on iturin A production in SSF were examined. All the tested carbon sources enhanced iturin A production by RB14-CS after 96 h incubation (Figure 7.33a). Among them, glucose showed the highest level of enhancement, increasing iturin A production by 45% compared with production without an additional carbon source. Although inorganic nitrogen sources had negative effects on iturin A production (Figure 7.33b), organic nitrogen sources, except Mieki, enhanced iturin A production. A high concentration of salt in Mieki (20.5% (w/v); based on a suggestion from the producer, caused by its neutralization after acid hydrolyzation of protein, may have the adverse effect on iturin A

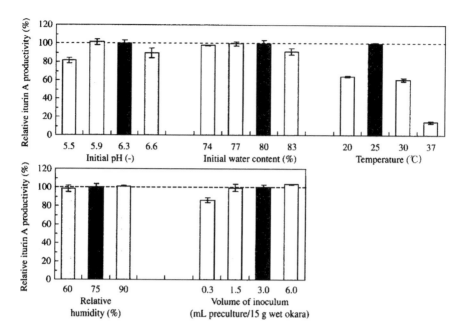

FIGURE 7.32 Relative iturin A production after 4 days of incubation under the various culture conditions. The black bar indicates the basal condition.

production. Among organic nitrogen sources, soybean meal was the most effective nitrogen source, increasing iturin A production by 38% compared with production without an added nitrogen source.

7.7.2.2 Medium Optimization Using Response Surface Methodology

Response surface methodology was introduced to determine the optimal concentrations of glucose and soybean meal for maximizing iturin A production over a 4-day cultivation. Based on the results shown in Figure 7.32, the temperature and the inoculum volume of preculture were set at 25°C and 3 ml/flask, respectively. Relative humidity was not controlled. The initial pH was not controlled because the initial pH of the solid material was 6.3. The experiment was carried out with two independent variables using a 22 full factorial design experiment with four star points ($\alpha = \pm 1.414$) and four replicates at the center point. The design of the experiment and results are presented in Tables 7.12 and 7.13, respectively. Each experiment was repeated twice.

Regression analysis was performed to fit the response function with the experimental data. As shown in Table 7.14, the F and P values were 91.49 and <0.0001, respectively. The statistical significance of the second-order model equation was checked, and the coefficient of determination (R^2) of the model was calculated to be 0.96, indicating that 96% of the variability in the response could be explained by the model. This indicates that the response equation provided a suitable model for the

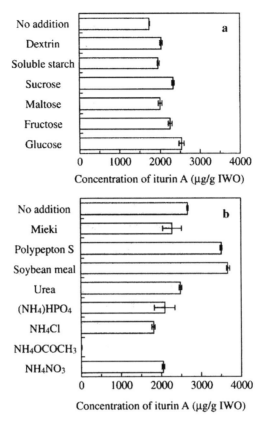

FIGURE 7.33 Iturin A production by *B. subtilis* RB14-CS cultured with different kinds of (a) carbon and (b) nitrogen sources (*N* = 3). (a) Cultures contained 0.375 g of each carbon source per 15 g IWO (initial wet okara). Dextrin and starch were sterilized simultaneously with okara, while the others were sterilized separately. (b) Each nitrogen source was added to a flask containing 15 g of IWO and 0.375 g of glucose to the C/N ratio of 11.

TABLE 7.14

Statistical Analysis for Models of Iturin A Production at Different Levels of Concentrations of Glucose and Soybean Meal $R^2 = 0.96$

Source	Sum of Squares	Degrees of Freedom	Mean Square	F Value	P Value
Model	2.289×10^7	5	4.578×10^6	91.49	< 0.0001
Residual	9.008×10^5	18	50043.20		
Total	2.379×10^7	23			

response surface of the experiment of iturin A production. The response equation obtained was as follows:

$$y = 5,386.49 + 687.65x_1 + 416.17x_2$$
$$-874.44x_1^2 - 630.43x_2^2 + 9.30x_1x_2$$

where x_1 and x_2 are the code values corresponding to the amount of glucose and soybean meal, respectively.

Figure 7.34 shows a three-dimensional diagram and a contour plot of the calculated response surface. The optimum values of glucose and soybean meal concentrations obtained for iturin A production were coded as $x_1 = 0.376$ and $x_2 = 0.330$, respectively. For calculation purposes, the normalized coded values x_1 and x_2 were defined as $x_1 = 2 (X_1 - 0.8)$, and $x_2 = X_2 - 1.5$, where X_1 is the concentration of glucose and X_2 is the concentration of soybean meal. According to these equations, the optimal concentrations of glucose and soybean meal for iturin A production were calculated to be 0.998 and 1.83 g per 15 g wet okara, respectively. The maximum value of iturin A production predicted from this model was 5591 μg/g IWO.

To confirm this result, we conducted a SSF culture experiment using these optimal conditions. The time course profiles of iturin A production and reducing sugar consumption are shown in Figure 7.35. The highest iturin A production of 5454 μg/g IWO was obtained after 4 days of cultivation under the optimal conditions. In SSF cultures lacking glucose and soybean meal, the maximum iturin A concentration after 4 days of cultivation was only 1924 μg/g IWO. When either glucose or soybean meal was added, iturin A production did not increase to the level of the optimum condition. In our previous study, iturin A concentration in SSF in the basal medium was 2633 μg/g IWO, and the reducing sugar was exhausted within 2 days of cultivation (42). The optimized conditions maintained the amount of reducing sugar up to 4 days, resulting in enhanced iturin A production. Viable cell number under the optimized condition rapidly increased, and during 1–3 days, it was 5–10 times higher than those in other conditions. The strong correlation between the experimental and statistical results confirms the validity of the response model and the existence of an optimal point.

7.7.2.3   Suppressive Effects of the Optimum SSF Product on the Plant Disease

The suppressive effects of the optimally cultured SSF product on the plant disease were examined in the pot test as shown in Table 7.15 and Figure 7.36. When the soil was infested only by *R. solani* K-1, 84% of tomato seedlings suffered from damping-off. The SSF product generated in basal medium showed a reduced disease occurrence of 52%. In contrast, SSF product generated under the optimized conditions showed a disease occurrence of only 19%. The iturin A concentration in the soil decreased immediately after addition of SSF product into soil, and iturin A was not detectable 17 days later (data not shown), indicating a short persistence of iturin A in soil, in sharp contrast to a long persistence of chemical pesticides.

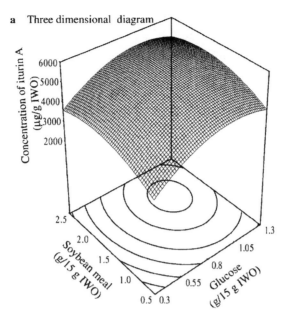

a Three dimensional diagram

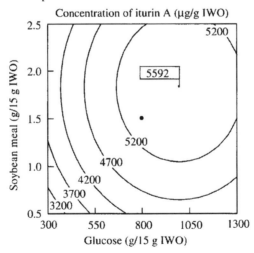

b Contour plot

FIGURE 7.34 (a) Three-dimensional diagram and (b) contour plot of the calculated response surface.

7.7.3 Discussion

Several previous studies used submerged fermentation to produce high concentrations of lipopeptide antibiotics from *B. subtilis* (19, 35, 60), but few reports have described the use of SSF to produce lipopeptide antibiotics from bacteria for use as a biological control agent. In Section 7.6, the advantages of SSF over submerged

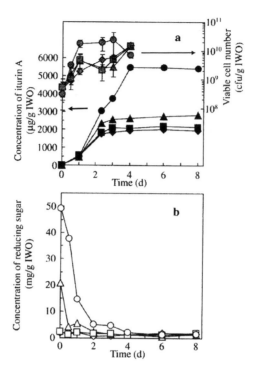

FIGURE 7.35 Time course of changes in concentrations of (a) iturin A (black symbols) and viable cell number (gray symbols) and (b) reducing sugar under the optimized conditions. Circles, cultures containing 1.83 g of soybean meal and 0.998 g of glucose per 15 g IWO. Squares, cultures containing no additional nutrients. Triangles, cultures containing 0.375 g of soybean meal per 15 g IWO. Diamonds, cultures containing 0.375 g of glucose per 15 g IWO.

TABLE 7.15

Suppressive Effects of Okara Cultured with *B. subtilis* RB14-CS on Damping-Off of Tomato Caused by *R. solani* K-1

Run	*R. solani*	Okara Cultured with RB14-CS (2 g/pot)	Disease Occurrence (%)
1	—	—	3[a]
2	+	—	84[d]
3	+	Basal medium	52[c]
4	+	Optimized medium	19[b]
		LSD *p* = 0.05	14

Notes: For each treatment, each datum is an average of the results from experiments repeated nine times. Means in any column with different letters are significantly different (*p* = 0.05) according to Fisher's protected least significant difference (LSD) analysis.

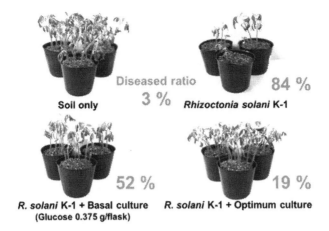

Diseased ratio

Soil only   3 %   Rhizoctonia solani K-1   84 %

R. solani K-1 + Basal culture   R. solani K-1 + Optimum culture
(Glucose 0.375 g/flask)   52 %   19 %

FIGURE 7.36 Photos of suppressive effects of solid culture under the optimum condition on damping-off of tomato caused by *R. solani* K-1. (*N* = 10) (Values show the diseased ratios.)

fermentation were shown when soybean curd residue was used as a solid nutrient for iturin A production by RB14-CS (42).

In this study, a statistical method with significant factors was applied to further enhance production of iturin A by RB14-CS grown on soybean curd residue in SSF. Previous reports on the effects of culture conditions on microbial growth and secondary metabolite production in SSF have mostly focused on fungi and the identified relevant conditions included initial water content (61, 62), relative humidity (63, 64), and volume of inoculum (65, 66). In contrast, our present findings on RB14-CS indicated that initial pH, temperature, and carbon and nitrogen sources were the most relevant factors for iturin A production. The optimum pH and temperatures were 5.9–6.3 and 25°C, respectively. As the initial pH of the intact solid material was almost 6.3, no specific control was conducted for this factor. The optimum temperature was 25°C, which was consistent with previous findings in the parent strain of *B. subtilis* RB14 (15). However, the addition of carbon and nitrogen sources, especially glucose and soybean meal, could effectively enhance the iturin A production in SSF.

The exact reasons why initial water content, relative humidity, and volume of inoculum had no significant effect of the production of iturin A by *B. subtilis* are not clear. However, the bacterial growth on soybean curd residue was significantly faster compared with fungal growth, and this resulted in shorter cultivation time and independence of initial water content or volume of inoculum in iturin A production. Furthermore, the fast growth may not induce evaporation of water from solid substrate. Actually, the cell growth of *B. subtilis* RB14-CS in solid-state fermentation and in submerged fermentation was similar as shown in Section 7.6.

The maximum concentration of iturin A produced experimentally under the statistically predicted optimum conditions was 5454 mg/kg IWO, which was close to the 5591 mg/kg IWO predicted by the model. This production was 2.8 times higher than that observed in solid culture without glucose or soybean meal, and 2.1 times higher than obtained in the basal medium (42). This increased production was likely due to the high cell density during the iturin A production period and the continuous availability of reducing sugars in the solid culture throughout the incubation under the optimized condition, which allowed continuous production of iturin A. When the response function was fit to the raw experimental data, the coefficient of determination (R^2) of the model was 0.96. This value was very close to 1, indicating that the statistical method overcame the difficulty of approximation due to the experimental error in SSF. This confirms the validity of the statistical method. This method corrects data variations in determining the effect of parameters and provides the optimum calculated values by a small number of experimental runs particularly in inhomogeneous cultures like SSF, which often cause large margins of error of experimental values.

Last, the results showed that solid culture produced under optimized conditions exhibited enhanced suppression of tomato seedling damping-off caused by *R. solani* K-1. This indicates that higher iturin A production under optimum culture conditions was associated with increased efficacy for RB14-CS as a biological control agent.

In scale-up in SSF, some problems such as difficult control of the rise in temperature and diffusion limitation of nutrients and oxygen occur (67, 68). However, several reactors have been proposed to solve these problems (69). A large-scale reactor is to be constructed based on this result in a tofu manufacturer to supply the okara directly and immediately after its production to minimize the energy utilization for sterilization and transport.

7.8   ASSOCIATION OF BIOFILM FORMATION WITH ITURIN PRODUCTION (70)

The higher iturin A production by strain RB14-CS in SSF may be due to biofilm formation on the surface of the soybean curd residue, okara. In natural environments, most bacterial cells grow in the situation of adhering to surfaces or interfaces, in the form of multicellular aggregates embedded in matrices commonly referred to as biofilms (71). Biofilms provide several benefits to their member cells, such as protection from environmental insults and assaults (72). Recently, *Bacillus* sp., including *B. subtilis*, has become a model organism for the study of biofilm formation by gram-positive bacteria, and these studies have shown that different genes are transcribed by biofilm-associated phases (73–78).

The production of poly-γ-DL-glutamic acid by *B. subtilis*, which is an extracellular polymer and contributes to formation of three-dimensional architecture of biofilm matrix, is dependent on *degQ* (79). This gene is a pleiotropic regulatory gene that controls the production of degradative enzymes, such as intracellular proteases and several secreted enzymes including levansucrase, alkaline proteases and metalloproteases, α-amylase, β-glucanase, and xylanase (80). In Chapter 6, it was reported that introduction of *degQ* into *B. subtilis* RM/iS2, a derivative of standard strain 168 carrying a complete iturin A operon and the 4′-phosphopantethienyl transferase *sfp*, increases iturin A production six- to eight-fold, indicating that *degQ* is also responsible for the high level of iturin A production by *B. subtilis* (81).

In this section, in order to clarify the relationship between the higher iturin A production by strain RB14-CS in SSF using soybean curd residue and biofilm formation, an air-membrane surface (AMS) bioreactor was used, where the cells grow as a biofilm on the surface of a semipermeable membrane and receive nutrients from the liquid medium below the membrane (48, 70). By comparing the cells from the AMS reactor and submerged fermentation, the biochemical and genetic differences were investigated.

7.8.1 Materials and Methods

7.8.1.1 Strain

B. subtilis RB14-CS was used as a sole producer of the antifungal iturin A and the characteristics of strain RB14-CS was described in Section 7.7.

7.8.1.2 Preparation of Seed Culture

The L medium was used for the growth of strain RB14-CS. The no. 3S medium was used for the growth of the bacterium and iturin A production both by SSF using the AMS and by SmF using a flask.

7.8.1.3 Solid-State Fermentation and Submerged Fermentation Using Soybean Curd Residue, Okara, as a Substrate

For SSF, 15 g of okara were placed in a 100 ml Erlenmeyer flask and autoclaved twice at 120°C for 20 min with an interval of 8–12 h to kill spore-forming microorganisms inhabiting the material. For SmF, the okara suspension medium was prepared by homogenizing 8 g of okara in 20 ml distilled water at 11,000 rpm for 1 min using a homogenizer (type ACE, Nihonseiki, Tokyo, Japan) and the medium suspension was transferred to a 200 ml Erlenmeyer flask and autoclaved twice. Okara in SSF and okara suspension in SmF were mixed with 25 mg of glucose, 0.67 mg of KH_2PO_4, and 2.5 mg of $MgSO_4 \cdot 7H_2O$ per g okara. The total volume of SmF was adjusted to 40 ml by adding sterile distilled water. To prepare a seed for SSF and SmF, preculture grown in L medium was inoculated into 100 ml of no. 3S medium and incubated at 30°C for 24 h in a shaking flask at 120 spm. Then, 0.2 ml of the preculture was added per gram of okara for both SSF and SmF. After mixing the okara with a stainless steel spatula, SSF was carried out statically in a water incubator at 28°C. SmF was carried out with shaking at 120 spm and 28°C.

7.8.1.4 Extraction and Quantitation of Iturin A from SSF and SmF

For the extraction of iturin A from SSF, 45 ml of methanol was added to the whole fermented solid product in the flask, and the mixture was shaken at 150 spm for 1 h. For the extraction of iturin A from SmF culture, 200 µl of culture broth were suspended in 800 µl of an extraction solution consisting of (1:1 [vol/vol]) acetonitrile–distilled water in a microtube, and shaken for 20 min. Then, these two extracts were centrifuged at $18,000 \times g$ for 10 min at 4°C, and filtered through a 0.2 µm pore PTFE membrane (JP020; Advantec Ltd., Tokyo, Japan). The filtrates were applied to reversed-phase HPLC with a column (Chromolith performance RP-18e 100-4.6, Merck KGaA, Germany). The HPLC system (LC-800 system, JASCO Co., Ltd., Tokyo, Japan) was operated at a flow rate of 2.0 ml/min with (35:65 [vol/vol]) acetonitrile–10 mM ammonium acetate as the eluent (27). The concentration of iturin A was determined using authentic iturin as a standard (Sigma, St. Louis, Missouri, USA).

7.8.1.5 Biofilm Fermentation in the AMS Reactor and Submerged Fermentation Using No. 3S Medium

A preculture grown in L medium after 16 h cultivation at 120 spm at 30°C was used as a seed for biofilm fermentation in an AMS reactor and for SmF. Some modifications were made to the AMS reactor described by Yan et al. (48). The procedures of two experimental reactors are shown in Figure 7.37.

Cellulose ester membranes (diameter, 47 mm; pore size, 0.2 µm; Toyo Roshi Co., Tokyo, Japan) were autoclaved at 121°C for 20 min and dried in the oven dryer at 70°C for at least 2 days. Fresh no. 3S medium (8.8 ml) was added to a sterile petri dish (diameter, 35 mm; height, 9 mm; Becton Dickinson), and the dish was covered with the membrane. In this system, the membrane was in contact with the medium on one side and with air on the other side and was held in place by surface tension. The liquid medium below the membrane was referred as spent medium. Then 88 µl of preculture (1 %) were inoculated and spread on the membrane surface. The petri

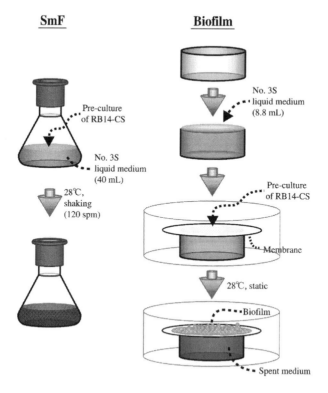

FIGURE 7.37 Experimental procedures for submerged fermentation (SmF) and biofilm fermentation (Biofilm) with AMS reactor using the same liquid no. 3S medium and *B subtilis* RB14-CS.

dish with the inoculated membrane was placed in a sterilized deep glass petri dish (diameter, 100 mm; height, 20 mm) to maintain sterility, and incubated at 28°C in an incubator.

For SmF, 40 ml of no. 3S medium were inoculated with 400 µl of RB14-CS pre-culture in a 200 ml Erlenmeyer flask and shaken at 120 spm and 28°C.

7.8.1.6 Scanning Electron Microscopy (SEM)

After a 2-day incubation, okara culture from SSF and biofilm from the AMS reactor were collected, fixed with 2.5% glutaraldehyde, and dehydrated in ethanol. The samples were then observed with a scanning electron microscope (S-5200; Hitachi, Tokyo, Japan).

7.8.1.7 Determination of Growth and Iturin A Production in the AMS Reactor and SmF

For the extraction of iturin A from the AMS reactor, the whole membrane covered with biofilm and the spent medium were mixed with an extraction solution of (1:1 [vol/vol]) acetonitrile–10 mM ammonium acetate and homogenized with a handy micro homogenizer (Physcotron NS-310E; Niti-on Co., Chiba, Japan). For SmF, the same volume of culture broth with that of the spent medium in AMS reactor was mixed with extraction solution. Sample preparation and HPLC analysis for measurement of the iturin A concentration were carried out as described earlier.

The samples prepared for determination of the iturin A concentration were also used for measurement of the glucose concentration, which was measured by HPLC with an ODS column (Shodex NH2P-50 4E; Showa Denko, Tokyo, Japan). The system was operated at a flow rate of 1.0 ml/min with (3:1 [vol/vol]) acetonitrile/ultrapure water as an eluent and at a column temperature of 40°C. The elution was monitored with a differential refractive index detector (RID-300; Jasco, Tokyo, Japan).

For the determination of the cell number, cell suspensions were prepared as described earlier except that extraction solution was not added. For the samples taken at or after 24 h, cell suspensions of biofilm were mildly sonicated by five 60 s pulses using a sonicator at 10 W (Model 5201; Ohtake, Tokyo, Japan) to disrupt the extracellular matrix, which can cause the difficulties in correctly determining the viable cell and spore numbers in the biofilm. Then, the sonicated mixture was serially diluted and spread onto L-agar plates (L medium containing 20 g/L agar). After 18 h at 37°C, the number of colonies was determined and expressed as colony-forming units (cfu). The suspension was also heated at 80°C for 15 min and spread onto L agar plates, and the number of colonies including only spores was expressed as cfu. The cell number after five cycles of mild sonication at 60 s was three times higher than if no sonication was performed and it did not change after more than six cycles of sonication (data not shown). The intensity level of sonication was not lethal to the cells because the suspended cell number in samples from SmF was almost the same with and without mild sonication (data not shown).

Biofilm and spent media from the AMS reactor and 8.8 ml of culture broth from SmF were collected in centrifuge tubes. Samples were homogenized with a handy

micro homogenizer (Physcotron NS-310E; Niti-on, Chiba, Japan) and centrifuged at
$12,000 \times g$ for 20 min at 4°C. After the precipitate was washed with cold distilled
water, precipitates were dried at 105°C for 24 h, and the weights of dried biomass
were measured.

7.8.1.8   Whole-Cell Protein Extraction for Proteome Analysis

Biofilm on the membrane and spent media from the AMS reactor and approximately
10 ml of culture broth from SmF were collected in centrifuge tubes. Samples were
homogenized with a handy micro homogenizer. The cell suspensions were centrifuged at $12,000 \times g$ at 4°C for 20 min, and the precipitates were transferred into
microtubes and washed twice with buffer A consisting of 2 mM ethylenediamine-
tetraacetic acid (EDTA), 2 mM dithiothreitol (DTT), and 25 mM Tris-HCl (pH 8.0).
After resuspension in 400 μl of buffer A, the cells were disrupted by sonication with
three 80 s pulses at 40 W with a sonicator (Model 5201; Ohtake, Tokyo, Japan).
Then, 400 μl of solubilization solution consisting of 8 M urea, 2 M thiourea, 2% (vol/
vol) Triton X-100, and 15 mM DTT was added to the sample, and the mixture was
centrifuged at $18,000 \times g$ for 10 min at 4°C. The protein extracts were then concentrated as described by Wassel and Flugge (82).

7.8.1.9   Two-Dimensional Electrophoresis

The protein content of each extract was determined by the Bradford method (83)
with a Protein Assay Kit II (Bio-Rad, Tokyo, Japan) using bovine serum albumin as
a standard. Rehydration solution (8 M urea, 2 M thiourea, 2% [vol/vol] Triton X-100,
12 μl/ml DeStreak reagent [GE Healthcare, Uppsala, Sweden], and 0.5% [vol/vol]
Pharmalyte 3–10 [GE Healthcare]) was added to each extract to adjust the protein
content to 1 μg/μl. For isoelectric focusing, pH 4–7 immobilized pH gradient (IPG)
gel strips (11 cm; GE Healthcare) were rehydrated overnight in 200 μl of protein
solution in an Immobiline DryStrip Reswelling Tray (GE Healthcare). Gels were
focused by a three-phase program using a Multiphor II: 300 V for 4 h, followed by a
linear gradient from 300 V to 3500 V for 5 h, and finally 3500 V for 7 h.

   After placing the IPG gels in equilibration buffer A (50 mM Tris-HCl, pH 6.8,
containing 6 M urea, 1.5 M thiourea, 30% [vol/vol] glycerol, and 1% [vol/vol] sodium
dodecyl sulfate [SDS], and 0.25% [vol/vol] DTT) for 10 min followed by buffer B
(50 mM Tris/HCl, pH 6.8, containing 6 M urea, 1.5 M thiourea, 30% [vol/vol] glycerol, 1% SDS, 4.5% [vol/vol] iodoacetamide, and a trace of bromophenol blue) for
10 min, the isoelectric focusing gels were embedded in a 8% to 18% (wt/vol) gradient
polyacrylamide separating gel. Electrophoresis was performed at 20 mA for 35 min
and then 50 mA for 75 min at a constant temperature of 15°C. After electrophoresis,
proteins were visualized by staining in 0.5% (vol/vol) Coomassie brilliant blue R250
solution for 1 h and destained in 50% (vol/vol) methanol, with 10% (vol/vol) acetic
acid. M. W. Marker "Daiichi"·II (Daiichi Pure Chemicals, Tokyo, Japan) was used
as a molecular mass marker.

7.8.1.10   Analysis of Proteome Patterns

Whole-cell protein extracts prepared from SmF and biofilm fermentation in AMS
reactor were analyzed on at least two separate occasions by electrophoresis. Gel

images were resized, matched, and compared by using PDQuest 2-D Gel analysis software (Version 6.1; Bio-Rad, Tokyo, Japan).

7.8.1.11 N-Terminal Amino Acid Sequencing and Protein Identifications

Two-dimensional electrophoresis was performed as described above and then the proteins were electrophoretically transferred to a commercial membrane (Immobilon-P; Millipore, Tokyo, Japan) using a horizontal blotting apparatus (ATTO Co., Tokyo, Japan). The regions of the membrane containing protein spots of interest were cut out and amino acid sequencing analysis was performed using an amino acid sequencing apparatus (PPSQ-21; Shimadzu, Kyoto, Japan) according to the standard method (84). Searches for homologous amino acid sequences were performed in the *B. subtilis* database BSORF (http://bacillus.genome.jp/) and the nonredundant database at the National Center for Biotechnology Information (www.ncbi.nlm.nih.gov/) using BLASTP.

7.8.1.12 RNA Isolation and Real-Time RT-PCR

Cell suspension (300–500 µl) prepared as described earlier and 600–1000 µl of RNA protect bacteria reagent (Qiagen, Tokyo, Japan) were mixed in a microtube and the following RNA stabilization was performed according to manufacturer's standard protocols.

For RNA extraction, an RNeasy mini kit (Qiagen) and an RNase-free DNase set (Qiagen) were used. Cell pellets obtained by centrifugation from the AMS reactor and from SmF were resuspended in 200 µl of 15 mg/ml lysozyme in TE buffer (pH 8.0) and incubated for 10 min. The suspension was mixed with 10 µl of 10% (vol/vol) SDS, incubated for 5 min, and mixed with 0.7 ml of RLT buffer provided in the RNeasy mini kit. The lysate was centrifuged at $18,000 \times g$ for 2 min, and the supernatant was treated following the protocol provided with the RNeasy mini kit. The isolated RNA was further treated with RNase-free DNase I (Takara, Tokyo, Japan) to remove contaminating DNA. After removal of DNase I by phenol/chloroform extraction, the integrity of the RNA samples was monitored by gel electrophoresis and the measurement of absorbance at 260 nm (A_{260}) and 280 nm (A_{280}).

The RNA (1 µg) was used for cDNA synthesis. Complementary DNA was synthesized using Superscript III (Invitrogen, Tokyo, Japan) and random octomer primers, according to the manufacturer's instructions, in a final volume of 20 µl. Real-time reverse transcription polymerase chain reaction (RT-PCR) was performed using SYBR Green master mix (Toyobo, Tokyo, Japan) and an ABI PRISM 5700 sequence detection system according to the manufacturer's instructions (Applied Biosystems, California, USA). The following primers (all 5′ to 3′) were used: for *ituB* (*ituB*-F, GGACAGAGCAGAGTATAAAGC; *ituB*-R, GCGATTTCTTCATTCTGTGCC), for *degQ* (*degQ*-F, CAGATCAAATACCTAGGACTCG; *degQ*-R, CATTGCATA ATTGTATTTATCGAGTTG), and for 16S ribosomal RNA (16SrRNA-F, CGTA GAGATGTGGAGGAACACC; 16SrRNA-R, CCCAACATCTCACGACACGAGC). PCR conditions were as follows: 94°C for 2 min and 40 cycles of 94°C for 30 sec, 59°C for 1 min and 72°C for 30 sec in 20 µl volumes.

Since all of the amplified fragments used to generate data points had melting curves identical to those of the positive controls and formed single bands of the expected sizes on an agarose gel, my team confirmed that the measured increase in fluorescence was due to amplification of only the specific targeted gene fragments. Relative RNA equivalents for each sample were obtained by standardization of 16S ribosomal RNA levels. Values reported are the average under the basal conditions (i.e., samples from biofilm fermentation after 24 h).

7.8.2 RESULTS

7.8.2.1 Comparison of Iturin A Production between SSF and SmF Using Soybean Curd Residue, Okara, as a Substrate

Figure 7.38 shows the iturin A production in SSF and in SmF using the same amounts of okara and nutrients. In SSF, okara was used as a solid substrate and in SmF, okara was used as a suspension. The concentrations of iturin A in the two systems are expressed as mg/g initial wet okara. The maximum concentration of iturin A in SSF per weight of initial wet okara was 30% higher than that in SmF.

7.8.2.2 Growth and Iturin A Production by Biofilm Fermentation in an AMS Reactor and in SmF Using No. 3S Liquid Medium

Biofilm fermentation was carried out in the AMS reactor using no. 3S medium as a model of SSF (see Figure 7.37). RB14-CS formed a biofilm with a complex architecture on the membrane at 28°C. It was confirmed that no cells existed in the spent medium, which is the liquid phase below the semipermeable membrane, because no colonies were detected when the spent medium was incubated on the L agar plate. In parallel, SmF was carried out using no. 3S medium in a 200 ml Erlenmeyer flask at a shaking speed of 120 spm and at 28°C. Iturin A production and growth in the two systems for 96 h are shown in Figure 7.39A and B, respectively. Figure 7.39C and D show the results after 48 h cultivation.

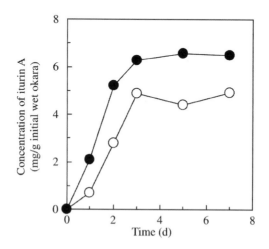

FIGURE 7.38 Changes in iturin A concentration when *B. subtilis* RB14-CS is grown by solid-state fermentation (SSF) (solid circles) and by submerged fermentation (SmF) (open circles) using soybean curd residue okara as substrate.

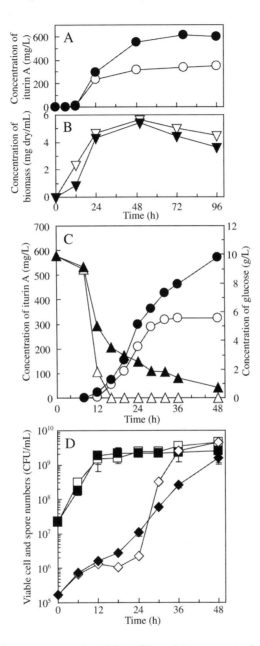

FIGURE 7.39 Changes in growth of RB14-CS, and the concentrations of iturin A and glucose in biofilm fermentation in an AMS reactor and in SmF using no. 3S medium. (A) Concentration of iturin A at 96 h; (B) concentration of dry biomass weight at 96 h; (C) concentration of iturin A (circles) and glucose (triangles) at 48 h; (D) viable cell number (squares) and spore number (diamonds) for 48 h. Solid symbols, biofilm fermentation; open symbols, SmF.

Iturin A production started in both systems after 12 h of incubation. Iturin A production by biofilm fermentation was higher than by SmF during all operation times (Figure 7.39A and C). The concentration of iturin A produced by biofilm fermentation increased continuously up to 48 h. The spore number also increased in a similar manner. However, the iturin A production in SmF leveled off after 28 h of cultivation, when the quick sporulation caused the majority of viable cells to become spores. After 48 h, the iturin A concentrations in the two systems did not change. The change in numbers of viable cells including vegetative cells and spores were similarly increased and maintained at approximately 2–3 × 10^9 CFU/ml (Figure 7.39D) in both systems. However, in biofilm fermentation, sporulation was initiated earlier than in SmF where enough glucose remained, and the number of spores increased continuously until 48 h of incubation. In SmF, spore number increased rapidly after 24 h and all cells became spores after 36 h (Figure 7.39D).

The biomass including the dry weights of cells and insoluble substances of okara reflected the growth of cells indirectly but the growth and maximum value of the biomass were almost the same in both systems (Figure 7.39B). The biomass decreased after 3 days probably due to degradation of the extracellular matrix or cell lysis, which may occur when the nutrients in the biofilm were exhausted (73, 85) or solubilization of the solid part of okara.

In SmF, glucose was consumed rapidly and exhausted after 16 h. In contrast, the concentration of glucose in biofilm fermentation decreased at a much slower rate and was almost exhausted after 48 h (Figure 7.39C). This suggests that the glucose supply to the cells in biofilm fermentation is limited by natural diffusion, in contrast to the free supply by forced mixing in SmF. The supply of other nutrients may be controlled in similar ways. Iturin A production was evaluated by three parameters as shown in Table 7.16. According to all three parameters, iturin production by biofilm fermentation was significantly higher than by SmF. Iturin A production per maximum viable cell number by biofilm fermentation was twofold higher than by SmF, and iturin A production per maximum biomass was 80% higher by biofilm fermentation than by SmF.

7.8.2.3 Proteome Analysis and Identification of Differentially Expressed Proteins

To characterize the differences in protein expression between the biofilm fermentation and SmF systems, two-dimensional electrophoresis of whole-cell protein extracts was performed after 12 h and 24 h as shown in Figure 7.40.

TABLE 7.16

Comparison of Iturin A Production between Biofilm Fermentation and SmF

	SmF	Biofilm	Biofilm/SmF
Per volume	349 (mg/l)	614 (mg/l)	1.76 (–)
Per maximum viable cell number	8.2 [µg/(10^8 × CFU)]	16.6 [µg/(10^8 × CFU)]	2.02 (–)
Per maximum biomass weight	62 (µg/mg)	113 (µg/mg)	1.82 (–)

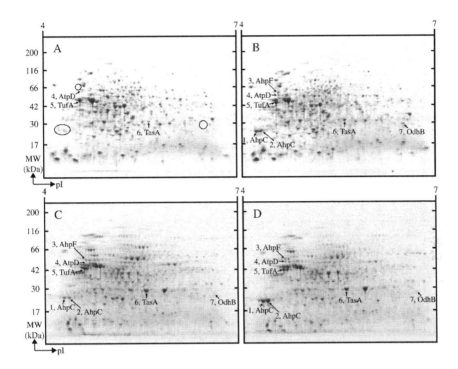

FIGURE 7.40 Two-dimensional electrophoresis profiles of whole-cell protein extracts obtained from *B. subtilis* RB14-CS. Profiles of cells at 12 h (A, B) and 24 h (C, D) from SmF (A, C) and biofilm fermentation in the AMS reactor (B, D). Unique protein spots are indicated by arrows. The pI values are plotted on the horizontal axes, and the molecular masses (MW) are plotted on the vertical axes.

The patterns in both systems were almost similar for cells after 24 and 36 h (data not shown). More than 300 distinct protein spots were observed with highly reproducible locations. By matching and comparing the patterns, seven protein spots, whose expression was reproducibly specific to or enhanced in biofilm fermentation, were detected. Then, N-terminal amino acid sequencing for these proteins was conducted and the results are summarized in Table 7.17.

In biofilm fermentation, protein spots 1–3 were detected after 12 h of incubation and the levels were higher in biofilm fermentation than in SmF at 24 h (Figure 7.40B and C). Spots 1 and 2 had the same molecular masses, but their pI values were different. As the amino acid sequences derived from the two protein gel spots were identical to those of a single protein, alkyl hydroperoxide reductase small subunit (AhpC), these spots may represent different posttranscriptionally modified versions of the same protein. The amino acid sequence derived from protein spot 3 was determined to be that of alkyl hydroperoxide reductase large subunit (AhpF).

The levels of protein spots 4–6 were higher in biofilm fermentation than in SmF at 12 h, but almost the same at 24 h. The amino acid sequence derived from protein spot 4 corresponded to ATP synthase beta chain (AtpD), which is a key enzyme in bacterial energy metabolism and produces ATP from ADP (86). The amino acid

TABLE 7.17

The Proteins Specific or Enhanced in Biofilm Fermentation Identified from Two-Dimensional Electrophoresis Gels of Whole-Cell Protein Extracts of *B. subtilis* RB14-CS

Spot	This Work		Database					
	Expression characteristics in biofilm fermentation	Sequence	Sequence	Gene	Protein Identity	Size (kDa)	pI	Identity (%)
1	Specific at 12 and enhanced at 24 and 36 h	MLIGKEVLPF	LIGKEVLPF	*ahpC*	Alkyl hydroperoxide reductase small subunit	20	4.28	90 [9/10]
2	Specific at 12 and enhanced at 24 and 36 h	MLIGKEVLPF	LIGKEVLPF	*ahpC*	Alkyl hydroperoxide reductase small subunit	20	4.28	90 [9/10]
3	Specific at 12 and enhanced at 24 and 36 h	MVLDANIKAQ	MVLDANIKAQ	*ahpF*	Alkyl hydroperoxide reductase large subunit	55	4.70	100 [10/10]
4	Enhanced at 12 h	VLGPVVDVR	VLGPVVDVR	*atpD*	ATPase beta chain	51	4.61	100 [9/9]
5	Enhanced at 12 h	VLHKKSGKG	VLHKKSGKG	*tufA*	Elongation factor Tu	43	4.72	100 [9/9]
6	Enhanced at 12 h	AFNDVKSTDA	AFNDIKSKDA	*tasA*	Translocation-dependent spore component	28	5.44	80 [8/10]
7	Specific at 12 and enhanced at 24 and 36 h	KRLVEVQQT	KRLVEVQQT	*odhB*	2-Oxoglutarate dehydrogenase	46	4.86	100 [9/9]

sequence derived from protein spot 5 corresponded to the sequence of the elongation factor Tu (TufA), which is a member of the GTPase family that has a function as molecular switches in various pathways. In prokaryotic cells, elongation factor Tu plays a central role in the elongation cycle of protein synthesis by forming a high-affinity complex with aminoacyl-tRNA and mediates its binding to the A site of the ribosome (87).

Expression of AtpD (spot 4) and TufA (spot 5) that play pivotal roles in energy metabolism and protein synthesis, respectively, is reduced by stringent response when amino acids and glucose were starved (88–90). The low expression of these two proteins at 12 h in SmF suggests that the cells in SmF suffered from a shortage of amino acids or glucose.

The N-terminal sequence of protein spot 6 was 80% identical to that of a translocation-dependent spore component (TasA). The 80% homology seemed to be insufficient to identify this protein as TasA, but the sequence obtained in this work (AFNDVKSTDA) was located in the same locus of the N-terminus of mature TasA after processing with signal peptidase (AFNDIKSKD) (91). Moreover, the molecular mass and pI in the database were consistent with those estimated from the gels. Considering these results and homology, it is concluded that protein spot 6 is TasA. The level of protein spot 7 was specifically found at 12 h in biofilm fermentation and higher in biofilm fermentation than in SmF at 24 h. The amino acid sequence derived from protein spot 7 corresponded to 2-oxoglutarate dehydrogenase (OdhB). Although the sequence (KRLVEVQQT) was 100% identical with that in the database, it is located far from the N-terminus of the full-length OdhB sequence (residues 201–417), and the molecular mass and pI in the database were different from those estimated from the gel. Therefore, it is not clear whether this protein is OdhB or not.

7.8.2.4 Real-Time RT-PCR

The expression of *ituB* and *degQ* by real-time RT-PCR is shown in Figure 7.41. The transcription of *ituB*, a member of the *itu* operon, which encodes the enzyme complex catalyzing the synthesis of antibiotic iturin A (34), was higher in biofilm fermentation than in SmF from 12 to 48 h. Transcription of *degQ* was also higher in biofilm fermentation from 24 to 48 h of incubation. Furthermore, these two genes were scarcely transcribed after 30 h in SmF, indicating that the higher production of iturin A in biofilm fermentation is due to difference at the level of transcription.

7.8.2.5 Scanning Electron Microscopy

Scanning electron micrographs of the cells from solid-state fermentation on okara and a biofilm from the AMS reactor are shown in Figure 7.42, indicating a similar biofilm formation in the two systems.

7.8.3 Discussion

B. subtilis RB14-CS produced a higher amount of antibiotic iturin A in SSF using a solid material of soybean curd residue, okara, than in SmF using suspended okara as nutrients. A previous study (42) also showed higher iturin A production by SSF on okara than by SmF in no. 3S medium, which is an efficient liquid medium for

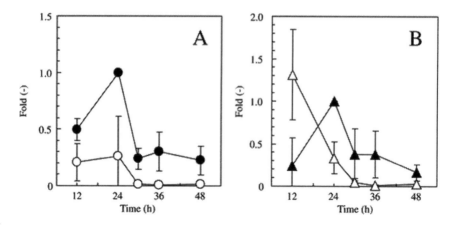

FIGURE 7.41 Expression of (A) *ituB* and (B) *degQ* in biofilm fermentation in the AMS reactor and SmF as determined by real-time RT-PCR. Symbols: solid symbols, biofilm fermentation; open symbols, SmF.

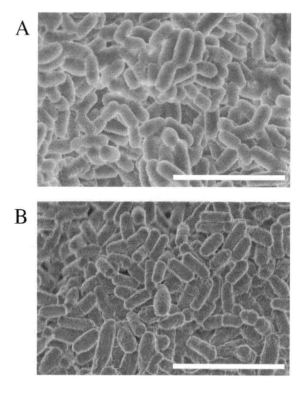

FIGURE 7.42 Scanning electron micrographs of (A) the surface of SSF using okara as a solid nutrient and (B) the biofilm in the AMS reactor. The specimens were photographed at a magnification 10,000. White bars represent 5 μm.

iturin A production. Thus, it appears that the higher production by SSF is independent of the medium compositions as long as the carbon or nitrogen contents are similar between the two systems. The AMS reactor used in this study had the following features: (i) growth of the bacterium as a biofilm on the surface of the membrane is similar to SSF in that cells grow on the surface of a solid substrate, (ii) the same medium as used for SmF is available, and (iii) cells are easily recovered from the membrane without contaminating the solid substrate. Thus, the biofilm formed in the AMS reactor can be used as a model biofilm system for SSF to elucidate higher production of iturin A in SSF.

At nearly the same viable cell numbers in the AMS and SmF cultures, iturin A productions per volume of medium and per weight of biomass were higher in biofilm fermentation than in SmF (Table 7.16). Therefore, this higher level of iturin A production by the biofilm was not due to an increase in cell number, but rather an increase in the production per cell. This means that the cells in the form of biofilm have a higher activity for production of iturin A than suspended cells, when the same nutrients are provided.

Because glucose of carbon and energy sources of *B. subtilis* (92) were exhausted earlier by SmF than by biofilm fermentation (Figure 7.39 C), it appears that glucose depletion results in lower production of iturin A by SmF. On the other hand, the glucose consumption rate was apparently lower in biofilm fermentation than in SmF, and residual glucose remained in the spent medium (Figure 7.39C). However, sporulation, which is mainly triggered by depletion of nutrients (93), occurred earlier in biofilm fermentation than in SmF during 12–24 h (Figure 7.39D). During this period, iturin A production also occurred more rapidly in biofilm fermentation than in SmF (Figure 7.39A). This suggests that some cells in the biofilm suffer from a local shortage of nutrients because of slow mass transfer through the membrane and the accumulated cells and partial depletion of nutrients that occurred in biofilm were associated with the higher iturin A production in biofilm fermentation.

Proteome analysis represented some specific characteristics of the cells in biofilms. Comparison of proteomes from the cells grown for 12 h revealed that proteins related to the stringent response (TufA and AtpD) were repressed in SmF (Figure 7.40, Table 7.17). This indicates that the cells in SmF were starved for amino acids or carbon sources. This was not observed in cells in the biofilm. However, in biofilm fermentation, proteins (AhpCF and OdhB) that are induced by starvation and stresses were expressed at a higher level, indicating that the cells also suffered from a shortage of nutrients, presumably due to the heterogeneity of the cells in biofilm structure. AhpC and AhpF are the two subunits of the same alkyl hydroperoxide reductase, and are encoded by the *ahpC* and *ahpF* genes, which are adjacent to each other and are most likely cotranscribed (94, 95). Alkyl hydroperoxide reductase is the best characterized bacterial enzyme involved in the metabolism of organic peroxide, and it detoxifies alkyl hydroperoxides, which are oxidizing agents generated *in vivo* under aerobic conditions that damage DNA (96). Transcription of *ahp* is induced by salt stress and heat stress, nutrient starvation, and H_2O_2 (95). The 2-oxo-glutarate dehydrogenase multienzyme complex, of which OdhB is a component, catalyzes the oxidative decarboxylation of 2-oxo-glutarate to succinyl coenzyme A, which is a reaction in the citric acid cycle (97). Transcription of the *odhAB* operon is strongly repressed by the presence of glucose in the medium (98). In SSF and AmF, nutrient depletion occurred, but the genes or proteins induced by nutrient depletion were not the same. This reflects the biochemical differences in the two systems.

Among differentially expressed proteins, the most intriguing is TasA, which is a major matrix in a major component of the *B. subtilis* biofilm matrix along with exopolysaccharides encoded by the *eps* operon and that expression of TasA is essential for formation of normal biofilm architecture (99). The early accumulation of TasA in biofilm in the AMS reactor suggests that the cells adapt to growth on a solid surface by producing extracellular matrix components.

The master regulator SinR represses the expression of the *eps* and *yqxM-sipW-tasA* operons, thereby controlling the synthesis of the major components of the biofilm matrix (99–101). The gene *sinI*, a direct antagonist of *sinR*, is positively regulated by the sporulation regulatory proteins Spo0A and σ^H (102, 103), and SinR-mediated repression of genes involved in biofilm formation is modulated by Spo0A (101). Additionally, transcription of *tasA* is positively regulated by *spo0A* (104). These findings indicate that production of TasA is related to sporulation. Thus, the early sporulation in biofilm fermentation observed here (Figure 7.40D) agrees with the high expression of TasA at 12 h as a typical characteristic of biofilms.

Expression of *itu* operon was higher and longer in biofilm fermentation than in SmF. This was also true for *degQ*, a gene that is essential for higher production of iturin A in SmF (81) (Figure 7.41). The *degQ* gene is also responsible for the production of a matrix component of *B. subtilis* biofilm, poly-γ-DL-glutamic acid (79), and it is positively regulated by two two-component regulatory systems, namely the ComP–ComA and the DegS–DegU systems (80). The ComQXPA signaling pathway, which includes the ComP–ComA system, is a major quorum response pathway in *B. subtilis* that regulates gene expression in response to fluctuations in cell density (122). ComQXPA regulates the transcription of genes including *srfA* (106, 107), *rapACEF* (108–111), and *degQ* (80). From the apparent volume of biofilm formed on the surface of the membrane, my team estimated that the cell density in the biofilm is at least eightfold higher than that in SmF (data not shown). Therefore, it is certain that regulation by ComQXPA occurred preferentially in biofilm fermentation.

DegS–DegU is a two-component system that regulates many cellular processes including extracellular protease production, competence development, and motility (112). Phosphorylated DegU (DegU-P) stimulates transcription of *aprE* and *nprE*, which encode the major extracellular proteases, and inhibit expression or the activity of an alternative sigma factor SigD (113, 114). Regulation of gene expression via DegS–DegU is affected by salt stress (115, 116). Biofilm may form a gradient of organic and inorganic ions existing at the solid surface, and the various exopolymers excreted by bacteria could concentrate ionic molecules from the bulk phase as the biofilm develops (117). These effects cause cells in biofilms to encounter higher salt stress (osmolarity) than cells in the liquid phase. During biofilm formation by *Escherichia coli*, the intracellular concentration of potassium ions is 1.6-fold higher in cells in the biofilm than in cells in liquid phase (117). Therefore, it is possible that *degQ* was enhanced by salt stress via DegS–DegU in biofilm fermentation and that expression of AhpCF, which is known to be induced by salt stress, was increased (94).

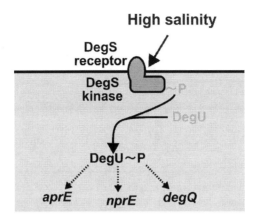

FIGURE 7.43 The summary of speculation.

The expression of *degQ* was induced by limitation of carbon, nitrogen, and phosphate (80), and was strongly repressed by glucose (118, 119). As mentioned earlier, in biofilm fermentation, cells are likely to encounter a long-term local depletion of nutrients including glucose due to limitation of diffusion, even though nutrients still remain in the water phase. This suggests that a slower supply of nutrients to cells prolongs the exposure of cells to nutrient limitation in biofilm fermentation, and contributes to the higher iturin A production via increased expression of *degQ*.

Consequently, our results showed that the higher production of iturin A by *B. subtilis* RB14-CS in biofilm fermentation is triggered by typical characteristics of biofilm, namely, a high cell density, high salinity, and/or prolonged period of local limitation of nutrients, via elevated expression of *degQ* and *itu* operon. The summary of the speculation is shown in Figure 7.43.

7.9 MAXIMIZATION OF HIGH SPORE DENSITY IN LIQUID CULTIVATION BY *B. SUBTILIS* RB14 (120)

In the direct treatment of *B. subtilis* culture into soil or on plant leaves or fruits, a higher spore density of the bacterium is preferable for the effective suppression of plant disease. To attain a high spore density culture, increased vegetative cell concentration must be established. For this purpose, the culture condition for obtaining high cellular mass and induction factors to convert vegetative cells to spores must be identified. In submerged fermentation in particular, oxygen limitation is inevitable in order to obtain a high cell concentration. In this section, the focal points are the efficient oxygen supply and metal concentrations for spore production.

7.9.1 MATERIALS AND METHODS

7.9.1.1 Strain Used

B. subtilis RB14, which produces the lipopeptide iturin and surfactin that contribute to the suppression of plant pathogens, was used.

7.9.1.2 Reactor

A stainless steel jar fermentor (5-liter volume, 10-liter nominal volume, BMS-10PI; ABLE Corp., Tokyo) was used. A basket-shaped unit for agitation was used (Figure 7.10) because the oxygen transfer coefficient ($K_L a$) of the fermenter was five times higher with this unit (2).

7.9.1.3 Medium

The basic SP medium consisted of a commercially available nitrogen source, peptone SE50M (Oriental Yeast Co., Tokyo) at 40 g/l, and a carbon source of sucrose syrup (Iwata Chemical Co., Tokyo) at 100 ml/l and K_2HPO_4 at 5g/l. Sucrose syrup is a waste product from the sugar production industry that contains 92% sucrose. SP agar medium contained SP medium and agar, 15 g/l.

7.9.1.4 Operation Condition

The temperature was controlled at 37°C, and the dissolved oxygen concentration was controlled at 30% of the saturation concentration. pH was maintained at 7.5 with 5N NaOH.

7.9.1.5 Cell Concentration

The total *B. subtilis* concentration, which contained vegetative cells and spores, was determined on the SP agar medium. The culture was heated at 70°C for 10 min, and the cells that appeared on the SP agar medium after the heat treatment were counted as spore number.

7.9.1.6 Oxygen Supply

As the cell concentration increased, the supply of air from a compressor was not effective because the dissolved oxygen (DO) concentration became almost 0, leading to oxygen limitation. Thus, an Airsep oxygen generator (Pressure Swing Adsorption system; Chart Japan Co., Ltd., Tokyo) was used to supply enriched oxygen.

7.9.2 RESULTS

7.9.2.1 The Effect of Metal Salts

As spore induction components, 1 M K_2HPO_4, 0.1 M $MgCl_2$, and 1 M $Ca(NO_3)_2$, were reported (121), 10 ml of each or different mixtures of these components were added during operation when the DO concentration started to decline and DO decline was stabilized. This stabilized decline reflected the decreased growth of the bacterium. Then, spore formation from vegetative cells began due to adverse effects on the cells, and the addition of the metal salts accelerated the spore formation. The result is shown in Figure 7.44.

The maximum spore concentration was highest with the mixture of $MgCl_2$ and $Ca(NO_3)_2$, and the value of the spore concentration was threefold higher compared with that of the control. It is interesting that the maximum cell concentration, which contained both vegetative cells and spores, was not necessarily related to the maximum spore concentration. In these experiments, the DO concentration decreased to almost 0 because the air supply was not sufficient to maintain the high DO concentration.

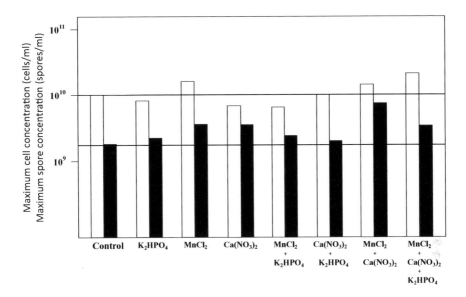

FIGURE 7.44 The effect of metal salts on the spore formation. The total cell concentration (□) includes vegetative cells and spores (■) of *B. subtilis* RB14.

The Airsep oxygen generator that supplies enriched oxygen by the adsorption of nitrogen from the air was then introduced. This generator introduction was also effective in preventing foaming during cultivation of *B. subtilis*. The introduction of the Airsep oxygen generator allowed maintaining DO concentration at 30% of the maximum concentration during cultivation operation and avoiding the limitation of DO.

7.9.2.2 Effect of Other Factors

The effect of the added volumes of the mixture of 0.1 M $MgCl_2$ and 1 M $Ca(NO_3)_2$ on the spore formation is shown in Figure 7.45. The 50 ml volume condition corresponded with the maximum spore concentration. When the volume of added peptone was varied, 4% peptone concentration was better for spore formation than both 1% and 8% (data not shown). The pH value was changed from 6.5 to 8, and pH 7.5 was optimal for the spore concentration (data not shown).

7.9.2.3 Maximizing Operation of Spore

The operation conditions for maximizing spore density are as follows:

The mixture of 50 ml 0.1 M $MgCl_2$ and 50 ml of 1 M $Ca(NO_3)_2$ was added at 9 h from the start of cultivation when the growth started to decrease.
The DO concentration was fixed at 30% of saturation by the Airsep oxygen generator.
The pH was maintained at 7.5 by adding 5 N NaOH solution.

The total cell number and spore concentration are shown in Figure 7.46.

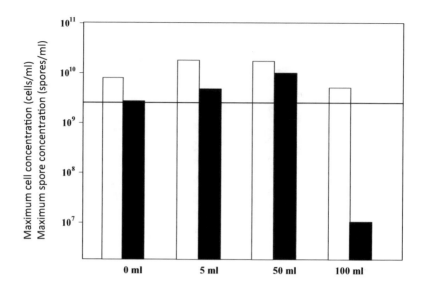

FIGURE 7.45 The effect of volumes of a mixture of 0.1 M MnCl$_2$ and 1 M Ca(NO$_3$)$_2$ on spore formation. The total cell concentration ($\square$) includes vegetative cells and spores ($\blacksquare$) of *B. subtilis* RB14.

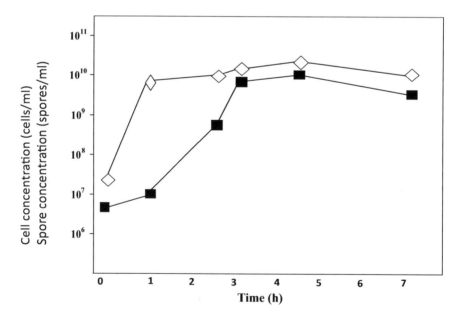

FIGURE 7.46 Change in cell concentration ($\lozenge$) and spore concentration ($\blacksquare$) during cultivation of *B. subtilis* RB14 under optimum operation condition. The cell concentration includes vegetative cells and spores.

When the total cell number reached nearly 10^{10} cells/ml and the DO was sufficient at 30% of saturation, the growth started to plateau, which was presumably because of product inhibition or because some minor nutrients became insufficient. At around this time, the mixture of 50 ml 0.1 M $MgCl_2$ and 50 ml of 1 M $Ca(NO_3)_2$ was added and the spore formation became active, leading to the achievement of the maximum spore density of 9.7×10^9 spores/ml. This value indicates that more than 90% of the culture consisted of spores. The culture obtained in this operation can be stably stored before use, and the treated suspension of the bacteria can be stably maintained in soil for a long period of time (Chapter 4).

REFERENCES

1. Phae, C. C., and Shoda, M. Investigation of optimal conditions for foam separation of iturin, an antifungal peptide produced by *Bacillus subtilis. J. Ferment. Bioeng.*, 71, 118–121 (1991).
2. Hayashi, T., Ohishi, T., and Shoda, M. Cultivation of baker's yeast by a fermenter with a basket-shaped unit for agitation. *Biotechnol. Techniq.*, 3, 381–384 (1989).
3. Rubin, A. J., Cassel, E. A., Henderson, O., Johnson, J. D., and Lamb, III, J. C. Microfloatation: new low gas-flow rate foam separation technique for bacteria and algae. *Biotechnol. Bioeng.*, 8, 135–151 (1966).
4. Grieves, R. B., and Wang, S. L. Foam separation of *Escherichia coli* with a cationic surfactant. *Biotechnol. Bioeng.*, 8, 323–336 (1966).
5. Wang, D. I. C., and Sinskey, A. Collection of microbial cells, pp. 121–152. In Perlman, D. (ed.), *Advances in Applied Microbiology*, vol. 12. Academic Press, New York (1970).
6. Effler, W. T., Tanner, R. D., and Malaney, G. W. Dynamic in-situ fractionation of extracellular proteins produced in a baker's yeast cultivation process, pp. 235–255. In Fiechter, A., Okada, H., and Tanner, R. D. (eds.), *Bioproducts and Bioprocesses.* Springer-Verlag, Berlin, (1989).
7. Ohno, A., Ano, T., and Shoda, M. Production of the antifungal peptide antibiotic, iturin by *Bacillus subtilis* NB22 in solid state fermentation. *J. Ferment. Bioeng.*, 75, 23–27 (1993).
8. Cannel, E., and Moo-Yang, M. Solid-state fermentation systems. *Proc. Biochem.* June/July, 2–7 (1980).
9. Sato, K., Miyazaki, S., Matsumoto, N., Yoshizawa, K., and Nakamura, K. Pilot-scale solid-state ethanol fermentation by inert gas circulation using moderately thermophilic yeast. *J. Ferment. Technol.*, 66, 173–178 (1988).
10. Ramesh, M. V., and Lonsane, B. K. Regulation of alpha-amylase production in *Bacillus licheniformis* M27 by enzyme end-products in submerged fermentation and its overcoming in solid state fermentation system. *Biotechnol. Lett.*, 13, 355–360 (1991).
11. Ohno, A., Ano, T., and Shoda, M. Use of soybean curd residue, okara, for the solid state substrate in the production of lipopeptide antibiotic, iturin A, by *Bacillus subtilis* NB22. *Process Biochem.*, 8, 801–806 (1996).
12. Phae, C.G., Shoda, M., and Kubota, H. Suppressive effect of *Bacillus subtilis* and its products on phytopathogenic microorganisms. *J. Ferment. Bioeng.*, 69, 1–7 (1990).
13. Phae, C.G., Shoda, M., Kita, N., Nakano, M., and Ushiyama, K. Biological control of crown and root rot and bacterial wilt of tomato by *Bacillus subtilis* NB22. *Ann. Phytopathol. Soc. Jpn.*, 58, 329–339 (1992).
14. Phae, C.G., Sasaki, M., Shoda, M., and Kubota, H. Characteristics of *Bacillus subtilis* isolated from composts suppressing phytopathogenic microorganisms. *Soil Sci. Plant Nutr.*, 36, 575–586 (1990).

15. Ohno, A., Ano, T., and Shoda, M. Effect of temperature on production of lipopeptide antibiotics, iturin A and surfactin by a dual producer, *Bacillus subtilis* RB14, in solid-state fermentation. *J. Ferment. Bioeng.*, 80, 517–519 (1995).

16. Sandrin, C., Peypoux, F., and Michel, G. Coproduction of surfactin and iturin A, lipopeptides with surfactant and antifungal properties, by *Bacillus subtilis*. *Biotechnol. Appl. Biochem.*, 12, 370–375 (1990).

17. Ohno, A., Ano, T., and Shoda, M. Production of a lipopeptide antibiotic surfactin with recombinant *Bacillus subtilis*. *Biotechnol. Letts.*, 14, 1165–1168 (1992).

18. Sheppard, J. D., and Mulligan, N. The production of surfactin by *Bacillus subtilis* grown on peat hydrolysate. *Appl. Microbiol. Biotechnol.*, 27, 110–116 (1987).

19. Cooper, D. G., MacDonald, C. R., Duff, S. T. B., and Kosaric, N. Enhanced production of surfactin from *Bacillus subtilis* by continuous product removal and metal cation additions. *Appl. Environ. Microbiol.*, 42, 408–412 (1981).

20. Tsuge, K., Ano, T., and Shoda, M. Characterization of *Bacillus subtilis* YB8, coproducer of lipopeptides and surfactin and plipastatin B1. *J. Gen. Appl. Microbiol.*, 41, 541–545 (1995).

21. Hiraoka, H., Ano, T., and Shoda, M. Molecular cloning of a gene responsible for the biosynthesis of the lipopeptide antibiotics iturin and surfactin. *J. Ferment. Bioeng.*, 74, 323–326 (1992).

22. Huang, C. C., Ano, T., and Shoda, M. Nucleotide sequence and characteristics of the gene, *lpa-14*, responsible for biosynthesis of the lipopeptide antibiotics iturin A and surfactin from *Bacillus subtilis* RB14. *J. Ferment. Bioeng.*, 76, 445–450 (1993).

23. Hiraoka, H., Asaka, O., Ano, T., and Shoda, M. Characteristics of *Bacillus subtilis* RB14, coproducer of peptide antibiotics iturin A and surfactin. *J. Gen. Appl. Microbiol.*, 38, 635–640 (1992).

24. Kaneda, T. *Iso-* and *anteiso*-fatty acids in bacteria: biosynthesis, function, and taxonomic significance. *Microbial. Rev.*, 55, 288–302 (1991).

25. Chan, M., Himes, R. H., and Akagi, J. M. Fatty acid composition of thermophilic, mesophilic, and psychrophilic *Clostridia*. *J. Bacteriol.*, 106, 876–881 (1971).

26. Mizumoto, S., Hirai, M., and Shoda, M. Enhanced iturin A production by *Bacillus subtilis* and its effect on suppression of the plant pathogen *Rhizoctonia solani*. *Appl. Microbiol. Biotechnol.*, 75, 1267–1274 (2007).

27. Asaka, O., and Shoda, M. Biocontrol of *Rhizoctonia solani* damping-off of tomato with *Bacillus subtilis* RB14. *Appl. Environ. Microbiol.*, 62, 4081–4085 (1996).

28. Ushiyama, K., Nishimura, J., and Aono, N. Damping-off of feather cockscomb (*Celosiaargentea* L. var. *cristata* O. Kuntze) caused by *Rhizoctonia solani* Kuhn. *Bull. Kanagawa Hortic. Exp. Stn.*, 34, 33–37 (1987).

29. Katayama, A., Hirai, M., Shoda, M., and Kubota, H. Factors affecting the stabilization period of sewage sludge in soil with reference to the gel chromatographic pattern. *Soil Sci. Plant Nutr.*, 32, 383–395 (1986).

30. Tokuda, Y., Ano, T., and Shoda, M. Survival of *Bacillus subtilis* NB22 and its transformant in soil. *Appl. Soil Ecol.*, 2, 85–94 (1995).

31. Leifert. C., Li, H., Chidburee, S., Hampson, S., Workman, S., Sigee, D., Epton, H. A., and Harbour, A. Antibiotic production and biocontrol activity by *Bacillus subtilis* CL27 and *Bacillus pumilus* CL45. *J. Appl. Bacteriol.*, 78, 97–108 (1995).

32. McCarter, S. M. Rhizoctonia diseases, pp. 21–22. In Jones, J. B., Jones, J. P., Stall, R. E., and Zitter, T. A. (eds.), *Compendium of Tomato Diseases*. APS Press, St. Paul, MN (1991).

33. Akpa, E., Jacques, P., Wathelet, B., Paquot, M., Fucks, R., Budzikiewicz, H., and Thonart, P. Influence of culture conditions on lipopeptide production by *Bacillus subtilis*. *Appl. Biochem. Biotechnol.*, 91, 551–561 (2001).

34. Tsuge, T., Akiyama, T., and Shoda, M. Cloning, sequencing, and characterization of the iturin A operon. *J. Bacteriol.*, 183, 6265–6273 (2001).

35. Leclere, V., Bechet, M., Adam, A., Guez, J. S., Wathelet, B., Ongena, M., Thonart, P., Gancel, F., Chollet-Imbert, M., and Jacques, P. Mycosubtilin overproduction by *Bacillus subtilis* BBG100 enhances the organism's antagonistic and biocontrol activities. *Appl. Environ. Microbiol.*, 71, 4577–4584 (2005).

36. Reddy, M. S., and Rahe, J. E. Growth effects associated with seed bacterization not correlated with populations of *Bacillus subtilis* inoculant in onion seedling rhizospheres. *Soil Biol. Biochem.*, 21, 373–378 (1989).

37. Safiyazov, J. S., Mannanov, R. N., and Sattarova, R. K. The use of bacterial antagonists for the control of cotton diseases. *Field Crop Res.* 43, 51–54 (1995).

38. Sailaja, P. R., Podile, A. R., and Reddanna, P. Biocontrol strain of *Bacillus subtilis* AF 1 rapidly induces lipoxygenase in groundnut (*Arachis hypogaea* L.) compared to crown root pathogen *Aspergillus niger. Eur. J. Plant Pathol.*, 104, 125–132 (1997).

39. Ryder, M. H., Yan, Z., Terrace, T. E., Rovira, A. D., Tang, W., and Correll, R. L. Use of *Bacillus* strains isolated in China to suppress take-all and rhizoctonia root rot, and promote seedling growth of glasshouse-grown wheat in Australian soils. *Soil Biol. Biochem.*, 31, 19–29 (1999).

40. Zheng, X. Y., and Sinclair, J. B. The effects of traits of *Bacillus megaterium* on seed and root colonization and their correlation with the suppression of *Rhizoctonia* root rot of soybean. *BioControl*, 45, 223–243 (2000).

41. Kondo, M., Hirai M., and Shoda, M. Integrated biological and chemical control of damping-off caused by *Rhizoctonia solani* using *Bacillus subtilis* RB14-C and flutolanil. *J. Biosci. Bioeng.*, 91, 173–177 (2001).

42. Mizumoto, S., Hirai, M., and Shoda, M. Production of lipopeptide antibiotic iturin A using soybean curd residue cultivated with *Bacillus subtilis* in solid-state fermentation. *Appl. Microbiol. Biotechnol.*, 72, 869–875 (2006).

43. Robinson, T., Singh, D., and Nigam, P. Solid state fermentation: a promising microbial technology for secondary metabolite production. *Appl. Microbiol. Biotechnol.*, 55, 284–289 (2001).

44. Acea, M. J., Moore, C. R., and Alexander, M. Survival and growth of bacteria introduced into soil. *Soil Biol. Biochem.*, 20, 509–515 (1988).

45. Conover, R. A. *Rhizoctonia* canker of tomato. *Phytopathology*, 39, 950–951 (1949).

46. Jones, J. B., Jones, J. P., Stall, R. E., and Zitter, T. A. *Compendium of Tomato Diseases*. APS Press, St. Paul, MN, pp. 21–22 (1991).

47. Ohno, A., Ano, T., and Shoda, M. Effect of temperature change and aeration on the production of antifungal peptide antibiotic iturin by *Bacillus subtilis* NB22 in liquid cultivation. *J. Ferment. Bioeng.*, 75, 463–465 (1993).

48. Yan, L., Boyd, K. G., Adams, D. R., and Burgess, J. G. Biofilm-specific cross-species induction of antimicrobial compounds in *Bacilli. Appl. Environ. Microbiol.*, 69, 3719–3727 (2003).

49. Stein, T. *Bacillus subtilis* antibiotics: structure, syntheses and specific functions. *Mol. Microbiol.* 56, 845–857 (2005).

50. Mah, T. F., Pitts, B., Pellock, Walker, G. C., Stewart, P. S., and O'Toole Pitts, G. A. A genetic basis for *Pseudomonas aeruginosa* biofilm antibiotic resistance. *Nature*, 426, 306–310 (2003).

51. Adams, T. T., Eiteman, M. A., and Hanel, B. M. Solid state fermentation of broiler litter for production of biocontrol agents. *Bioresour. Technol.*, 82, 33–41 (2002).

52. Katayama, A., Gomez, M. M., Ker, K. C., Hirai, M., Shoda, M., and Kubota, H. Decomposition process of various organic wastes in soil with reference to gel chromatography. *Soil Sci. Plant Nutr.*, 33, 471–486 (1987).

53. Szczech, M., and Shoda, M. The influence of *Bacillus subtilis* RB14-C on the development of *Rhizoctonia solani* and indigenous microorganisms in the soil. *Can. J. Microbiol.*, 51, 405–411 (2005).

54. Van Elsas, J. D., Dijkstra, A. F., Govaert, J. M., and Van Veen, J. A. Survival of *Pseudomonas fluorescens* and *Bacillus subtilis* introduced into two soils of different texture in field microplots. *FEMS. Microbiol. Ecol.*, 38, 151–160 (1986).

55. Mizumoto, S., and Shoda, M. Medium optimization of antifungal lipopeptide, iturin A, production by *Bacillus subtilis* in solid-state fermentation by response surface methodology. *Appl. Microbiol. Biotechnol.*, 76, 101–108 (2007).

56. Souza, M. C. O., Robetro, I. C., and Milagres, A. M. F. Solid-state fermentation for xylanase production by *Thermoascus aurantiacus* using response surface methodology. *Appl. Microbiol. Biotechnol.*, 52, 768–772 (1999).

57. Park, Y. S., Kang, S. W., Lee, J. S., and Hong, S. I. Xylanase production in solid state fermentation by *Aspergillus niger* mutant using statistical experimental designs. *Appl. Microbiol. Biotechnol.*, 58, 761–766 (2002).

58. Tarocco, F., Lecuona, R. E., Couto, A. S., and Arcas, J. A. Optimization of erythritol and glycerol accumulation in conidia of *Beauveria bassiana* by solid-state fermentation, using response surface methodology. *Appl. Microbiol. Biotechnol.*, 68, 481–488 (2005).

59. Xiong, C., Shouwen, C., Ming, Z., and Ziniu, Y. Medium optimization by response surface methodology for poly-γ-glutamic acid production using dairy manure as the basis of a solid substrate. *Appl. Microbiol. Biotechnol.*, 69, 390–396 (2005).

60. Jacques, P., Hbid, C., Destain, J., Razfindralambo, H., Paquot, M., De Pauw, E., and Thonart, P. Optimization of biosurfactant lipopeptide production from *Bacillus subtilis* S449 by Plackett-Burman design. *Appl. Biochem. Biotechnol.*, 77, 223–233 (1999).

61. Kolicheski, M. B., Roccol, C. R., Marin, B., Medeiros, E., and Raimbault, M. Citric acid production on three cellulosic supports in solid state fermentation, pp. 449–462. In Roussos, S., Lonsane, B. K., Raimbault, M., and Viniegra-Gonzalez, G. (eds.), *Advances in Solid State Fermentation: Proceedings of the 2nd International Symposium on Solid State Fermentation FMS-95*. Montpellier, France (1997).

62. Castilho, L. R., Medronho, R. A., and Alves, T. L. M. Production and extraction of pectinases obtained by solid state fermentation of agroindustrial residues with *Aspergillus niger*. *Bioresour. Technol.*, 71, 45–50 (2000).

63. Ortiz-Vazouez, E., Granados-Baeza, M., and Rivera-Munoz, G. Effect of culture conditions on lipolytic enzyme production by *Penicillium candidum* in a solid state fermentation. *Biotechnol. Adv.*, 11, 409–416 (1993).

64. Sugano, Y., Matsuo, C., and Shoda, M. Efficient production of a heterologous peroxidase, DyP from *Geotrichum candidum* Dec I, on solid-state culture of *Aspergillus oryzae* RD005. *J. Biosci. Bioeng.*, 92, 594–597 (2001).

65. Gawande, P. V., and Kamat, M. Y. Production of *Aspergillus* xylanase by lignocellulosic waste fermentation and its application. *J. Appl. Microbiol.*, 87, 511–519 (1999).

66. Elinbaum, S., Ferreyra, H., Ellenrieder, G., and Cuevas, C. Production of *Aspergillus terreus* β-L-rhamnosidase by solid state fermentation. *Lett. Appl. Microbiol.*, 34, 67–71 (2002).

67. Raimbault, M. General and microbiological aspects of solid substrate fermentation. *Electron. J. Biotechnol.*, 1, 174–188 (1998).

68. Raghavarao, K. S. M. S., Ranganathan, T. V., and Karanth, N. G. Some engineering aspects of solid-state fermentation. *Biochem. Eng. J.*, 13, 127–135 (2003).

69. Durand, A. Bioreactor designs for solid state fermentation. *Biochem Eng. J.*, 13, 113–125 (2003).

70. Shoda, M., and Mizumoto, S. Biofilm formed on the soybean curd residue is associated with efficient production of a lipopeptide by *Bacillus subtilis*, pp. 333–352. In Maxell, J. E. (ed.), *Soybeans Cultivation, Uses and Nutrition*, Nova Scientific Publishers, New York (2011).

71. Branda, S. S., Vik, A., Friedman, L., and Kolter, R. Biofilms: the matrix revisited. *Trends Microbiol.*, 13, 20–26 (2005).

72. Davey, M. E., and O'Toole, G. A. Microbial biofilms: from ecology to molecular genetics. *Microbiol. Mol. Biol. Rev.*, 64, 847–867 (2000).

73. Waite, R. D., Papakonstantinopoulou, A., Littler, E., and Curtis, M. A. Transcriptome analysis of *Pseudomonas aeruginosa* growth: comparison of gene expression in planktonic cultures and developing and mature biofilms. *J. Bacteriol.*, 187, 6571–6576 (2005).

74. Morikawa, M. Beneficial biofilm formation by industrial bacteria *Bacillus subtilis* and related species. *J. Biosci. Bioeng.*, 101, 1–8 (2006).

75. Oosthuizen, M. C., Steyn, B., Theron, J., Cosette, P., Lindsay, D., von Holy, A., and Brozel, S. Proteomic analysis reveals differential protein expression by *Bacillus cereus* during biofilm formation. *Appl. Environ. Microbiol.*, 68, 2770–2780 (2002).

76. Ren, D., Bedzyk, L. A., Setlow, P., Thomas, S. M., Ye, R. W., and Wood, T. K. Gene expression in *Bacillus subtilis* surface biofilms with and without sporulation and the importance of *yveR* for biofilm maintenance. *Biotechnol. Bioeng.*, 86, 344–364 (2004).

77. Resch, A., Rosenstein, R., Nerz, C., and Gotz, F. Differential gene expression profiling of *Staphylococcus aureus* cultivated under biofilm and planktonic conditions. *Appl. Environ. Microbiol.*, 71, 2663–2676 (2005).

78. Stanley, N. R., Britton, R. A., Grossman, A. D., and Lazazzera, B. A. Identification of catabolite repression as a physiological regulator of biofilm formation by *Bacillus subtilis* by use of DNA microarrays. *J. Bacteriol.*, 185, 1951–1957 (2003).

79. Stanley, N. R., and Lazazzera, B. A. Defining the genetic differences between wild and domestic strains of *Bacillus subtilis* that affect poly-γ-DL-glutamic acid production and biofilm formation. *Mol. Microbiol.*, 57, 1143–1158 (2005).

80. Msadek, T., Kunst, F., Klier, A., and Rapoport, G. DegS-DegU and ComP-ComA modulator-effector pairs control expression of the *Bacillus subtilis* pleiotropic regulatory gene *degQ*. *J. Bacteriol.*, 173, 2366–2377 (1991).

81. Tsuge, K., Inoue, S., Ano, T., Itaya, M., and Shoda, M. Horizontal transfer of iturin A operon, *itu*, to *Bacillus subtilis* 168 and conversion into an iturin A producer. *Antimicrob. Agents Chemother.*, 49, 4641–4648 (2005).

82. Wassel, D., and Flugge, U. I. A method for the quantitative recovery of protein in dilute solution in the presence of detergents and lipids. *Anal. Biochem.*, 138, 141–143 (1984).

83. Bradford, M. M. A rapid and sensitive method for the quantification of microgram quantities of protein utilizing the principle of protein-dye binding. *Anal. Biochem.*, 72, 248–254 (1976).

84. Edman, P. A method for the determination of the amino acid sequence in peptides, *Arch. Biochem. Biophys.*, 11, 475–476 (1949).

85. Allison, D. G., Ruiz, B., SanJose, C., Jape, A., and Gilbert, P. Extracellular products as mediators of the formation and detachment of *Pseudomonas fluorescens* biofilms. *FEMS Microbiol. Lett.*, 167, 179–184 (1998).

86. Santana, M., Ionescu, M., Vertes, A., and Longin, R. *Bacillus subtilis* F₀F₁ ATPase: DNA sequence of the *atp* operon and characterization of *atp* mutants. *J. Bacteriol.*, 176, 6802–6811 (1994).

87. Krasny, L., Mesters, J. R., Tieleman, L. N., Kraal, B., Fucik, V., Hilgenfeld, R., and Jonak, J. Structure and expression of elongation factor Tu from *Bacillus stearothermophilus*. *J. Mol. Biol.*, 283, 371–381 (1998).

88. Bernhardt, J., Weibezahn, J., Scharf, C., and Hecker, M. *Bacillus subtilis* during feast and famine: visualization of the overall regulation of protein synthesis during glucose starvation by proteome analysis. *Genome Res.*, 13, 224–237 (2003).

89. Eymann, C., Homuth, G., Scharf, C. C., and Hecker, M. *Bacillus subtilis* functional genomics: global characterization of the stringent response by proteome and transcriptome analysis. *J. Bacteriol.*, 184, 2500–2520 (2002).

90. Inaoka, T., Takahashi, K., Ohnishi-Kameyama, M., Yoshida, M., and Ochi, K. Guanine nucleotides guanosine 5′-diphosphate 3′-diphosphate and GTP co-operatively regulate the production of an antibiotic bacilysin in *Bacillus subtilis*. *J. Biol. Chem.*, 278, 2169–2176 (2003).

91. Stover, A., and Driks, A. Secretion, localization, and antibacterial activity of TasA, a *Bacillus subtilis* spore-associated protein. *J. Bacteriol.*, 181, 1664–1672 (1999).

92. Stulke, J., and Hillen, W. Regulation of carbon catabolism in *Bacillus* species. *Annu. Rev. Microbiol.*, 54, 849–880 (2000).

93. Errington, J. Regulation of endospore formation in *Bacillus subtilis*. *Nat. Rev. Microbiol.*, 1, 117–126 (2003).

94. Antelmann, H., Engelmann, S., Schmid, R., and Hecker, M. General and oxidative stress responses in *Bacillus subtilis*: cloning, expression and mutation of the alkyl hydroperoxide reductase operon. *J. Bacteriol.*, 178, 6571–6578 (1996).

95. Bsat, N., Chen, L., and Helmann, J. D. Mutation of the *Bacillus subtilis* alkyl hydroperoxide reductase (*ahpCF*) operon reveals compensatory interactions among hydrogen peroxide stress genes. *J. Bacteriol.*, 178, 6579–6586 (1996).

96. Imlay, J. A., and Linn, S. DNA damage and oxygen radical toxicity. *Science*, 240, 1302–1309 (1988).

97. Carlsson, P., and Hederstedt, L. Genetic characterization of *Bacillus subtilis odhA* and *odhB*, encoding 2-oxoglutarate dehydrogenase and dihydrolipoamide transsuccinylase, respectively. *J. Bacteriol.*, 171, 3667–3672 (1989).

98. Resnekov, O., Melin, L., Carlsson, P., Mannerlov, M., von Gabain, A., and Hederstedt, L. Organization and regulation of the *Bacillus subtilis odhAB* operon, which encodes two of the subenzymes of the 2-oxoglutarate dehydrogenase complex. *Mol. Gen. Genet.*, 234, 285–296 (1992).

99. Branda, S. S., Chu, F., Kearns, D. B., Losick, R., and Kolter, R. A major protein component of the *Bacillus subtilis* biofilm matrix. *Mol. Microbiol.*, 59, 1229–1238 (2006).

100. Chu, F., Kearns, D. B., Branda, S. S., Kolter, R., and Losick, R. Targets of the master regulator of biofilm formation in *Bacillus subtilis*. *Mol. Microbiol.*, 59, 1216–1228 (2006).

101. Kearns, D. B., Chu, F., Branda, S. S., Kolter, R., and Losick, R. A master regulator for biofilm formation by *Bacillus subtilis*. *Mol. Microbiol.*, 55, 739–749 (2005).

102. Gaur, N. K., Cabane, K., and Smith, I. 1988. Structure and expression of the *Bacillus subtilis* sin operon. *J. Bacteriol.*, 170, 1046–1053.

103. Shafikhani, S. H., Mandic-Mulec, I., Strauch, M. A., Smith, I., and Leighton, T. Postexponential regulation of *sin* operon expression in *Bacillus subtilis*. *J. Bacteriol.*, 184, 564–571 (2002).

104. Stover, A., and Driks, A. Regulation of synthesis of the *Bacillus subtilis* transition-phase, spore-associated antibacterial protein TasA. *J. Bacteriol.*, 181, 5476–5481 (1999).

105. Koumoutsi, A., Chen, X.-H., Henne, A., Licsegang, H., Hitzeroth, G., Franke, P., Valer, J., and Borriss, R. Structural and functional characterization of gene clusters directing nonribosomal synthesis of bioactive cyclic lipopeptides in *Bacillus amyloliquefaciens* strain FZB42. *J. Bacteriol.*, 186, 1084–1096 (2004).

106. Nakano, M. M., Xia, L., and Zuber, P. Transcription initiation region of the *srfA* operon, which is controlled by the ComP-ComA signal transduction system in *Bacillus subtilis*. *J. Bacteriol.*, 173, 5487–5493 (1991).

107. Nakano, M. M., and Zuber, P. Mutational analysis of the regulatory region of the *srfA* operon in *Bacillus subtilis*. *J. Bacteriol.*, 175, 3188–3191 (1993).

108. Jarmer, H., Larsen, T. S., Krogh, A., Saxild, H. H., Brunak, S., and Knudsen, S. Sigma-A recognition sites in the *Bacillus subtilis* genome. *Microbiology*, 147, 2417–2424 (2001).

109. Jiang, M., Grau, R., and Perego, M. Differential processing of propeptide inhibitors of Rap phosphatases in *Bacillus subtilis*. *J. Bacteriol.*, 182, 303–310 (2000).

110. Lazazzera, B. A., Kurtser, I. G., McQuade, R. S., and Grossman, A. D. An autoregulatory circuit affecting peptide signaling in *Bacillus subtilis*. *J. Bacteriol.*, 181, 5193–5200 (1999).

111. Mueller, J. P., Bukusoglu, G., and Sonenshein, A. L. Transcriptional regulation of *Bacillus subtilis* glucose starvation-inducible genes: control of *gsiA* by the ComP-ComA signal transduction system. *J. Bacteriol.*, 174, 4361–4373 (1992).

112. Ogura, M., Yamaguchi, H., Yoshida, K., Fujita, Y., and Tanaka, T. DNA microarray analysis of *Bacillus subtilis* DegU, ComA and PhoP regulons: an approach to comprehensive analysis of *B. subtilis* two-component regulatory systems. *Nucleic Acids Res.*, 29, 3804–3817 (2001).

113. Ogura, M., and Tanaka, T. Transcription of *Bacillus subtilis degR* is σ_D dependent and suppressed by multicopy *proB* through σ_D. *J. Bacteriol.*, 178, 216–222 (1996).

114. Tokunaga, T., Rashid, M. H., Kuroda, A., and Sekiguchi, J. Effect of *degS-degU* mutations on the expression of *sigD*, encoding an alternative sigma factor, and autolysin operon of *Bacillus subtilis*. *J. Bacteriol.*, 176, 5177–5180 (1994).

115. Kunst, F., and Rapoport, G. Salt stress is an environmental signal affecting degradative enzyme synthesis in *Bacillus subtilis*. *J. Bacteriol.*, 177, 2403–2407 (1995).

116. Steil, L., Hoffman, T., Budde, I., Volker, U., and Bremer, E. Genome-wide transcriptional profiling analysis of adaptation of *Bacillus subtilis* to high salinity. *J. Bacteriol.*, 185, 6358–6370 (2003).

117. Prigent-Combaret, C., Vidal, O., Dorel, C., and Lejeune, P. Abiotic surface sensing and biofilm-dependent regulation of gene expression in *Escherichia coli*. *J. Bacteriol.*, 181, 5993–6002 (1999).

118. Msadek, T., Kunst, F., Henner, D., Klier, A., Rapoport, G., and Dedonder, R. Signal transduction pathway controlling synthesis of a class of degradative enzymes in *Bacillus subtilis*: expression of the regulatory genes and analysis of mutations in *degS* and *degU*. *J. Bacteriol.*, 172, 824–834 (1990).

119. Yang, M., Ferrari, E., Chen, E., and Henner, D. J. Identification of the pleiotropic *sacQ* gene of *Bacillus subtilis*. *J. Bacteriol.*, 166, 113–119 (1986).

120. Ueda, T. A procedure for high spore density cultivation by *Bacillus subtilis* RB14, master's theses, Tokyo Institute of Technology (2007).

121. Japan patent, No. 4-234981. Mitui-Toatu Chemical Co., Ltd. High-density cultivation system of *Bacillus subtilis* (1992).

122. Comella, N., and Grossman, A. D. Conservation of genes and processes controlled by the quorum response in bacteria: characterization of genes controlled by the quorum-sensing transcription factor ComA in *Bacillus subtilis*. *Mol. Microbiol.*, 57, 1159–1174 (2005).

8 Surfactin Production and Plasmid Stability

8.1 SURFACTIN PRODUCTION BY A RECOMBINANT *B. SUBTILIS* IN LIQUID CULTIVATION (1)

Bacillus subtilis MI113 is a derivative of the standard strain Marburg 168. *B. subtilis* RB14 is a strain that exhibits suppressibility against several plant pathogens *in vitro* and *in vivo* by producing lipopeptide antibiotics iturin and surfactin. A genetic locus that conferred the ability to produce surfactin was cloned from RB14 and reduced in size in *B. subtilis* MI113 in this laboratory (2). The productivity of surfactin by the recombinant strain, B. subtilis MI113(pC112) was one and half times larger than that by RB14. A recombinant strain, *B. subtilis* MI113(pC194), was used to confirm no surfactin production of the host strain holding vector alone.

8.1.1 MATERIALS AND METHODS

8.1.1.1 Strain and Plasmid
B. subtilis RB14 and MI113 and plasmids pC112 and pC194 (3) were used.

8.1.1.2 Cultivation of *B. subtilis* MI113(pC112)
Each culture of *B. subtilis* MI113(pC194), MI113(pC112), and RB14 was grown in no. 3S medium in submerged fermentation (SmF). Main cultivation was carried out under the same conditions except that the medium contained 30 g/l Polypepton S. For the cases of MI113(pC112), cultivations were carried out with or without chloramphenicol (Cm) selection.

8.1.1.3 Measurement of Cell Concentration
Cell concentration was determined by viable cell count on no. 3 medium agar plates, where Polypepton S was replaced by Polypepton.

8.1.1.4 Evaluation of Plasmid Stability
Plasmid stability of the recombinant strain was evaluated by replica plating of the 100 individual colonies onto no. 3 medium agar plates supplemented with or without 5 μg/ml chloramphenicol followed by the incubation at 30°C for 18 h before counting the colonies.

8.1.1.5   Extraction and Measurement of Surfactin and Iturin

Cultivations were carried out in individual flasks, each containing 40 ml of the culture. One culture in a flask at a time was terminated at 7, 24, 32, 48, 72, 96, and 120 h, and transferred to 100 ml conical flasks, acidified with concentrated HCl to pH 2.0, and kept at 4°C for peptide precipitation. Then, the culture broths were centrifuged for 10 min at $11,000 \times g$ at 4°C. The precipitate was extracted with 10 ml methanol at room temperature for more than 1.5 h. The crude extract was filtered with a 0.20 μm polytetrafluoroethylene (PTFE) filter (JP020; Advantec Toyo Ltd., Tokyo). The filtrate was injected into high-performance liquid chromatography (HPLC) (JASCO 880 PU) using an ultraviolet (UV) detector (JASCO 870 UV) to determine the concentration of surfactin at 205 nm. A calibration curve was prepared by using pure surfactin (Wako Pure Chemicals Co., Osaka).

Conditions for HPLC were as follows: mobile phase, acetonitrile: 1% (w/w) acetic acid = 4:1 (v/v); column, 4.6 mm Φ × 250 mm ODS-2 (GL Sciences, Tokyo); column temperature, 30°C; flow rate, 1.5 ml/min.

Concentration of iturin was determined by HPLC with modification to the method previously described in Chapter 2 as follows: mobile phase, acetonitrile: 10 mM ammonium acetate = 3:4 (v/v); column, 4.6 mm Φ × 250 mm ODS-2 (GL Sciences, Tokyo); column temperature, 30°C; flow rate, 1.0 ml/min.

8.1.2   Results

8.1.2.1   Time Course of Surfactin Production by B. *subtilis* MI113(pC112) and RB14

Figure 8.1 shows the time course of surfactin production by *B. subtilis* MI113(pC112) and RB14. MI113(pC112) was grown in the medium with or without chloramphenicol. Both RB14 and MI113(pC112) attained maximum surfactin concentration at 24 h, and thereafter the concentration remained constant throughout cultivation. Maximum surfactin concentration was larger for MI113(pC112) than for RB14. No surfactin was produced by the host strain with the vector plasmid, MI113(pC194).

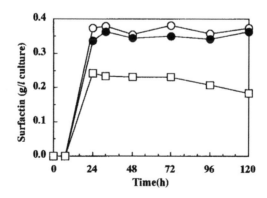

FIGURE 8.1   Time courses of surfactin production by *B. subtilis* MI113(pC112) and RB14. Symbols: o, MI113(pC112)(+Cm); ●, MI113(pC112)(–Cm); □, RB14.

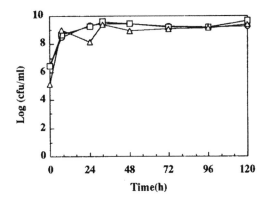

FIGURE 8.2 Time courses of cell concentration in viable cell count. Symbols: o, MI113(pC112)(+Cm); □, MI113(pC112)(−Cm); Δ, RB14.

8.1.2.2 Time Course of Cell Concentration

Cell concentrations of these strains in the number of viable cells are shown in Figure 8.2. As viable cell numbers were almost the same in MI113(pC112) with/ without Cm and RB14, the enhanced productivity of surfactin was obvious as shown in Figure 8.1.

8.1.2.3 Plasmid Stability

Time course of plasmid stability is shown in Figure 8.3. During the 120 h of the experiment, plasmid stability was 62%–82%. Since the accumulation of surfactin reached maximum in 24 h as shown in Figure 8.1 in the recombinant strain with/ without Cm, plasmid stability during this period had little effect on the production of surfactin. Therefore, surfactin production by MI113(pC112) is possible without selective pressure of chloramphenicol.

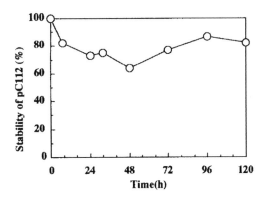

FIGURE 8.3 Time course of plasmid stability of pC112 in no. 3S medium with 3% Polypepton S. Plasmid stability was measured by replica plating.

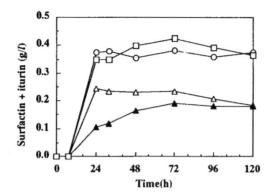

FIGURE 8.4 Time courses of total of surfactin and iturin concentrations by *B. subtilis* MI113(pC112) and RB14. Symbols: o, surfactin by MI113(pC112)(+Cm); △, surfactin by RB14;▲, iturin by RB14; □, surfactin + iturin by RB14.

The productivity of a lipopeptide antibiotic surfactin by the recombinant strain *B. subtilis* MI113(pC112) was larger than that by *B. subtilis* RB14, in which the surfactin producing gene originated.

Actually, RB14 produces another lipopeptide antibiotic iturin, but the recombinant strain was only a surfactin producer. The total amount of lipopeptides, surfactin and iturin, by RB14 was nearly equal to the amount of surfactin produced by the recombinant strain (Figure 8.4). As these two compounds have similar structures (Figure 3.7), competitive distribution of the components, such as fatty acids used for the side chain, between the two lipopeptides, by RB14 may have disappeared in the recombinant strain.

Surfactin is the most effective biosurfactant so far (4), and as such may have many potential industrial applications, which include uses related to oil recovery, emulsion stabilization, etc. Biosurfactants provide an advantage over synthetic surfactants because most are biodegradable and generally less toxic.

Our recombinant strain is suitable for surfactin production because similar by-products are not produced along with surfactin.

8.2 SURFACTIN PRODUCTION AND PLASMID STABILITY IN SOLID-STATE FERMENTATION (SSF) (5)

Production of a lipopeptide antibiotic, surfactin, in solid-state fermentation (SSF) on soybean curd residue, okara, as a solid substrate was carried out using *B. subtilis* MI113 with a recombinant plasmid pC112, which contains *lpa-14*, a gene related to surfactin production cloned from a wild-type surfactin producer, *B. subtilis* RB14. The optimal moisture content and temperature for the production of surfactin were 82% and 37°C, respectively. The amount of surfactin produced by MI113 (pC112) was as high as 2.0 g/kg wet weight, which was eight times as high as that of the original *B. subtilis* RB14 at the optimal temperature for surfactin production, 37°C.

Although the stability of the plasmid showed a similar pattern in both SSF and sub-merged fermentation (SmF), production of surfactin in SSF was four to five times more efficient than in SmF.

8.2.1 MATERIALS AND METHODS

8.2.1.1 Strains and Recombinant Plasmid Used

B. subtilis MI113 (*arg*-15 *trpC2 hsmM hsrM*) is a derivative of Marburg 168 which produces no surfactin. *B. subtilis* RB14 is a strain that exhibits suppressibility against several plant pathogens *in vitro* by producing lipopeptide antibiotics iturin A and surfactin (6). A genetic locus, *lpa-14*, which is related to the production of surfactin and iturin A, was cloned from RB14 and reduced in size in *B. subtilis* MI113 (6). The structure of the resulting recombinant plasmid, pC112, is shown in Figure 8.5 (6, 7).

The recombinant plasmid holds two *Hind*III fragments with sizes 2700 and 590 bp as indicated in the figure, between a unique *Hind*III site of the vector plasmid pC194 which expresses chloramphenicol resistance (Cm'). Both its exact location and its sequence were determined by Huang et al. (7). The plasmid, pC112, was transformed into *B. subtilis* MI113 and the transformant, *B. subtilis* MI113 (pC112), gained an ability to produce surfactin.

8.2.1.2 Preparation of Seeding Culture

Each culture of *B. subtilis* MI113(pC112) and RB14 grown in no. 3S medium was mixed with glycerol at 1:1 (w/w) and stored at −40°C. A 0.1 ml portion of the glycerol

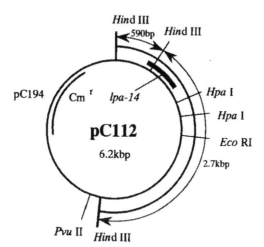

FIGURE 8.5 Structure of a recombinant plasmid pC112. The *Hind*III fragment shown in a solid arc is the vector, pC194, and Cm' indicates the chloramphenicol resistance gene of the vector. The largest *Hind*III fragment shown in a white box is pIB111, obtained by Hiraoka et al. (6), from *B. subtilis* RB14 chromosome. The exact location (shown as a black bar) and the sequence of a gene, *lpa-14*, associated with the production of surfactin, was determined by Huang et al. (7).

suspension was inoculated into a 500 ml shaken flask containing 100 ml of no. 3S medium (0.1% inoculation) and cultivated at 30°C at 120 strokes/min as preculture. When MI113(pC112) was inoculated, additionally chloramphenicol was added at 5 µg/ml.

8.2.1.3 Submerged Fermentation (SmF) in No. 3S Medium Modified to 3% Polypepton S

After 20–24 h cultivation of preculture, 40 µl was inoculated into each of the 200 ml conical flasks containing 40 ml of 3S medium, modified to 3% (30 g/l) Polypepton S. They were shaken at 30°C at 120 rpm and the whole culture of a flask was terminated at a desired time for the evaluation of surfactin production and plasmid stability in SmF.

8.2.1.4 Solid-State Fermentation

Soybean curd residue, okara, consists mostly of water-insoluble components of soybean, since the water-soluble portion of soybeans is separated from okara in aqueous suspension known as soy milk, which is subjected to the tofu manufacturing process. Table 8.1 summarizes the compositions of Polypepton S and okara by elemental analysis. Polypepton S was used for the seeding culture for the better growth and productivity of surfactin (data not shown).

Okara (150–200 g, initial moisture content of 77.6%) was placed in a glass petri dish (diameter: 25 cm, depth: 6 cm) and autoclaved twice at 120°C for 30 min at an interval of 8–12 h. Interval sterilization was necessary to kill the spore-forming microorganisms inhabiting the material. After autoclaving, 15 g each of okara was placed into a 100 ml sterile conical flask with a cotton plug. For the fortification of nutrients (8), 833 µl aqueous solution of 0.45 g/ml glucose, 75 µl of 1 M KH_2PO_4, and 225 µl of 1 M Mg SO_4 were added. Moisture content of the okara after this treatment was 82%. Sterile distilled water was added to vary the moisture content of the sterile okara up to 95% according to experimental design. Partially dried okara with the moisture content of 39% provided by Fukuyama Technical Exchange Center (Fukuyama, Hiroshima) was used likewise to investigate the effect of moisture content at 37°C after soaking overnight with an appropriate amount of distilled water before autoclaving. The preculture was mixed for seeding thoroughly with a sterile stainless spatula at a ratio of 3 ml to every 15 g of okara with the moisture content of 77.6%. Cultivation was carried out for 48 h at various temperatures. Viable cell

TABLE 8.1
Elemental Analysis of Lyophilized Okara as Compared with Polypepton S

	Carbon (%)	Hydrogen (%)	Nitrogen (%)	Sulfur (%)	Ash (%)
Okara	46.34	6.99	3.99	0.25	3.59
Polypepton S	34.68	6.10	7.63	0.47	13.02

number, plasmid stability, and surfactin production were measured as described in the following paragraphs at the specified time.

8.2.1.5   Measurement of Viable Cells

For SmF, viable cells were counted at 8, 12, 32, and 48 h by sampling 1 ml of the culture at a time and plating onto no. 3S medium agar plates followed by serial dilution. For SSF, viable cell number was estimated by colony counting at 12 h intervals up to 48 h as follows. One gram each of the solid cultures was placed in a sterile 18 mm Φ test tube, mixed thoroughly with 9 ml of sterile distilled water with a vortex, diluted serially up to 10^{-6}, and spread onto the same agar plate. After 18 h of incubation at 30°C, the number of colonies appeared was counted.

8.2.1.6   Evaluation of Plasmid Stability

Plasmid stability of the recombinant strain at different temperatures of main cultivation was evaluated by replica plating in both SSF and SmF. One hundred individual colonies that appeared on the no. 3S agar medium described earlier were randomly selected and replica-plated with wooden toothpicks onto no. 3S medium agar plates supplemented with or without 5 μg/ml chloramphenicol and incubated for 18–20 h at 30°C. Then the number of colonies was counted and those grown on the agar plates containing chloramphenicol were regarded to hold the plasmid.

8.2.1.7   Extraction and Measurement of Surfactin

Crude extract of surfactin from SMF culture was prepared as follows: whole culture (40 ml) was harvested at 24 and 48 h of cultivation, acidified with concentrated HCl to adjust pH to 2, and cooled to 4°C overnight for peptide precipitation. Then, the precipitate was collected by centrifugation at 15,500 × g at 4°C and extracted with 10 ml methanol for 1.5 h at room temperature. In the case of SSF, crude extract of surfactin was isolated by adding 45 ml of methanol to the fermented product of 15 g of intact okara, i.e., 3.36 g of initial dry weight of the intact okara, and the mixture was shaken at 92 strokes/min for 60 min with a reciprocal shaker (Eyelashaker SS-8, Tokyo Rikakikai Co., Tokyo). In either case, the crude extract was filtered with a 0.20 μm PTFE membrane filter (JP020; Advantec Toyo Ltd., Tokyo) and 20 μl of the filtrate was injected into an HPLC column (880 PU; Japan Spectroscopic Co., Tokyo) using a UV detector (870 UV; Japan Spectroscopic Co., Tokyo) to determine the concentration of surfactin at 205 nm. A calibration curve was prepared by using the commercial product of surfactin (Wako Pure Chemicals Co., Osaka, Japan) purified to more than 95% to exhibit a single peak by our HPLC system. Conditions for HPLC were as follows: mobile phase, acetonitrile: 1% (w/w) acetic acid = 4:1 (v/v); column, 4.6 mm Φ × 250 mm ODS-2, a reverse-phase, hydrophobic column (GL Sciences, Tokyo); column temperature, 30°C; flow rate, 1.5 ml/min. The mobile phase was modified later with 3.8 mM trifluoroacetic acid in place of 1% acetic acid to improve the quantification method.

8.2.1.8   Estimation of the Soluble Nutrients

Soluble carbon and nitrogen in SSF were estimated to be 24.4 and 1.72 g/kg wet weight, respectively, by the following procedures: The water-soluble fraction of okara was obtained after the homogenization of okara at 10,000 rpm for 10 min with a homogenizer (Excel Auto, Nissei Seisakusho, Tokyo) and centrifugation at 15,500 × g for 10 min. The supernatant was dried and the carbon and nitrogen contents were measured by elemental analysis.

8.2.2 RESULTS

8.2.2.1 Effect of Moisture Content on Surfactin Production by *B. subtilis* MI113(pC112)

Figure 8.6 shows the effect of initial moisture content on relative surfactin productivity with okara and partially dried okara at 37°C, where the maximum value of each on a dry and wet basis was set at 100%. It turned out that 70% was optimal on a wet basis. However, 82% was adopted for further experiments because of the following two reasons. One is that moisture content of the intact okara produced by the tofu factory was 77%, which is close to the moisture content of 70%. The other is that relative surfactin productivity on a dry basis was highest with the moisture content of 82%, as shown in Figure 8.6, which implies highest per-unit dry mass of the substrate. Little change in moisture content was observed during the 48 h period. The wheat bran in the range of 45%–90% of moisture content showed only 70%–80% of surfactin productivity in comparison with that of okara (data not shown).

8.2.2.2 Effect of Temperature on Surfactin Production by *B. subtilis* MI113(pC112) and RB14

Effect of temperature on surfactin production of the transformant *B. subtilis* MI113 (pC112) and the original strain RB14 is shown in Figure 8.7. Initial moisture content was fixed at 82% in accordance with the former result. A maximum of 1.8–2.0 g

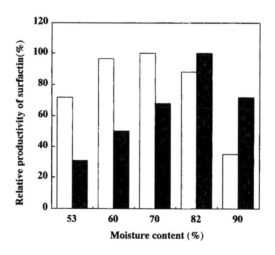

FIGURE 8.6 Effect of initial moisture content on relative productivity of surfactin in the solid-state fermentation of okara after 48 h of cultivation of a recombinant *B. subtilis* MI113 (pC112) at 37°C. White bar, wet basis; shaded bar, dry basis.

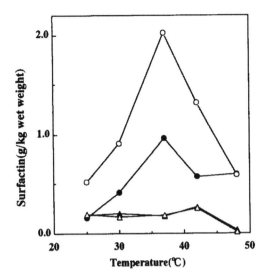

FIGURE 8.7 Effect of temperature on surfactin production in solid-state fermentation of okara by a recombinant strain *B. subtilis* MI113(pC112) at (●) 24 h, (○) 48 h; and by the original strain, *B. subtilis* RB14, at (▲) 24 h and (△) 48 h, in this order.

surfactin/kg wet weight was attained by MI113(pC112) at 37°C at 48 h, but the yield significantly decreased at other temperatures. The productivity of RB14 was maintained at a relatively lower level in the temperature ranges tested.

8.2.2.3 Effect of Temperature on Viable Cell Count of *B. subtilis* MI113(pC112) and RB14 and Plasmid Stability in MI113(pC112)

Figures 8.8a and b show the viable cell counts of transformant, MI113(pC112) and RB14, respectively, in SSF. Figure 8.8c is the plasmid stability of MI113(pC112) in SSF after 12, 24, and 48 h of cultivation at temperatures 30°C, 37°C, and 42°C, while Figure 8.9 shows the plasmid stability in SmF at 42°C. Figure 8.8a indicates that the maximum cell population reached in the 24 h period of SSF was around 10^{10} colony-forming units (cfu)/g wet weight at 37°C–42°C and, in 48 h, decreased to the order of 10^9 cfu/g wet weight, mainly due to autolysis. The cell number at 30°C showed a gradual increase to 7×10^9. Since the viable count of cells of MI113(pC112) at 37°C is nearly one order of magnitude higher than that at 25°C (data not shown), the amount of surfactin produced is associated with the cell number. The cell number at 42°C was similar to that at 37°C, but loss of productivity at 42°C is partly due to the plasmid loss from 30% to 60% as the cultivation temperature rises from 37°C to 42°C (Figure 8.8c). A wild-type strain, RB14, grew quickly in 12 h at 30°C, followed by a lytic decrease later on, while at 37°C–42°C, the constant increase in cell number until 48 h from 10^9 to 10^{10} cfu/g wet weight was observed, as shown in Figure 8.8b. These cell numbers in RB14 were not directly associated with the productivity of surfactin, as shown in Figure 8.7, although the reason for the lower productivity of RB14 in SSF is unknown at present.

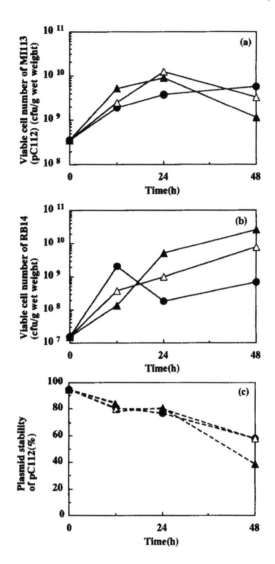

FIGURE 8.8 Count of viable cells of *B. subtilis* (a) MI113(pC112) and (b) RB14, and (c) plasmid stability of pC112 at 30°C, 37°C, and 42°C in the solid-state fermentation of okara: (•) 30°C; (Δ) 37°C; (▲) 42°C.

8.2.3 DISCUSSION

It was demonstrated that a recombinant *B. subtilis*, an obligate aerobe, produced a biosurfactant, surfactin, on a solid-state fermentation of food processing waste, okara, with higher productivity in comparison with the wild strain, RB14.

Surfactin production was maximum at 48 h in spite of the maximal cell number being achieved at 24 h because surfactin, a typical secondary metabolite (4, 9), is deliberately produced at the late stationary or the declining phase.

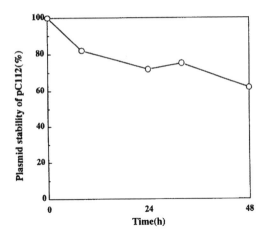

FIGURE 8.9 Time course of plasmid stability of pC112 in no. 3S medium with 3% Polypepton S in submerged fermentation at 42°C.

Several reasons why the productivity in SSF was higher than that in SmF of the same recombinant can be given. In SmF, the nutrients and oxygen are homogeneously distributed in the liquid medium, and this accelerates the growth of the cells rather than production of surfactin as a secondary metabolite. Actually, the logarithmic growth phase in SmF lasted for about 10 h. However, the growth rate in SSF for 10 h was much lower than that in SmF. This indicates that the bacteria in SSF are left to consume only the portion of nutrients gradually dissolving into water from the solid substrate and the cell division occurs on the surface of solid. This may easily cause limitation of nutritional supply, and thus this triggers rather slower growth and the more accelerated synthesis of the secondary metabolites than that in SmF.

Contents of soluble carbon and nitrogen were compared in both SSF and SmF. Comparison between SmF and SSF is based on the fact that 1 g of okara is almost equivalent to 1 ml in volume, hence, 1 g surfactin/l in SmF is approximately 1 g surfactin/kg wet weight in SSF. In SmF, no. 3S medium modified to 3% Polypepton S was used as a standard for SmF. The soluble fractions of carbon and nitrogen in the modified no. 3S medium were calculated from the data in Table 8.1 to be 14.4 and 2.29 g/l, respectively. Although 70% more carbon was found in SSF, the amount of surfactin produced by MI113 (pC112) in SSF was sixfold of that by the same strain in SmF in Chapter 5 (10). The amount of surfactin produced, 1.8–2.0 g/kg wet weight at this experiment, was 10–20 times as much as that reported in SmF as shown in Table 8.2. Such enhanced productivity may be partly due to the overcoming (11) of end-product inhibition of surfactin (4) in SSF, although the detail of the mechanism is not clear.

It is known that the gene *lpa-14* plays a role as a dual controller to the production of surfactin and iturin A, another lipopeptide antibiotic (8, 15, 16), in RB14. This gene works as a single controller of surfactin in *B. subtilis* MI113, which lacks at least the function of iturin A gene (6, 7). This may result in the enhanced production of surfactin as compared with RB14.

TABLE 8.2

Productivity of Surfactin in Various Media

Strain	Medium	Type of Fermentation	Surfactin Production (mg/l)	Reference
B. subtilis ATCC21332	Semi-synthetic	SmF	100	(9)
B. subtilis ATCC21332	Synthetic	SmF	250	(4)
B. subtilis ATCC21332	Synthetic	SmF	800 (only in the foam)	(4)
B. subtilis ATCC21332	Peat hydrolysate	SMF	160	(12)
Suf-1, a mutant of ATCC21332	Synthetic	SmF	550	(13)
B. subtilis ATCC21332	Synthetic	SmF, aqueous two-phase	350	(14)
ATCC55033, a mutant of ATCC21332	Semi-synthetic	SmF	3500–4300	(5)
B. subtilis RB14	Semi-synthetic	SmF	250	(10)
B. subtilis RB14	Okara	SSF	200–250 (mg/kg wet weight)	This work
B. subtilis MI113 (pC112)	Semi-synthetic	SMF	350	(10)
B. subtilis MI113 (pC112)	Okara	SSF	2000 (mg/kg wet weight)	This work

Notes: Quantification in this work is on a single homologue of surfactin. SmF, submerged fermentation; SSF, solid-state fermentation.

The effect of temperature on the surfactin production was significant, as shown in Figure 8.7. Rising the temperature accelerated the plasmid loss, as shown in Figure 8.8c, in SSF. A similar pattern of instability of the plasmid was also observed in SmF. About 60% of viable cells held the plasmid at 48 h in SmF (Figure 8.9) and the loss increased as the temperature rose (data not shown). This indicates the stable maintenance of plasmid is a factor in increasing the productivity of surfactin in SSF.

Conclusively, the recombinant strain *B. subtilis* MI113(pC112) surpassed the original strain *B. subtilis* RB14 in the production of surfactin, with productivity of 10 times or higher than the values reported in previous papers as shown in Table 8.2. As there is no nuisance of foaming in SSF, the production of biosurfactant will be promising if proper design of reactors and operation are materialized.

8.3 NEW VARIANTS OF SURFACTIN IN SOLID-STATE FERMENTATION BY A RECOMBINANT *BACILLUS SUBTILIS* (17)

8.3.1 MATERIALS AND METHODS

8.3.1.1 Strains and Medium Used

Plasmid pC115, which carries *lpa-14*, a gene related to surfactin and iturin A production, was cloned from a wild-type surfactin and iturin A coproducer, *B. subtilis* RB14-C. This plasmid was transformed into *B. subtilis* MI113, a nonproducer of surfactin, by the competent-cell method as described in Chapter 5 (6, 7), and *B. subtilis* MI113(pC115) gained the ability to produce surfactin.

The medium used for precultivation of the transformant was no. 3S medium as described in Section 8.2. As the plasmid pC115 holds a chloramphenicol-resistance site, 5 µg chloramphenicol/ml ethanol solution was added to 100 ml medium to select growth of only the recombinant MI113(pC115). Precultivation was performed at 30°C at 120 strokes/min, for 24 h.

Okara, which is a solid waste of tofu production, was used as a solid substrate, and elemental analysis of the okara (dry basis) showed C 46.3, H 6.99, N 3.99, S 0.25, and ash 3.59%. A 15 g sample (wet weight; initial moisture content of 84%) of okara was placed in a 100 ml conical flask with a cotton plug and autoclaved twice at 120°C for 20 min with an interval of 7–8 h. A 3 ml sample of the preculture was mixed with the sterilized okara and incubated for 48 h at 37°C statically.

8.3.1.2 Extraction and Analysis of Surfactin Variants

Three equivalent volumes of methanol were lidded to fermented okara and the mixture was shaken at 92 shakes/min for 1 h with a reciprocal shaker to extract surfactin. The crude extract was then centrifuged at 4000 rpm for 20 min, the methanol was evaporated, the residue was suspended in acetone, and surfactin was eluted by acetone through the activated carbon column. Distilled water was added to the acetone eluate in the ratio of 15:13 (v/v), and the solution was cooled in a refrigerator to crystallize crude surfactin. After evaporation of acetone, crude white powdered surfactin was obtained. Nine surfactin variants were fractionated by HPLC (Type l-4000W; Yanako Co. Ltd. Kyoto, Japan) column (Prep-Ods, 20 mm diameter × 250 mm long; GL Science Co. Ltd, Yokohama, Japan) and detected by a UV monitor (Shodex M-315; Showadenko Co. Ltd., Tokyo, Japan) at 205 nm. Conditions for HPLC were as follows: mobile phase: 1.8 mM trifluoroacetic acid:acetonitrile = 1:4 (v/v); flow rate: 7.5 ml/min (7). Each fraction as applied to a Sephadex LH-20 column (10 mm diameter × 510 mm long, flow rate 0.1 ml/min) and eluted by acetone to remove small amounts of contaminants. Then each fraction was subjected to fast atom bombardment mass spectrometry (FAB-MS) and amino acid analysis.

8.3.2 RESULTS

A yield of 4.4 g crude surfactin was obtained from 1 kg wet okara (84% moisture content). Figure 8.10 shows a preparative HPLC chart with the designated peak numbers

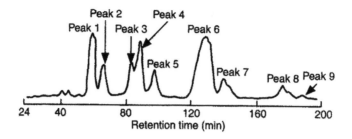

FIGURE 8.10   Preparative HPLC chart of crude surfactin.

of nine variants. Table 8.3 summarizes the results of FAB-MS and amino acid analysis for each purified peak sample fractionated by HPLC. The molar ratio relative to Asp (which is denoted A in Table 8.3) is calculated for each peak by putting Asp equal to 1. The peak 6 sample in Figure 8.10 showed the same infrared (IR) spectrum as authentic surfactin (Wako Chemicals,Tokyo) with characteristic bands of peptides (3300 cm^{-1} NH, 1650 cm^{-1} CO, and 1520 cm^{-1} CN), aliphatic chains (2960, 2930, and 2850 cm^{-1} CH$_2$ and CH$_3$), and the ester carbonyl group (1730 cm^{-1} CO) (data not shown). With reference to the IR data, amino acid composition, and FAB-MS results, peak 6 is judged to be the standard surfactin consisting of seven peptides and a lipophilic part comprising *iso*-, *anteiso*-, or *n*-C$_{15}$-β-hydroxy fatty acid. As the IR spectrum of crude surfactin containing all variants showed the same bands as peak 6, peaks 1–5 and 7–9 were considered as surfactin variants. The presence of two kinds of peptide rings in peaks 2, 5, and 7 (Figure 8.10) was suggested because Ile was detected by amino acid analysis and two molecular ion peaks appeared in their FAB-MS spectra. On the basis of the molar ratio relative to Ile, the amino acid compositions of two surfactin variants in those peaks were estimated. The detailed estimation in the case of peak 2 is as follows. The observed molar ratio of the amino acid components is shown as A in Table 8.3. The amino acid composition of a surfactin variant containing Ile has been reported to be Glu:Leu:Val:Asp:Ile = 1:3:1:1:1 (19), therefore, the amino acid composition of the surfactin variant containing Ile, which is denoted B in Table 8.3, should be Glu:Leu:Val:Asp:Ile = 0.3:0.9:0.3:0.3:0.3. The residual amino acid composition, namely, A minus B, is Glu:Leu:Val:Asp = 0.7:2.1:1.3:0.7, which is shown as C in Table 8.3. Values denoted in brackets under B and C are the relative molar ratios of surfactin variants B and C when Asp is set equal to 1. Since C contains additional Val instead of the Ile in B, and the values for *m/z* of B and C were 1008 and 994, respectively, the difference of 14 between these values is explained by the difference of molecular mass between Val(117) and Ile(131). This indicates that B and C have the same carbon number in the aliphatic side chain. The relative ratios of B/C in peak 2 can be estimated as 3/7 from the ratio of Asp in this peak. The same procedure was applied to peaks 5 and 7, and the amino acid compositions of each mixture are shown in Table 8.3. Peak 5 consisted of B and D. B contains three units of Leu and one unit of Ile, while D contains four units of Leu. As the molecular masses of the amino acid rings in B and D are the same, the two *m/z* values of 1008 and 1022 are judged to be due to the different numbers of aliphatic side chains.

TABLE 8.3

Fast Atom Bombardment Mass Spectrometry and Amino Acid Analysis of Isolated Surfactin Variants

Peak Number in Figure 8.10		Glu	Leu	Val	Asp	Ile	FAB-MS Spectra (m/z)	
		\textit{Amino Acid Components (Molar Ratio)}					\textit{FAB-MS Spectra (m/z)}	
1	A	1.0	4.0	1.0	1.0		$1008(C_{13})^b$	
2	A	1.0	3.0	1.6	1.0	0.3		
B:C = 3:7[a]	B	0.3	0.9	0.3	0.3	0.3	$1008(C_{13})$	
		(1.0	3.0	1.0	1.0)	1.0)		
	C	0.7	2.1	1.3	0.7		$994(C_{13})$	
		(1.0	3.0	1.9	1.0)			
3	A	1.1	3.0	1.1	1.0	1.0	$1008(C_{13})$	
4	A	1.0	4.0	1.0	1.0		$1022(C_{14})$	
5	A	1.0	3.4	1.3	1.0	0.4		
	B	0.4	1.2	0.52	0.4	0.4	$1008(C_{13})$	$1022(C_{14})$
		(1.0	3.0	1.3	1.0	1.0)		
	D	0.6	2.2	0.78	0.6		$1008(C_{13})$	$1022(C_{14})$
		(1.0	3.7	1.3	1.0)			
6	A	1.0	4.2	1.0	1.0		$1036(C_{15})$	
7		1.2	3.3	1.9	1.0	0.3		
B:C = 3:7[a]	B	0.36	0.96	0.36	0.3	0.3	$1036(C_{15})$	
		(1.2	3.2	1.2	1.0	1.0)		
	C	0.84	2.34	1.54	0.7		$1002(C_{15})$	
		(1.2	3.3	2.2	1.0)			
8	A	1.1	3.0	1.0	1.0		$1036(C_{15})$	$1050(C_{16})$
9	A	1.2	3.3	1.2	1.0		$1050(C_{16})$	

Parentheses indicate the relative molar ratio of amino acids in surfactin variants at Ile = 1 or Asp = 1.

[a] Relative ratio of surfactin variants B/C in the mixture.

[b] The carbon number of the fatty acid assumed from amino acid components and *m/z* value in each surfactin variant is shown in parenthesis in the final column.

A–Data by amino acid analysis carried out with each peak sample as shown in Figure 8.1.

B–Surfactin variant with amio acid ratio of Glu:Leu:Val:Asp:Ile = 1:3:1:1:1.

C–Surfactin variant with amino acid ratio of Glu:Leu:Val:Asp = 1:3:2:1.

D–Surfactin variant with amino acid ratio of Glu:Leu:Val:Asp = 1:4:1:1.

8.3.3 DISCUSSION

The structures of surfactin variants produced by solid-state fermentation of recombinant *B. subtilis* were determined. As peak 6 is a standard surfactin, peaks 2, 3, 5, and 7–9 contained variants where the Leu moiety of the peptides ring was replaced by Ile, while peaks 2 and 7 had variants where Leu was substituted by Val. Peak 9 is a newly isolated single peak that has one Ile and α- C_{16} fatty acid chain. The existence of a surfactin variant having a fatty acid moiety of C_{16} was suggested by the fatty acid analysis of the products of *Bacillus licheniformis* BAS 50 (20), but it had not actually been isolated. The reason why these variants are produced in solid-state fermentation is mainly the relatively high cell density (10^{10}–10^{11} cfu/g solid) and the high accumulation of surfactin compared with liquid culture that were attained. It is not known which Leu in the surfactin structure is replaced by Ile or Val. Recently studies on the surfactin synthetases (22–26) have shown that three subunits—SrfAORFl, SrfAORF2, and SrfAORF3—in the multienzyme complex for surfactin synthesis recognize L-Glu$^{(1)}$-L-Leu$^{(2)}$-D-Leu$^{(3)}$, L-Val$^{(4)}$-L-Asp$^{(5)}$-D-Leu$^{(6)}$, and L-Leu$^{(7)}$, respectively. The amino acid–binding sites in SrfAORFl, SrfAORF2, and SrfAORF3 can recognize more than one amino acid with different affinity, and a slight change of activity of SrfAORFl or SrfAORF2 with Val or Ile was measured (23). SrfAORF3 was purified without cross-contamination and changes in the kinetic data for SrfAORF3 with Leu, Ile, and Val were reported as Ks = 0.84, 2.91, and 1.89, and V_{max} = 10058, 4051, and 936 cpm/min, respectively (23). These data indicate that the effect of amino acid substitution is mainly due to L-Leu$^{(7)}$, and this reasonably explains the amount of the Ile variant being larger than that of the Val variant.

8.4 PLASMID STABILITY IN *B. SUBTILIS* NB22 (27)

Although *B. subtilis* plasmid stabilities in nutrient-rich liquid culture have been studied intensively and genetic elements are indicated to be required for stable maintenance (33, 34, 36), no reports have appeared on plasmid stability under conditions of nutritional limitation and sporulation. The elucidation of plasmid stability during sporulation, as well as during vegetative growth, will be important in field application of genetically manipulated microorganisms by furthering our understanding of the partitioning of plasmid in spore-forming bacteria.

8.4.1 MATERIALS AND METHODS

8.4.1.1 Strain Used

B. subtilis NB22-l, a spontaneous streptomycin-resistant mutant of NB22 (8), was used. The growth rate and iturin productivity of NB22-1 were confirmed to be the same as those of the parental strain NB22 (data not shown). NB22-1 was transformed with the plasmids pC194 and pUB110 by the alkali metal ion treatment method (Chapter 5) (18). The stability of the plasmid in no. 3 medium was investigated over 100 generations at 30°C in successive cultivations by diluting the culture 10^{-6} times every 12 h.

8.4.2 RESULTS

8.4.2.1 Stability of pC194

Plasmid pC194 was kept stably in NB22-1 in the presence or absence of chloramphenicol in no. 3 medium, as in other *B. subtilis* strains (28). The fraction of plasmid-carrying cells was 98% of the total population after 100 generations: The plasmid loss rate was calculated to be 0.02% per cell generation. Plasmid stability in a prolonged cultivation with nutritional limitation was investigated. Vegetative cells, confirmed to be without spores, were inoculated into no. 3 medium at a ratio of 1%, and cultivated for 10 days. The plasmid was stably maintained on the 1st day, but gradual curing was then observed, and the plasmid stability decreased to 28% by the 10th day, which was remarkably lower than the stability in the successive cultivations with logarithmic growth described earlier (Figure 8.11). As sporulation was observed in the later period of the logarithmic or stationary phase, the number of spores and the stability of the plasmid in the spores were determined. Spores appeared after one day and increased in number to 1.6×10^8 cfu/ml by the 10th day, as shown in Figure 8.11. The plasmid stability in spores was only 57%, even on the first day, suggesting that the stability of the plasmid in vegetative cells and in spores is controlled in different ways.

8.4.2.2 Stability in Schaeffer's Medium

To analyze this phenomenon in more detail, plasmid stability in Schaeffer's sporulation medium (29) was investigated (Table 8.4). Sporulation was initiated at around 12 h and proceeded significantly between 12 h and 15 h; the plasmid stability in spores was low, being about 50% to 60% throughout the 10-day incubation period. No further curing was observed after the sporulation. This instability of the plasmid

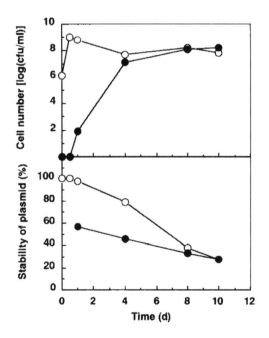

FIGURE 8.11 Stability of pC194 in *B. subtilis* NB22-1 in prolonged incubation in no. 3 medium at 30°C. The number of spores was counted as the number of colonies appearing on no. 3 agar medium after the culture broth was treated for 15 min at 80°C. All the vegetative cells were confirmed to have been killed after 5 min heat treatment at 60°C, while no changes in the numbers of spores were observed, even after 20-min treatment at 80°C. The number of spores and the total number of viable cells, consisting of spores and vegetative cells, are both expressed as colony-forming units (cfu). Open and closed circles represent the data for total cells and spores, respectively.

TABLE 8.4

Stability of pC194 in *B. subtilis* NB22-1 Cultivated in Schaeffer's Sporulation Medium at 30°C

	Total Cells		Spores[a]	
Time	Total Cell Number [log(cfu/ml)]	Plasmid Stability (%)	Number of Spores [log(cfu/ml)]	Plasmid Stability in Spores (%)
0 h	4.6	100	0	—
12 h	8.9	88	1.3	—
15 h	8.7	82	7.1	56
24 h	8.6	63	9.0	65
2 days	7.4	69	8.5	56
4 days	8.2	66	8.7	62
10 days	8.3	52	8.7	48

[a] Spores were counted as described in the text.

was not circumvented by the addition of chloramphenicol for selective pressure (data not shown). Six independent colonies were chosen from the 10th-day culture, and each colony was recultivated in the sporulation medium with or without chloramphenicol for 10 days. The plasmid stabilities of the six colonies were low, at about 50% to 60%. These results mean that not even a single molecule of the plasmid was partitioned into half of spore cells made in the medium, although the copy number of pC194 was about 15 per chromosomal equivalent in vegetative cells (28).

To ascertain the effect of the sporulation rate on plasmid stability, cultures in the sporulation medium were carried out at 15°C and 23°C. Although the growth rate and the spore-forming rate in the sporulation medium at 23°C and 15°C were significantly lower than those at 30°C, the plasmid stabilities in spores were the same values and thus no improvement in stability was obtained by changing the cultivation temperature (data not shown).

8.4.2.3   Stability of pUB110

As unexpected instability of pC194 was observed during sporulation, another stable plasmid pUB110 was used. As shown in Table 8.5, spores became detectable on the 2nd day, and a plasmid stability of almost 100% was observed on day 8. This result was quite different from that of pC194. The presence of kanamycin had no effect on the stability of the plasmid during sporulation (data not shown). This result with plasmid pUB110 indicates that sporulation does not necessarily cause instability of the plasmid. As pUB110 was stably maintained, this plasmid may have a gene for stable maintenance during sporulation, or the higher copy number (about 50 per chromosomal equivalent) (30) of this plasmid than that of pC194 (about 15 per chromosomal equivalent) (28) may raise the probability of plasmid distribution into the spore cells.

The stability of pUB110 in prolonged incubation in no. 3 medium is shown in Figure 8.12. One hundred percent of plasmids were maintained for 4 days. Significant curing was, however, observed after the 8th day; only 20% of the

TABLE 8.5

Stability of pUB110 in *B. subtilis* NB22-1 Cultivated in Schaeffer's Sporulation Medium at 30°C

Time (days)	Total cells		Spores[a]	
	Total Cell Number [log(cfu/ml)]	Plasmid Stability (%)	Number of Spores [log(cfu/ml)]	Plasmid Stability in Spores (%)
0	5.0	100	0	—
0.5	7.1	100	0	—
1	8.4	100	0	—
2	7.9	100	5.7	100
4	7.1	98	7.3	96
8	7.4	98	7.0	97

[a] Spores were counted by the method described in the text.

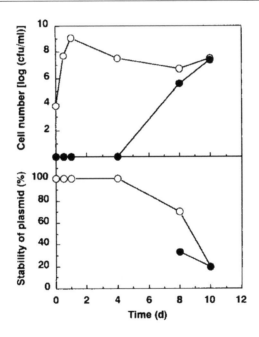

FIGURE 8.12 Stability of pUB110 in *B. subtilis* NB22-I in prolonged incubation in no. 3 medium at 30°C. Spores were counted as described in Figure 8.11. Open and closed circles represent the data for total cells and spores, respectively.

plasmids were maintained in spores by day 10. The different plasmid stabilities observed in spores made in prolonged incubation (Figure 8.11 for pC194 and Figure 8.12 for pUB110) and those in Schaeffer's medium may reflect different mechanisms of spore formation or of plasmid partition functioning in different media or cultivation conditions.

8.4.3 DISCUSSION

The mechanism for the instability of pC194 and pUB110 is not yet clear, but the following may be speculated for pC194: (i) Some exclusive function against plasmid DNA in the host cell might exist during the spore-forming period, or (ii) there may be nonhomogeneous distribution of plasmids in the cytosol. As curing was observed even in the presence of selective pressure and the cured cells could not grow in the medium with the antibiotic, curing should occur after cell division, and most probably at the stage of septum formation in sporulation. Although plasmid instabilities after the logarithmic growth phase has been reported (31, 32), the instability of *Bacillus* plasmids during sporulation has not. The *par* genes required for stable maintenance of plasmids in *B. subtilis* have been reported (33, 34, 36); the function of the *par* gene during sporulation, however, is unclear.

DNA rearrangement during endospore formation of *B. subtilis* has been elucidated (38–40). In the late stage of sporulation, the cells are divided into two compartments: mother cell and forespore. The DNA rearrangement is the result of site-specific recombination within a 5 bp sequence (AATGA) present identically in *spoIVCB* and *spoIIIC*, and this occurs specifically in the mother cell. As the 5 bp sequence is found both in pC194 and pUB110 (41, 42), these plasmids also may interact with the chromosomal DNA in the mother cell. However, the number and distribution of the sequence in the two plasmids are significantly different (41, 42), which might cause the different stability of the plasmids during sporulation.

The fact that two stable plasmids, pC194 and pUB110, became unstable during sporulation under nutritionally limited conditions indicates the necessity of the selection or development of stable vectors in oligotrophic conditions where the application of genetically engineered *Bacillus* spp. to the natural environment is concerned.

REFERENCES

1. Ohno, A., Ano, T., and Shoda, M. Production of a lipopeptide antibiotic surfactin with recombinant *Bacillus subtilis*. *Biotechnol. Letts.*, 14, 1165–1168 (1992).
2. Hiraoka, H., Ano, T., and Shoda, M. Rapid transformation of *Bacillus subtilis* using KCl-treatment. *J. Ferment. Bioeng.*, 74, 241–243 (1992).
3. Ehrlich, S. D. Replication and expression of plasmids from *Staphylococcus aureus* in *Bacillus subtilis. Proc. Natl. Acad. Sci. U.S.A.*, 74, 1680–1682 (1977).
4. Cooper, D. G., MacDonald, C. R., Duff, S. J. B., and Kosaric, N. Enhanced production of surfactin from *Bacillus subtilis* by continuous product removal and metal cation additions. *Appl. Environ. Microbiol.*, 42, 408–412 (1981).
5. Ohno, A., Ano, T., and Shoda, M. Production of a lipopeptide antibiotic, surfactin, by recombinant *Bacillus subtilis* in solid state fermentation. *Biotechnol. Bioeng.*, 47, 209–214 (1995).
6. Hiraoka, H., Ano, T., and Shoda, M. Molecular cloning of a gene responsible for the biosynthesis of the lipopeptide antibiotics iturin and surfactin. *J. Ferment. Bioeng.*, 74, 323–326 (1992).
7. Huang, C. C., Ano, T., and Shoda, M. Nucleotide sequence and characteristics of the gene, *lpa-14*, responsible for biosynthesis of the lipopeptide antibiotics iturin A and surfactin from *Bacillus subtilis* RB14. *J. Ferment. Bioeng.*, 76, 445–450 (1993).

8. Ohno, A., Ano, T., and Shoda, M. Production of the antifungal peptide antibiotic, iturin by *Bacillus subtilis* NB22 in solid state fermentation. *J. Ferment. Bioeng.*,75, 23–27 (1993).

9. Tsuge, K., Ano, T., and Shoda, M. Characterization of *Bacillus subtilis* YB8, coproducer of lipopeptides and surfactin and plipastatin B1. *J. Gen. Appl. Microbiol.*, 41, 541–545 (1995).

10. Ohno, A., Ano, T., and Shoda, M. Production of a lipopeptide antibiotic surfactin with recombinant *Bacillus subtilis*. *Biotechnol. Lett.*, 14, 1165–1168 (1992).

11. Sandrin, C., Peypoux, F., and Michel, G. Coproduction of surfactin and iturin A, lipopeptides with surfactant and antifungal properties, by *Bacillus subtilis*. *Biotechnol. Appl. Biochem.*, 12, 370–375 (1990).

12. Sheppard, J. D., and Mulligan, C. N. The production of surfactin by *Bacillus subtilis* grown on peat hydrolysate. *Appl. Microbiol. Biotechnol.*, 27, 110–116 (1989).

13. Mulligan, C. N., Chow, T. Y.-K., and Gibbs, B. F. Enhanced biosurfactant production by a mutant *Bacillus subtilis* strain. *Appl. Microbiol. Biotechnol.*, 31, 486–489 (1989).

14. Drauin, C. M., and Cooper, D. G. Biosurfactants and aqueous two-phase fermentation. *Biotechnol. Bioeng.*, 40, 86–90 (1992).

15. Uesugi, Y. *Methods in Pesticide Science*, vol. 2. Soft Science Pub. Co., Tokyo (1981).

16. Chang, I.-P., and Kommedahl, T. Biological control of seedling blight of corn by coating kernels with antagonistic microorganisms. *Phytopathology*, 58, 1395–1401 (1968).

17. Nakayama, S., Takahashi, S., Hirai, M., and Shoda, M. Isolation of new variants of surfactin by a recombinant *Bacillus subtilis*. *Appl. Microbiol. Biotechnol.*, 48, 80–82 (1997).

18. Ano, T., Kobayashi, A., and Shoda, M. Transformation of *Bacillus subtilis* with the treatment by alkali cations. *Biotechnol. Lett.*, 12, 99–104 (1990).

19. Jenny, K., Kappeli, O., and Fiechter, A. Biosurfactants from *Bacillus licheniformis*: structural analysis and characterization. *Appl. Microbiol. Biotechnol.*, 36, 5–13 (1991).

20. Yakimov, M. M., Timmis, K. N., Wray, V., and Fredrickson, H. L. Characterization of a new lipopeptide surfactant produced by thermotolerant and halotolerant subsurface *Bacillus licheniformis* BAS 50. *Appl. Environ. Microbiol.*, 61, 1706–1713 (1995).

21. Besson, F. Characterization of the surfactin synthetase isolated from the *Bacillus subtilis* strain producing iturin. *Biotechnol. Lett.*, 16, 1269–1274 (1994).

22. Besson, F., and Michel, G. Biosynthesis of iturin and surfactin by *Bacillus subtilis*: evidence for amino acid activating enzymes. *Biotechnical. Lett.*, 14, 1013–1018 (1992).

23. Galli, G., Rodriguez, F., Cosmina, P., Pratesi, C., Nogarotto, R., de Ferra, F., and Grandi, G. Characterization of the surfactin synthetase multi-enzyme complex. *Biochim. Biophys. Acta*, 205, 19–28 (1994).

24. Menkhaus, M., Ullrich, C., Kluge, B., Yater, J., Vollenbroich, D., and Kamp, R. M. Structural and functional organization of the surfactin synthetase multienzyme system. *J. Biol. Chem.*, 268, 7678–7684 (1993).

25. Ullrich, C., Kluge, B., Palacz, Z., and Yater, J. Cell-free biosynthesis of surfactin, a cyclic lipopeptide produced by *Bacillus subtilis*. *Biochemistry*, 30, 6503–6508 (1991).

26. Vollenbroich, D., Mehta, N., Zuber, P., Vater, J., and Kamp, R. M. Analysis of surfactin synthetase subunits in *srf* A mutants of *Bacillus subtilis* OKB 105. *J. Bacteriol.*, 176, 395–400 (1994).

27. Tokuda, Y., Ano, T., and Shoda, M. Characteristics of plasmid stability in *Bacillus subtilis* NB22, an antifungal-antibiotic iturin producer. *J. Ferment. Bioeng.*, 75, 319–321 (1993).

28. Alonso, J. C., and Trautner, T. A. A gene controlling segregation of the *Bacillus subtilis* plasmid pC194. *Mol. Gen. Genet.*, 198, 427–431 (1985).

29. Schaeffer, P., Millet, J., and Aubert J.-P. Catabolic repression of bacterial sporulation. *Proc. Natl. Acad. Sci. U.S.A.*, 54, 704–711 (1965).

30. Keggins, K. M., Lovett, P. S., and Duvall, E. J. Molecular cloning of genetically active fragments of *Bacillus* DNA in *Bacillus subtilis* and properties of the vector plasmid pUB110. *Proc. Natl. Acad. Sci. U.S.A.*, 75, 1423–1427 (1978).
31. Pinches, A., Louw, M. E., and Watson, T. G. Growth, plasmid stability and α-amylase production in batch fermentations using a recombinant *Bacillus subtilis* strain. *Biotechnol. Lett.*, 7, 621–626 (1985).
32. Harington, A., Watson, T. G., Louw, M. E., Rodel, J. E., and Thomson, J. A. Stability during fermentation of a recombinant α-amylase plasmid in *Bacillus subtilis*. *Appl. Microbiol. Biotechnol.*, 27, 521–527 (1988).
33. Bron, S., Bosma, P., van Belkum, M., and Luxen, E. Stability function in the *Bacillus subtilis* plasmid pTA1060. *Plasmid*, 18, 8–15 (1987).
34. Chang, S., Chang, S.-Y., and Gray, O. Structural and genetic analyses of a par locus that regulates plasmid partition in *Bacillus subtilis*. *J. Bacteriol.*, 169, 3952–3962 (1987).
35. Wei, D., Parulekar, S. J., Stark, B. C., and Weigand, W. A. Plasmid stability and α-amylase production in batch and continuous cultures of *Bacillus subtilis* TN106 (pAT5). *Biotechnol. Bioeng.*, 33, 1010–1020 (1989).
36. Devine, K. M., Hogan, S. T., Higgins, D. G., and McConnell, D. J. Replication and segregational stability of *Bacillus* plasmid pBAAI. *J. Bacteriol.*, 171, 1166–1172 (1989).
37. Leonhardt, H., and Alonso, J. C. Parameters affecting plasmid stability in *Bacillus subtilis*. *Gene*, 103, 107–111 (1991).
38. Sato, T., Samori, Y., and Kobayashi, Y. The *cisA* cistron of *Bacillus subtilis* sporulation gene *spoIVC* encodes a protein homologous to a site-specific recombinase. *J. Bacteriol.*, 172, 1092–1098 (1990).
39. Stragier, P., Kunkel, B., Kroos, L., and Losick, R. Chromosomal rearrangement generating a composite gene for a developmental transcription factor. *Science*, 243, 507–512 (1989).
40. Losick, R., and Stragier, P. Crisscross regulation of cell-type specific gene expression during development in *B. subtilis*. *Nature (London)*, 355, 601–604 (1992).
41. Lovett, P. S., and Ambulos, N. P. The nucleotide sequence and restriction site analysis of pC194, pp. 349–354. In Harwood, C. R. (ed.), *Bacillus (Biotechnology Handbooks)*, vol. 2. Plenum Press, New York (1989).
42. Lovett, P. S., and Ambulos, N. P. The nucleotide sequence and restriction site analysis of pUB110, pp. 333–340. In Harwood, C. R. (ed.), *Bacillus (Biotechnology Handbooks)*, vol. 2. Plenum Press, New York (1989).

9 Co-Use of *B. subtilis* with Chemical Pesticide

Over use of chemical pesticides causes several environmental problems and thus, the use of biological control agents has drawn attention and has been considered as a promising alternative to chemical pesticides. Many applications of fungus and bacteria have been attempted (1–5). However, it is difficult to assess 100% suppressibility of plant disease in soil only by application of microorganisms, mainly because of the complexity of the plant and soil system and relatively lower efficacies than those by applying chemicals under various environmental conditions. Therefore, the integrated use of chemical pesticides and microorganisms may be one of the practical methods applicable in the field. By this method, a reduction in the amount of chemical pesticides used in is anticipated. Some attempts of the integrated use of biological agents and chemical pesticides have been reported (6–8).

B. subtilis RB14-C was isolated as a microbial agent to suppress the growth of plant pathogens both *in vitro* and *in vivo* (Chapter 2). The characteristics of this bacterium were described in detail in the previous papers (9–12). In this chapter, this bacterium was used in combination with a chemical pesticide, flutolanil, which is a commercially available chemical pesticide against the damping-off caused by *Rhizoctonia solani*. First, *B. subtilis* RB14-C was cultivated in the solid and liquid media containing different concentrations of flutolanil, and resistance of RB14-C to flutolanil was confirmed. Then, mixtures of flutolanil solution and RB14-C culture broth were prepared and applied to tomato seedlings plant test *in vivo*.

9.1 MATERIALS AND METHODS (13)

9.1.1 STRAIN AND MEDIUM USED

Bacillus subtilis RB14-C is a spontaneous streptomycin-resistant mutant of strain RB14 and is a coproducer of antifungal substances, iturin A and surfactin (Chapter 3). The growth rate and the antifungal activity of RB14-C were the same as those of the parent strain RB-14 (12). The growth media of RB14-C and a plant pathogen, *Rhizoctonia solani*, were described in Chapter 3. The plant test was the same as that described in Chapter 3.

9.1.2 CHEMICAL PESTICIDE USED

A 25% flutolanil solution of α,α,α-trifluoro-3′-isopropoxy-o-toluanilide (Nihon Noyaku Co. Ltd., Tokyo) was used as a chemical pesticide.

9.1.3 GROWTH OF RB14-C ON AGAR PLATES CONTAINING FLUTOLANIL

After RB14-C was cultivated in the L medium with a shaking flask at 37°C for 16 h, and 50 μl of the properly diluted culture broth was spread on L agar (L medium plus 1.5% agar) plates containing 100 mg/l flutolanil. The number of colonies that appeared on the plates was counted after 24 h of incubation at 37°C and compared with that of the control plates containing no flutolanil.

9.1.4 GROWTH OF RB14-C AND PRODUCTIVITY OF ITURIN A AND SURFACTIN IN LIQUID MEDIUM CONTAINING FLUTOLANIL

A cell culture (400 μl) of RB14-C in the L medium was poured into each of the 10 shaking flasks containing 40 ml of the no. 3 medium with 10 mg/l flutolanil and all flasks were shaken at 38°C at 120 strokes per minute (spm). At each sampling time, one flask was taken, the optical density of which at 660 nm was measured and 20 ml of the entire culture broth was acidified with 12 N HCl at pH 2.0 and centrifuged at 12,000 × g for 25 min. The precipitate was resuspended in 10 ml of methanol for extraction and the suspension was centrifuged again. The supernatant was filtered through a 0.2 μm pore-sized polytetrafluoroethylene (PTFE) membrane (P020; Advantec, Tokyo) and the filtrate was injected into a high-performance liquid chromatography (HPLC) column (ODS [octyldecyl silanolate]-2, 4.6 mm in diameter by 250 mm; GL Sciences, Tokyo) (12). The concentration of 10 mg/l flutolanil used was prepared according to the supplier's manual, that is, the recommended dose of flutolanil is 1.7 mg per 150 g soil. The water content of the soil used here was approximately 120 ml, and the flutolanil concentration in the water in the pot was approximately 14 mg/l.

9.1.5 SUPPRESSIVE EFFECT OF THE MIXTURE OF A SOLUTION CONTAINING ITURIN A AND SURFACTIN AND FLUTOLANIL ON THE GROWTH OF *R. SOLANI IN VITRO*

RB-14C was grown in the no. 3 medium at 38°C for 5 days and culture broth was prepared in the same manner to measure the concentrations of iturin A and surfactin by HPLC. Iturin A (900 mg/l) and surfactin (2500 mg/l) were detected in the filtrate. Next, 5 ml of the potato dextrose agar (PDA) medium containing different concentrations of iturin A and surfactin and 0.2 mg/l flutolanil as the final concentration was poured into a plastic plate. As a control, a PDA plate containing only 0.2 mg/l flutolanil was prepared. The concentration of 0.2 mg/l flutolanil was determined to be a maximum concentration at which *R. solani* could grow on the PDA plate. Then 5 mm plugs were taken from a PDA petri dish culture of *R. solani* K-1 and placed at the center of the PDA plate prepared as described earlier. After incubation at 28°C for 8 days, the areas of the fungal growth on the plate containing a mixture of iturin A and surfactin and flutolanil relative to that on the plate containing only flutolanil were measured.

9.1.6 PLANT TEST

The soil used in this study and the characteristics of the soil were the same as those described in Chapter 3 (12). The soil was autoclaved four times at 12 h intervals for 60 min at 121°C. Sterilized soil (150 g) was put into a plastic pot with a diameter of about 90 mm, and the moisture content was maintained at 60% of the maximum water-holding capacity by the daily addition of sterilized water.

R. solani K-1 was incubated statically in the dark in the PDP medium at 28°C for 1 week. The mycelial mats that formed on the surface of the medium were homogenized at 4000 rpm for 2 min in sterile water and inoculated into the soil at a ratio of one piece mycelial mat to one pot 6 days before planting the germinated tomato seeds.

The plant test was performed according to the procedure described in Chapter 3 (12). Tomato seeds (Ponderosa) were disinfected with 80% ethanol and then with 0.5% sodium hypochlorite. After rinsing with sterile water, the seeds were germinated on a 2% agar plate at 30°C for 2 days. Each pot was sown with nine germinated seeds after the RB14-C culture and/or flutolanil were introduced into the soil and then placed in a growth chamber at 30°C with 90% relative humidity under 16 h of light (about 12,000 lux). After 2 weeks, seedlings that showed no or very weak growth were judged to be diseased. Then, the percentage of diseased seedlings divided by the nine seedlings in one pot was determined as the diseased ratio. Furthermore, the shoots were clipped off at their bases, and their lengths and dry weights were measured. A statistical analysis using Fisher's protected least significant difference was carried out.

9.1.7 PREPARATION OF **RB14-C** AND FLUTOLANIL FOR PLANT TEST

As a preliminary test, different concentrations of flutolanil in the range of 30–1000 µg/pot were added to the soil in each pot and the plant test was carried out. Then, 385 µg/pot flutolanil was found to be the concentration that reduces the occurrence of the disease to about 80%. Therefore, the culture broth of RB14-C was used together with flutolanil at the concentrations of 385 µg/pot and 94 µg/pot. Then, the following four samples of the culture broth of RB14-C grown in the no. 3 medium for 5 days and/or flutolanil were introduced into the soil in a pot 2 days after *R. solani* inoculation: (i) Flutolanil was mixed with the soil to a final concentration of 94 µg/pot. (ii) Flutolanil was mixed with the soil to a final concentration of 385 µg/pot. (iii) A 30 ml culture broth of RB14-C was mixed with the soil in a pot. (iv) Flutolanil (94 µg) and 30 ml of culture broth were mixed with the soil in a pot.

9.1.8 CELL RECOVERY FROM THE SOIL AND COUNTING
THE NUMBER OF VIABLE CELLS

Soil samples were analyzed before planting, and at day 3 and day 14 after planting. Three grams of soil was suspended in 8 ml of 0.85% NaCl solution (pH 8.0) in a 50 ml Erlenmeyer flask, and then shaken for 15 min at 140 spm at room temperature. The suspension was serially diluted in 0.85% NaCl solution and plated onto L agar plates containing 100 µg/ml streptomycin. The plates were incubated at 38°C for 12 h, and the number of viable cells was expressed as colony-forming units (cfu).

9.1.9   QUANTITATIVE ANALYSIS OF ITURIN A AND SURFACTIN RECOVERED FROM SOIL

Three grams of soil was suspended in 21 ml of a mixture of acetonitrile:3.8 mM trifluoroacetic acid (4:1 [v/v]) in a 50 ml Erlenmeyer flask and then shaken for 1 h at 140 spm at room temperature. The soil in the suspension was then removed using a filter paper (Toyo Roshi Co. Ltd., Tokyo) and the filtrate was evaporated. The precipitate was subjected to extraction using 2 ml of methanol for 2 h and the extract was applied to HPLC analysis as described in Chapter 3 (12).

9.2   RESULTS AND DISCUSSION

9.2.1   GROWTH OF RB14-C ON PLATE CONTAINING FLUTOLANIL AND PRODUCTIVITY OF ITURIN A AND SURFACTIN IN MEDIUM CONTAINING FLUTOLANIL

The numbers of RB14-C viable cells that appeared on the L agar plate containing 100 mg/l flutolanil and on the L agar plate without flutolanil were almost the same as those obtained after 24 h of incubation (data not shown). This suggests that RB14-C is resistant to a high concentration of 100 mg/l flutolanil. When the growth of RB14-C in the liquid L medium containing 10 mg/l flutolanil was compared with that in the same medium without flutolanil as shown in Figure 9.1, no difference in the growth pattern was observed. The production patterns of iturin

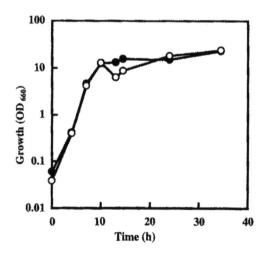

FIGURE 9.1   Growth of *B. subtilis* RB14-C in no. 3 medium with 10 mg/l flutolanil (•) and without flutolanil (o).

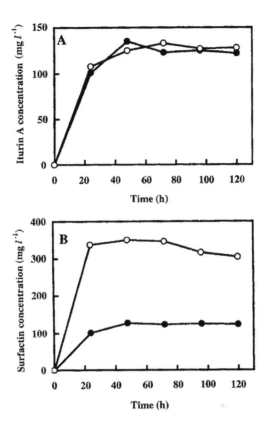

FIGURE 9.2 Production of (A) iturin A and (B) surfactin by RB14-C in no. 3 medium containing 10 mg/l flutolanil (●) and no flutolanil (○).

A and surfactin in the liquid L medium are shown in Figure 9.2. After 5 days of incubation, the final concentration of iturin A was about 122 mg/l in the medium containing 10 mg/l flutolanil, while it was 128 mg/l in the flutolanil-free medium. The surfactin concentrations were 123 mg/l in the medium containing flutolanil and 304 mg/l in the flutolanil-free medium. This suggests that the biosynthetic pathway of surfactin was adversely affected by 10 mg/l flutolanil, but that of iturin A of RB14-C was unaffected. However, since the antifungal effect of iturin A was about 100 times greater than that of surfactin (12), the deteriorated surfactin production may have a minor effect on the ability of RB14-C to suppress the growth of the plant pathogen.

The functional mechanism of flutolanil against plant pathogens was speculated to be an inhibitory effect on the activity of succinate dehydrogenase of plant pathogens. Therefore, the reduction in the surfactin production in the presence of flutolanil may be partly related to a similar inhibitory effect on *B. subtilis*. On the other hand, the flutolanil solution used in this study was a commercial product containing some other detergents that may have adverse effects on the surfactin production in *B. subtilis*.

9.2.2   COMPARISON OF GROWTH OF *R. SOLANI* ON PLATES CONTAINING A MIXTURE OF ITURIN A AND SURFACTIN FROM RB14-C AND FLUTOLANIL

Figure 9.3 shows the relative growth areas of *R. solani* on the plates after incubation for 7 days. The growth of *R. solani* decreased in proportion to the increase in the iturin A concentration. Although a mixture of iturin A and surfactin was used, iturin A concentration was used in the figure due to its strong antifungal effect. Flutolanil at a concentration of 0.2 mg/l reduced the growth of *R. solani* by 60% compared with that on the PDA plate containing no flutolanil (Figure 9.3A). When different concentrations of iturin A were mixed with 0.2 mg/l flutolanil, a significant increase

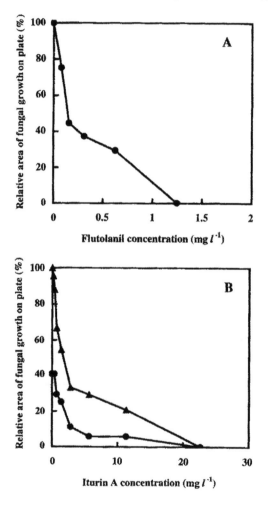

FIGURE 9.3   Relative growth areas of *R. solani* on the PDA plates containing (A) different concentrations of flutolanil and (B) different concentrations of iturin A. In (B), 0.2 mg/l flutolanil was added (•) and no flutolanil was added (▲). The abscissa is represented by the iturin A concentration due to its strong antifungal effect, although the mixture of iturin A and surfactin was applied to the experiment.

in the growth inhibitory effect on *R. solani* was observed. In the iturin A concentration range of 5–20 mg/l, the growth of *R. solani* was reduced by 95%, indicating a synergistic effect of iturin A and flutolanil (Figure 9.3B).

9.2.3 PLANT TEST

The results of the plant test are shown in Table 9.1. The data are average values of the results from experiments repeated four times. The diseased ratio of the pot containing only *R. solani* reached 90%. When flutolanil was introduced into the pot, the diseased ratio was reduced to 18%–65%, depending on the concentration. The culture broth of RB14-C suppressed the occurrence of the disease to 51%, and integrated use of the culture broth of RB14-C and flutolanil (94 µg/pot) reduced the diseased ratio to 23%, which is almost equivalent to that when 375 µg/pot flutolanil was introduced to the soil. The shoot weight and the shoot length significantly increased when the amount of flutolanil was increased from 94 µg/pot to 375 µg/pot. When only RB14-C was applied to the soil, the values of the shoot weight and the shoot length were larger than those of the single application of flutolanil at 94 µg/pot. The values obtained from the integrated use of RB14-C and flutolanil at 94 µg/pot were almost equivalent to those of the single use of flutolanil at 375 µg/pot, indicating a more stable growth of tomato seedlings by application of RB14-C. These data indicate that the integrated use of RB14-C and flutolanil can reduce the amount of flutolanil used to one-fourth of that of the single use of flutolanil, with the same effect of reducing disease occurrence.

The number of RB14-C viable cells in the soil after 14 days of plant testing is shown in Table 9.2. Although a slight decrease in the number of viable cells was

TABLE 9.1

Effects of Treatments of *B. subtilis* RB14-C Culture Broth and/or Flutolanil on the Suppression of Damping-Off in Tomato Plants Caused by *R. solani* 14 Days after Planting

Treatment					
R. solani (3.0 g/pot)	Flutolanil (µg/pot)	RB14-C (ml)	Shoot Length (mm)	Dry Weight of Shoots (mg/pot)	Diseased Ratio (%)
+	–	–	9.2[a]	41.4[a]	90.3[a]
+	94	–	29.6[b]	112[ab]	65.3[b]
+	375	–	67.4[cd]	192[bc]	18.1[cd]
+	0	30	47.7[bc]	227[cd]	51.4[b]
+	94	30	70.9[d]	286[d]	23.6[c]
–	–	–	106.7[e]	307[d]	0[d]
LSD *P* = 0.05			20.2	86.7	20.9

Notes: For each treatment, each datum is an average of the results from experiments repeated four times. Means in any column with different letters are significantly different (*P* = 0.05) according to Fisher's protected least significant difference (LSD) analysis.

TABLE 9.2

Populations of *B. subtilis* RB14-C in the Soil Treated with RB14-C Culture Broth and Flutolanil at 0, 3, and 14 Days after Planting

Treatment	Cell Number ($\times 10^8$ cfu/g dry soil)		
	0 Days	3 Days	14 Days
RB14-C	5.6	3.7	2.4
RB14-C + flutolanil 94 µg/pot	4.5	4.1	2.3

Note: Each value is an average of the results from two pots in one experiment.

TABLE 9.3

Concentrations of Iturin A and Surfactin Recovered from the Soil Samples at 0, 3, and 14 Days after Planting When a Mixture of RB14-C Culture Broth and Flutolanil Was Used

Treatment		Concentration of Antibiotics (µg/g dry soil)		
		0 Days	3 Days	14 Days
RB14-C	Iturin A	13	3.9	0.4
	Surfactin	76	60	19
RB14-C + flutolanil 94 µg/pot	Iturin A	16	3.0	0.3
	Surfactin	68	48	13

Note: Each value is an average of the results from two pots in one experiment.

observed, the same order of 10^8 cfu was detected at day 14, indicating that RB14-C was not influenced by the existence of flutolanil in the soil. Table 9.3 shows the concentrations of iturin A and surfactin extracted from the soil samples. When the culture broth of RB14-C and flutolanil were mixed into the soil, 16 µg of iturin A introduced into soil decreased to about 3 µg after 3 days of incubation, and almost no iturin A was detected after 14 days. Similarly, 68 µg of surfactin on day 0 decreased to 13 µg after 14 days. These degradation patterns were similar to those in the soil that contained only the culture broth of RB14-C (12). In the reports on integration of *Bacillus* spp. and chemicals against *Fusarium* or *Pythium* diseases (6, 7), neither the number of viable cells of *Bacillus* spp. nor that of antifungal substances were detected.

Although the persistence of flutolanil in the soil was not yet clarified, the biodegradability of iturin A and surfactin is supposed to be significantly higher than that of flutolanil. Therefore, the combined use of RB14-C will reduce the environmental problems due to the use of chemical pesticide. It was confirmed in a previous paper (12) that when the solution containing iturin A and surfactin was applied into the soil, both substances were not detected after 14 days of incubation. The causes of the reduction in the iturin A and surfactin concentrations in the soil may be due to leaching during watering of the plants, possible biodegradation by airborne microorganisms, or irreversible binding to soil materials or humic acids. Since concentrations of iturin A and surfactin in centrifuged culture supernatant were maintained for more than one month (data not shown), such decrease in the concentration of iturin A and surfactin in the soil was probably related to the environment.

According to the data on the amount of flutolanil actually used in the agricultural field, about 10–30 times higher flutolanil concentration was applied in the field mainly due to its overuse, which was about 100 times higher than the concentration used in this experiment. Therefore, the effect of the integrated use of RB14-C culture broth and flutolanil will result in a more significant reduction in the amount of chemical pesticides applied in the field.

REFERENCES

1. Cook, R. J. Twenty-five years of progress towards biological control, pp. 1–14. In Hornby, D. (ed.), *Biological Control of Soil-Borne Plant Pathogens*. C.A.B. International, Wallingford (1990).
2. Gutterson, N. Microbial fungicides: recent approaches to elucidating mechanisms. *Crit. Rev. Biotechnol.*, 10, 69–91 (1990).
3. Weller, D. M. Biological control of soilborne plant pathogens in the rhizosphere with bacteria. *Annu. Rev. Phytopathol.*, 26, 389–408 (1988).
4. Phae, C. G., Shoda, M., Kita, N., Nakano, M., and Ushiyama, K. Biological control of crown and root rot and bacterial wilt of tomato by *Bacillus subtilis* NB22. *Ann. Phytopath. Soc. Japan*, 58, 329–339 (1992).
5. Shoda, M. Bacterial control of plant diseases. *J. Biosci. Bioeng.*, 89, 515–521 (2000).
6. Hwang, S. F. Potential for integrated biological and chemical control of seedling rot and preemergence damping off caused by *Fusarium avenaceumn* in lentil with *Bacillus subtilis* and Vitaflo®-280. *Z. Pflanzenkr. Pflanzenschutz*, 101, 189–199 (1994).
7. Hwang, S. F., Chang, K. F., Howard, R. J., Deneka, B. A., and Turnbull, G. D. Decrease in incidence of *Pythium* damping-off of field pea by seed treatment with *Bacillus* spp. and metalaxyl. *Z. Pflanzenkr. Pflanzenschutz*, 103, 31–41 (1996).
8. Mandeel, Q. Integration of biological and chemical control of *Fusarium* wilt of radish. *Z. Pflanzenkr. Pflanzenschutz*, 103, 610–619 (1996).
9. Hiraoka, H., Asaka, O., Ano, T., and Shoda, M. Characterization of *Bacillus subtilis* RB14, coproducer of peptide antibiotics iturin A and surfactin. *J. Gen. Appl. Microbiol.*, 38, 635–640 (1992).
10. Hiraoka, H., Ano, T., and Shoda, M. Molecular cloning of a gene responsible for the biosynthesis of the lipopeptide antibiotics iturin and surfactin. *J. Ferment. Bioeng.*, 4, 323–328 (1992).
11. Huang, C.-C., Ano, T., and Shoda, M. Nucleotide sequence and characteristics of a gene, *lpa-14* responsible for the bio-synthesis of the lipopeptide antibiotics iturin A and surfactin from *Bacillus subtilis* RB14. *J. Ferment. Bioeng.*, 86, 445–450 (1993).
12. Asaka, O. and Shoda, M. Biocontrol of *Rhizoctonia solani* damping-off of tomato with *Bacillus subtilis* RB14. *Appl. Environ. Microbiol.*, 62, 4081–4085 (1996).
13. Kondo, M., Hirai M., and Shoda, M. Integrated biological and chemical control of damping–off caused by *Rhizoctonia solani* using *Bacillus subtilis* RB14-C and flutolanil. *J. Biosci. Bioeng.*, 91, 173–177 (2001).

10 Mixed Culture Effect on Biocontrol

In recent decades numerous microorganisms have been identified as a capable of suppressing plant pathogens (1). Inoculation of soil with a single strain of biocontrol agent only rarely leads to a level of protection obtained with chemicals, and a positive effect is often inconsistent (2, 3). The reason for this unstable efficacy is a complexity of the natural system, where microbial inoculum is affected by biotic and abiotic factors. The way to improve plant bioprotection may be the use of mixtures or combinations of biological agents, especially when they exhibit different or complementary modes of action. Under natural, changeable conditions one mechanism may compensate the other and result in an additive or synergistic effect. Pierson and Weller (4) showed that combining strains of fluorescent pseudomonads increased control of wheat take-all compared to the same strains applied individually. There was a general trend in all experiments toward greater suppression and enhanced consistency against the pathogens using strain mixtures. Control efficacy obtained by Guetsky et al. (5, 6), which used a mixture of *Pichia guilermondii* and *Bacillus mycoides*, was higher than that achieved by their separate inoculation, and the variability of suppression was reduced. Other reports demonstrated that combinations of microorganisms provided better plant protection than single strains (7–9).

In this study, two bacterial strains that exhibited strong antagonistic activity against *Rhizoctonia solani* were used. One strain was *B. subtilis* RB14-C, which is a producer of lipopeptide antibiotics iturin A and surfactin (10, 11). Iturin interacts with cytoplasmatic membrane increasing its permeability because of formation of ion-conducting pores (12). The advantage of using *Bacillus* in the mixture is its property to form spores resistant to unfavorable natural conditions and its tolerance to antimicrobial substances released by other microorganisms in soil and by co-inoculant.

The second strain used in this work was *Burkholderia cepacia* BY isolated from volcanic soil. This strain exhibited strong antagonism against *R. solani* in paired-culture tests on potato dextrose agar (PDA) and on nutrient agar, creating clear zones of inhibition between colonies, and when the cells were applied to the soil, damping-off of tomato plants caused by this pathogen was significantly decreased (Szczech, unpublished data). The mechanism of activity of this strain is not yet known. However, bacteria from *B. cepacia* complex (Bcc) are reported as producers of numerous antibiotics (cepacin, cepaciamide, cepacidine, pseulanes, phenylpyrroles, phenazine, pyrrolnitrin, antibiotic called AFC-BC11) and siderophores (13–17). Moreover, *B. cepacia* is reported as widely spread in the natural environment (18) and was studied as a successful biological control agent (17, 19, 20).

In this work, different combinations were used of *B. subtilis* RB14-C with *B. cepacia* BY to control *Rhizoctonia* damping-off of tomato plants under growth chamber conditions. The specific objectives were to test the potential of the combined treatments to enhance the efficacy of microbial inoculants and to study their effect on the population of the pathogen in soil. The effect of RB14-C on BY growth and BY on RB14-C population in *in vitro* experiments and in the soil environment were also investigated.

10.1 MATERIALS AND METHODS (1)

10.1.1 MICROORGANISMS AND INOCULUM PRODUCTION

The strain *B. subtilis* RB14-C, a stable mutant of *B. subtilis* RB14 resistant to streptomycin, was used. Fresh cells were obtained from stock culture stored at –80°C and grown in Luria broth (Polypepton 10 g, yeast extract 5 g, NaCl 5 g, agar 20 g, distilled water 1 liter, pH 7.0) for 24 h at 30°C. Inoculum was prepared in 200 ml flasks containing 40 ml of no. 3S medium (Polypepton S 30 g, glucose 10 g, KH_2PO_4 1 g, $MgSO_4 \cdot 7H_2O$ 0.5 g, distilled water 1 liter, pH 7.0) amended with 100 µg/ml of streptomycin. The flasks were shaken (120 strokes/min) for 5 days at 30°C. Bacterial culture broth was centrifuged (6000 × g for 15 min at 4°C) and the supernatant was discarded. The cell pellet was suspended in 0.85% NaCl solution and centrifuged again at the same conditions. The supernatant was removed and washed cells were resuspended in sterile distilled water. The concentration of the cells in the suspension was approximately 10^9 colony-forming units (cfu)/ml.

The strain BY, identified as *B. cepacia*, was isolated from volcanic soil and exhibited strong antagonism against *R. solani*. The inoculum of BY was prepared from stock culture stored at –80°C and grown on 25% PDA plates for 4 days at 28°C. After that, sterile 0.85% NaCl solution was added to the plates and cells were gently scratched from the surface of the medium. Bacterial suspension, collected from the plates, was adjusted with sterile 0.85% NaCl solution to obtain cell density approximately 10^9 cfu/ml.

Pathogenic fungus *R. solani* strain Kl, anastomosis group AG-4, was used for soil infestation. To prepare inoculum potato dextrose broth (PDB; potato infusion 200 g, glucose 20 g, Polypepton 1 g, distilled water 1 liter, pH 5.6) was used. Next, 50 ml of PDB was seeded in 200 ml flasks with mycelial disk of *R. solani* and then incubated in the dark at 28°C for 4 days. The mycelial mat was removed from the medium and homogenized in sterile distilled water.

10.1.2 EXAMINATION OF MUTUAL INHIBITION OF RB14-C AND BY IN *IN VITRO* TESTS

To check if metabolites produced by RB14-C can inhibit growth of BY on agar plates, the cell suspension of RB14-C in 0.85% NaCl solution at density approximately 10^7 cfu/ml, was dropped into four wells cut in nutrient agar and incubated for 24 h at 28°C. After that time a suspension of BY bacterium (10^7 cfu/ml) was sprayed over the plates. Then the plates were incubated for the next 48 h at 28°C, and the presence of zones of inhibition of BY growth, around the wells with RB14-C, was observed. A similar test was performed to check the antibiosis of BY against RB14-C. In this test, BY was dropped into wells cut in nutrient agar and then plates were sprayed with RB14-C.

To determine the sensitivity of BY to iturin A, produced by *B. subtilis* RBI4-C, a diffusion test was performed. The dilution of BY (10^6 cfu/ml) was spread on the surface of nutrient agar and then filter discs (5 mm) containing 5 µg of iturin A were placed. After 2 days of incubation at 28°C, the presence of clear zones of inhibition around filters was observed.

10.1.3 Inhibition of *R. solani* in Microcosm Studies

The soil used in this test was volcanic ash mixed with vermiculite in the ratio 4:1 (wt:wt). The mixture was sterilized at 120°C for 20 min in 1-liter containers. Then the mixture was brought to approximately 60% of the maximum water-holding capacity by addition of sterile water and fertilizers: N 0.04%, P 0.09%, K 0.06%, Ca 0.06%, Mg 0.06%, Fe 0.001%.

One part of the soil was mixed with cell suspension of RB14-C (10^9 cfu/ml) at a dose 12 ml of the suspension per 100 g. The second part of the soil was mixed with the same dose of BY suspension at the same density, and the third part with the mixture of RB14-C and BY (12 ml each, 10^9 cfu/ml). Soil not treated with bacteria was used as a control. Sterilized glass fiber filters (47 mm diameter) (GA-55; Advantec, Japan) were covered with a thin layer of water agar and placed in the center of petri plates. The filters were placed in the center with disk of *R. solani* on PDA. The plates were then filled with the soils, sealed with parafilm and incubated for 7 days at 28°C. Two plates were prepared for each treatment. After incubation, the filters were removed, carefully washed under a stream of tap water, dried on paper towels, and stained in lactophenol blue solution (Merck, Germany). The microscopic studies of *R. solani* mycelium were performed at 100× and 1000× magnifications. The test was repeated.

10.1.4 Biocontrol of *Rhizoctonia* Damping-Off
under Growth Chamber Conditions

The experiments were performed in plastic pots (9 cm diameter) containing volcanic soil prepared as described earlier, but not sterilized. The soil was infested with inoculum of *R. solani* (RS) at a dose of 1 g of homogenized mycelium per pot (220 g soil). The pots were incubated for 7 days (until seed planting) in a growth chamber at 30°C, humidity 90%, and under 12 h of light. During incubation the soil in pots was mixed with *B. subtilis* RB14-C and *B. cepacia* BY at a rate of 25 ml (10^9 cfu/ml) of the bacterial suspension per pot. Application of the bacteria was performed as follows: RB14-C 4 days before seed planting (RB); BY 4 days before seed planting (BY4); RB14-C together with BY 4 days before seed planting (RB + BY4); BY 2 days before seed planting (BY2); RB14-C 4 days and BY 2 days before seed planting

(RB + BY2); BY before seed planting (BY0); RB14-C 4 days and BY immediately before seed planting (RB + BY0). Soil without *R. solani* and bacterial applications served as a control (C).

Seeds of tomato cv. Ogata-Fukuju (Takei Seed Co., Ltd., Kyoto, Japan) were surface sterilized in 70% ethanol for 4 min and rinsed five times in sterile distilled water. Then, they were germinated on 2% water agar for 2 days at 30°C and planted into the soil (nine seeds per pot, three pots for treatment). The pots were kept in randomized blocks in the growth chamber under conditions described above. After 2 weeks the number of plants in each pot and dry weight of plant shoots were recorded as a measure of *Rhizoctonia* damping-off severity. The experiment was repeated three times.

10.1.5   NUMBER OF *R. SOLANI* IN SOIL

Six grams of the soil were collected in petri plates from each pot before seed planting. The soil was thoroughly mixed, and then 1 g was put into a new plate. Each sample was replicated three times. Sterile and cooled 2% agar (15 ml) was poured into plates and carefully mixed with the soil. Once the soil–agar mixture solidified, a layer (approximately 10 ml) of selective for *Rhizoctonia* Ko and Hora (KH) medium was poured on the surface (21). This medium was supplemented with gallic acid 400 mg, chloramphenicol 50 mg, streptomycin 50 mg, metalaxyl 10 mg, and benomyl 2 mg/1 liter. After 24 h of incubation at 28°C the colonies grown on the surface of KH medium were counted. The number of *Rhizoctonia* propagules was expressed per 1 g of dry weight of the soil.

10.1.6   NUMBER OF *B. SUBTILIS* RB14-C IN SOIL

One gram of the soil, from soil collected to determine number of *R. solani*, was placed into a tube containing 10 ml of 0.85% NaCl solution and agitated in a vortex for 60 s. The suspension was serially diluted and plated on Luria medium amended with 100 µg/ml of streptomycin. The plates were incubated for 18 h at 37°C, and the number of colonies was counted and defined as the total number of *B. subtilis*. The number of spores was estimated by plating aliquots of diluted samples after heating of soil dilutions in a water bath at 80°C for 10 min. The total number of bacteria and number of the spores were expressed as cfu/1 g of dry weight of the soil.

10.1.7   STATISTICAL ANALYSIS

The data were subjected to analysis of variance, and Student's *t*-test was used to estimate the significance of differences between the means ($P \leq 0.05$).

10.2   RESULTS

10.2.1   MUTUAL ANTAGONISM OF RB14-C AND BY

The growth of *B. subtilis* RB14-C on nutrient agar was slightly inhibited by metabolites produced by *B. cepacia* BY. There were clear zones of inhibition (approximately 2 mm) visible between inoculum of BY and colonies of RB14-C spread over the plates. Opposite to RB14-C, BY was not sensitive to antibiotics produced by RB14-C. BY was resistant to iturin A.

10.2.2 INHIBITION OF *R. SOLANI* IN MICROCOSM STUDIES

The mycelium of *R. solani* grown on the filters in soil mixed with cells of *B. subtilis* RB14-C was slightly less dense and less branched than control mycelium recovered from nonbacterized soil (Figure 10.1a and b). The addition of *B. cepacia* BY to the soil strongly suppressed growth of the fungus (Figure 10.1c) and only a few hyphae were observed in several fields of view. A combination of both bacteria gave the best inhibition of *R. solani* (Figure 10.1d). Mycelium was almost not visible on the filters.

10.2.3 BIOCONTROL OF *R. SOLANI* IN POT EXPERIMENTS

Both bacterial strains, used in growth chamber experiments, significantly protected tomato plants against damping-off caused by *R. solani* (Table 10.1). When single strains were used, the best effect was obtained with *B. cepacia* applied 4 days before seed planting (BY4). Then, the number of plants in RB, BY2 and BY0 treatments did not differ significantly. Combining RB14-C with BY gave various results relating to the time of BY application. The best effect was obtained when BY was added to the soil 2 days after RB14-C inoculation (Figure 10.2). Protection of plants in this case

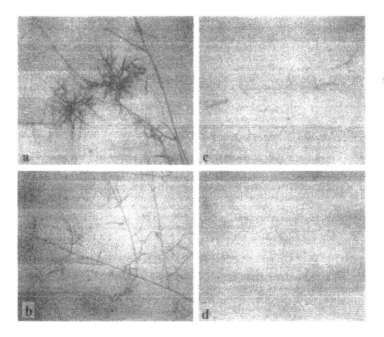

FIGURE 10.1 Light micrographs of the mycelial growth of *Rhizoctonia solani* in microcosms studies after 1 week of incubation in soil inoculated with *Bacillus subtilis* RB14-C and *Burkholderia cepacia* BY (magnification: 100 ×). Treatments: (a) control, (b) washed cells of RB14-C, (c) cells of BY, and (d) mixture of both bacteria.

TABLE 10.1

Infection of Tomato Plants Caused by *Rhizoctonia solani* in Soil Inoculated with *Bacillus subtilis* RB14-C and *Burkholderia cepacia* BY

Treatment	Number of Plants/Pot*	Dry Weight (%)
C	9.0[a]	100.0[ab]
RS	0.6[e]	59.9[c]
RB	3.4[d]	88.7[b]
BY0	2.3[d]	63.0[c]
BY2	2.9[d]	104.7[a]
BY4	4.6[bc]	103.5[a]
RB + BY0	2.9[d]	87.0[bc]
RB + BY2	5.1[b]	98.0[ab]
RB + BY4	3.7[bcd]	95.7[ab]

* Infection is expressed as a number of plants in pots and the percent of dry weight of the plants. Every pot was planted with nine germinated seeds. Dry weight was determined after 2 weeks of growth of tomato plants in the growth chamber. Data are the average of three experiments and they were analyzed using Student's *t*-test ($P \leq 0.05$). Letter "a" indicates the highest value. The same letters within a column mean no significant differences between the numbers.

was higher than when both strains were used separately, RB and BY2 (Table 10.1). Good effect was also obtained when RB and BY were added to the soil together (RB + BY4), but the number of plants in this combined treatment did not differ significantly compared to single treatments of RB and BY4. Application of BY immediately before seed planting did not improve the effect of RB14-C added 4 days before to the same soil.

R. solani strongly decreased dry weight of tomato plants (Table 10.1), whereas bacterial application significantly enhanced plant growth in soil infested with the pathogen. An exception was BY0, where the weight of plants was similar to plants infected with *R. solani*. Dry weight of plants grown in bacterized soil without *R. solani* did not differ significantly from that in control soil (data not presented). In infested soil, the weight of the plants was the highest in BY2 and BY4 treatments. Combining BY with RB14-C did not give any significant difference compared to control plants. RB14-C applied separately increased plant weight compared to RS, but it was lower than that of control plants. BY0 combined with RB14-C resulted in plant weight similar to that obtained in RB.

10.2.4 POPULATION OF *R. SOLANI* IN SOIL

Application of *B. subtilis* RB14-C decreased population of the pathogen in soil (Figure 10.3), but *B. cepacia* BY drastically reduced the population of *R. solani*.

FIGURE 10.2   Infection of tomato plants by *Rhizoctonia solani* in soil inoculated with *Bacillus subtilis* RB14-C and *B. cepacia* BY. (a) Control: Soil without the pathogen and bacteria. In treatments (b)–(e), *R. solani*, RB14-C, BY and both of RB14-C and BY were added to the soil before seed planting, respectively. The application method was described in detail in section 10.1.4.

The strongest reduction was obtained when BY was applied before seed planting. Introduction of this bacterium 4 and 2 days before seed planting (BY4 and BY2) also strongly decreased the population of the pathogen in soil, but the number of recovered colonies of *R. solani* was higher than in BY0 (Figure 10.3). Combining RB14-C with BY0 and BY4 did not vary from the effect indicated by single application of BY0 and BY4. Application of BY 2 days after RB14-C gave an intermediate effect compared to RB + BY0 and RB + BY4. However, in this combination the number of *R. solani* propagules in soil was significantly lower than in single treatments of RB and BY2.

10.2.5   NUMBER OF *B. SUBTILIS* RB14-C IN SOIL

B. subtilis RB14-C, 4 days after application, persisted in the soil at the approximate level of 7.0×10^8 cfu/g of dry weight of the soil and mostly in the form of spores (Table 10.2). Combining RB14-C with *B. cepacia* BY had no significant effect on

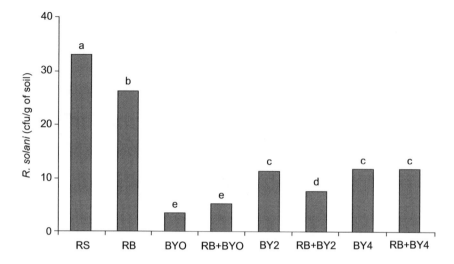

FIGURE 10.3   Population of *R. solani* in inoculated with *B. subtilis* RB 14-C and *B. cepacia* BY. The number of colonies expressed per gram of dry weight of the soil. Student's *t*-test ($P \leq 0.05$) was used to analyze the data. Letter "a" indicates the highest value. The same letters indicate no significant difference between bars.

TABLE 10.2
Number of *Bacillus subtilis* RB14-C in Soil
Treated with *Burkholderia cepacia* BY

Treatment	Total Number (cfu $\times$ 10⁸)	Spores (cfu $\times$ 10⁸)
RB	6.8[a]	6.4[a]
RB + BY0	5.5[a]	4.5[b]
RB + BY2	6.3[a]	4.0[b]
RB + BY4	6.5[a]	4.5[b]

Notes: Soil samples were taken before seed planting. The number of
bacteria was expressed as cfu/1 g of dry weight of the soil.
Data are the average of three experiments and they were analyzed using the Student's *t*-test ($P \leq 0.05$). Letter "a" indicates the highest value. The same letters within a column
mean no significant differences between the numbers.

the total population of RB14-C. The number of spores of *Bacillus* decreased significantly when BY was added.

10.3   DISCUSSION

Increasing the diversity of the biocontrol system, through the use of combined microorganisms with multiple modes of action, may be a means of achieving a higher level of disease control, because there is a greater probability that at least some of the mechanisms involved in biocontrol will be expressed over a wider range of environmental conditions. Both strains used in the present experiments, *B. subtilis* RB14-C and *B. cepacia* BY, belong to bacterial groups described as potential biocontrol agents with board spectrum of antifungal activity (22–29). However, authors did not find any report about combination of these bacteria to use in biocontrol. In microcosm studies, the mixture of RB14-C and BY applied to the soil almost completely inhibited growth of *R. solani* mycelium on filter disks buried in the soil (Figure 10.1). *B. cepacia* BY itself exhibited good fungal inhibition, but slow growth of the pathogen mycelium was still observed. *B. subtilis* only slightly reduced growth of *R. solani*.

In pot experiments, the best control of damping-off caused by *R. solani* was obtained only in combination, RB + BY2, where BY was applied to the soil 2 days after RB14-C (Table 10.1 and Figure 10.2). This combination of both agents gave significantly better plant protection compared to the effect of the bacteria used separately. Reduction of plant infection, similar to that in RB + BY2, was observed in BY4 (Table 10.1). However, when BY was mixed with RB14-C (RB + BY4) the additive control was not obtained, and plant protection was even lower than the effect of single BY4. A combination of RB14-C with BY added to the soil before seed planting (RB + BY0) also did not enhance plant protection compared to single bacterial application. There are reports that combined biocontrol agents did not always provide better control than the isolates used singly (9, 30). Negative interactions between introduced agents and also between agents and indigenous microflora can influence their performance in soil. Competition for nutrients or a detrimental effect of secondary metabolites produced by one organism on the growth of the other was reported as the cause of the mutual suppression (9, 31).

The *in vitro* test on agar plates showed that BY was not inhibited by RBI4-C and was resistant to pure iturin A. The strong resistance of Bcc bacteria to numerous antibiotics was reported in several papers (16, 24, 32). Oppositely to BY, RBl4-C was slightly inhibited by *B. cepacia* on the plates. However, in soil that kind of inhibition was not observed and the total number of RB14-C isolated from all soils was not significantly different (Table 10.2). Such discrepancy between *in vitro* and *in vivo* antagonism was also demonstrated (4, 9).

In microcosm tests, the mixture of RB14-C and BY almost completely reduced growth of *R. solani,* but strong inhibition of the pathogen could be because of lower density of *Rhizoctonia* in this test than in pot experiments. The soil in pots was infested with *R. solani* 3 days before bacterial application, allowing the pathogen to establish high population. In microcosms, a small agar disk of the fungus was placed on the filter and immediately buried in soil mixed with the bacteria. In this case all microbes were introduced to the system at the same time and the "starting" pathogen density was low. Moreover, in microcosms sterilized soil was used that eliminated the interactions of the agents with native soil microflora, whereas in pot experiments the soil was not autoclaved.

In the case of RB + BY2, RB14-C was introduced to the soil 2 days before BY. It was found that during 24 h after inoculation, RB14-C, added as a cell suspension, can produce antibiotic iturin A, and then active cells transform into spores (Szczech, unpublished data). After 2 days, co-inoculated BY could not be affected by not active RB14-C, and metabolites produced by *B. cepacia* probably did not affect resistant spores of *Bacillus*. Therefore, independent activity of both strains could be responsible for increased plant protection.

In RB + BY0 combination, where BY was added before seed planting, protection of the plants was at the level similar to that obtained by RB14-C only (Table 10.1). Single application of BY0 indicated the lowest control among other bacterial treatments, and dry weight of the plants was similar to that measured in RS. It shows that this application had only a slight effect on tomato protection, probably because of short-time between bacterial introduction and seed planting, whereas *R. solani* can infect plants within several hours (33). On the other hand, both single and combined introduction of BY before seed planting strongly reduced the number of *R. solani* in soil (Figure 10.2). In as much as this bacterium was applied as a cell suspension, it suggests that it is probably capable of releasing some antifungal substances almost immediately after introduction to the soil and reducing the population of the pathogen able to growth on the selective medium. A significant decrease of the pathogen density in soil by BY was also observed in other treatments, but it was not as strong like in the case of BY0. Perhaps the population of BY applied several days earlier could decrease and its effect on the pathogen could be lower.

The experiments did not show a clear relation between the reduced population of *R. solani* and the rate of disease control (Table 10.1 and Figure 10.3). This result suggests that inhibition of *R. solani* growth in soil is probably only partly responsible for the suppressive effect indicated by the bacteria, and the mechanism is more complex. RB14-C only slightly decreased the number of the pathogens in soil compared to BY, and it was confirmed in microcosm studies. In the case of *B. subtilis* RB14-C, its suppressiveness to *R. solani* depended on the production of the antibiotics iturin A and surfactin (34), but the exact mechanism still remains unclear. Bacteria Bcc are known to produce a wide range of secondary metabolites such as antibiotics and other nonvolatile and volatile compounds (16). It was found that *R. solani* is sensitive to pyrrolnitrin and phenazine produced by *B. cepacia* (14, 16), but another antibiotics can be also involved (15). Other modes of disease control can be competition for nutrients or promotion of plant growth (14, 19). This latter is under question, because the weight of tomato plants grown in volcanic soil and treated only with washed cells of RB14-C or BY did not indicate any significant differences compared to control plants (data not presented).

At present it is difficult to speculate the mode of action of used agents, and explanation needs future investigation. The current study provides evidence that some combinations of *B. subtilis* RB14-C with *B. cepacia* BY can lead to achieving greater damping-off suppression than that exhibited by these strains used separately. The protective effect was clearly related to the term of both agents' introduction. It is possible that by changing the system of bacterial application we can enhance the efficacy of that combination, but more complete comprehension of the ecology and dynamics of bacterial populations in soil is necessary for effective management of these biocontrol agents.

REFERENCES

1. Szczech, M., and Shoda, M. Biocontrol of *Rhizoctonia* damping-off of tomato by *Bacillus subtilis* combined with *Burkholderia cepacian*. *J. Phytopathol.*, 152, 549–556 (2004).
2. Weller, O. M. Biological control of soilborne pathogens in the rhizosphere with bacteria. *Annu. Rev. Phytopathol.*, 26, 379–407 (1998).
3. Koch, E. Evaluation of commercial products for microbial control of soil-borne plant diseases. *Crop Protect.*, 18, 119–125 (1999).
4. Pierson, E. A., and Weller, D. M. Use of mixtures of fluorescent pseudomonads to suppress take-all and improve the growth of wheat. *Phytopathology*, 84, 940–947 (1994).
5. Guetsky, R., Shtienberg, D., Elad, Y., and Dinoor, A. Combining bio-control agents to reduce the variability of biological control. *Phytopathology*, 91, 621–627 (2001).
6. Guetsky, R., Shtienberg, D., Elad, Y., Fischer, E., and Dinoor, A. Improving biological control by combining biocontrol agents each with several mechanisms of disease suppression. *Phytopathology*, 92, 976–985 (2002).
7. Duffy, B. K., and Weller, D. M. Use of *Gaeumannomyces graminis* var. *graminis* alone and in combination with fluorescent *Pseudomonas* spp. to suppress take-all of wheat. *Plant Dis.*, 79, 907–911 (1995).
8. Sung, K. C., and Chung, Y. R. Enhanced suppression of rice sheath blight using combination of bacteria which produce chitinases or antibiotics, pp. 370–372. In Ogoshi, A., Kobayashi, K., Homma, Y., Kodama, F., Kondo, N., and Akino, S. (eds.), *Plant Growth Promoting Rhizobacteria, Present Status and Future Prospects: Proceedings of the International Workshop on Plant Growth-Promoting Rhizobacteria*. Nakanishi Printing, Sapporo, Japan (1997).
9. De Boer, M., van der Sluis, I., van Loon, L. C., and Bakker, P. A. H. M. Combining fluorescent *Pseudomonas* spp. strains to enhance suppression of fusarium wilt of radish. *Eur. J. Plant Pathol.*, 105, 201–210 (1999).
10. Ohno, A., Ano, T., and Shoda, M. Production of a lipopeptide antibiotic surfactin with recombinant *Bacillus subtilis*. *Biotech. Letts.*, 14, 1165–1168 (1992).
11. Huang, C. C., Ano, T., and Shoda, M. Nucleotide sequence and characteristic of the gene, *lpa-14*, responsible for biosynthesis of lipopeptide antibiotics iturin A and surfactin from *Bacillus subtilis* RB14. *J. Ferment. Bioeng.*, 76, 445–450 (1993).
12. Maget-Dana, R., and Peypoux, F. Iturins, a special class of pore-forming lipopeptides: biological and physicochemical properties. *Toxicology*, 87, 151–174 (1994).
13. Burkhead, K. D., Schisler, D. A., and Slininger, P. J. Pyrrolnitrin production by biological control agent *Pseudomonas cepacia* B37w in culture and in colonized wounds of potatoes. *Appl. Environ. Microbiol.*, 60, 2031–2039 (1994).
14. Bevivino, A., Sarrocco, S., Dalmastri, C., Tabacchioni, S., Cantale, C., and Chiarini, L. Characterization of a free-living maize-rhizosphere population of *Burkholderia cepacia*: effect of seed treatment on disease suppression and growth promotion of maize. *FEMS Microbiol. Ecol.*, 27, 225–237 (1998).
15. Kang, Y., Carlson, R., Tharpe, W., and Schell, M. Characterization of genes involved in biosynthesis of a novel antibiotic from *Burkholderia cepacia* BCI 1 and their role in biological control of *Rhizoctonia solani*. *Appl. Environ. Microbiol.*, 64, 3939–3947 (1998).
16. Parke, J. L., and Gurian-Sherman, D. Diversity of the *Burkholderia cepacia* complex and implications for risk assessment of biological control strains. *Ann. Rev. Phytopathol.*, 39, 225–258 (2001).
17. Hwang, J., Chilton, W. S., and Benson, D. M. Pyrrolnitrin production by *Burkholderia cepacia* and biocontrol of *Rhizoctonia* stem rot of poinsettia. *Biol. Control*, 25, 56–63 (2002).

18. Miller, S. C. M., LiPuma, J. J., and Parke, J. L. Culture-based and non-growth-dependent detection of the *Burkholderia cepacian* complex in soil environments. *Appl. Environ. Microbiol.*, 68, 3750–3758 (2002).

19. Zaki, K., Misaghi, I. J., and Heydari, A. Control of cotton seedling damping-off in the field by *Burkholderia (Pseudomonas) cepacia*. *Plant Dis.*, 82, 291–293 (1998).

20. Heungens, K., and Parke, J. L. Postinfection biological control of *Oomycete* pathogens of pea by *Burkholdena cepacia* AMMDRI. *Phytopathology*, 91, 383–391 (2001).

21. Ko, W., and Hora, F. K. A selective medium for the quantitative determination of *Rhizoctonia solani* in soil. *Phytopathology*, 61, 707–710 (1971).

22. Cartwright, D. K., Chilton, W. S., and Benson, D. M. Pyrrolnitrin and phenazine production by *Pseudomonas cepacia*, strain 5.58, a biocontrol agent of *Rhizoctonia solani*. *Appl. Microbiol. Biotechnol.*, 43, 211–216 (1995).

23. Nemec, S., Datnoff, L. E., and Strandberg, J. Efficacy of biocontrol agents in planting mixes to colonize plant roots and control root diseases of vegetables and citrus. *Crop Protect.*, 15, 735–742 (1996).

24. Brannen, P. M., and Kenney, D. S. Kodiak—a successful biological control product for suppression of soil-borne plant pathogens of cotton. *J. Industr. Microbiol. Biotechnol.*, 19, 169–171 (1997).

25. Sfalanga, A., Di Cello, F., Mugnai, L., Tegli, S., Fani, R., and Surico, G. Isolation and characterisation of a new antagonistic *Burkholderia* strain from the rhizosphere of healthy tomato plants. *Res. Microbiol.*, 150, 45–59 (1999).

26. Li, W., Roberts, D. P., Dery, P. D., Meyer, S. L. F., Lohrke, S., Lumsden, R. D., and Hebbar, K. P. Broad spectrum anti-biotic activity and disease suppression by the potential biocontrol agent *Burkholderia ambifaria* BC–F. *Crop Protect.*, 21, 129–135 (2002).

27. Wulff, E. G., Mguni, C. M., Mortensen, C. N., Keswani, C. L., and Hockenhull, J. Biological control of black rot (*Xanthomonas campestris* pv. *campestris*) of brassicas with an antagonistic strain of *Bacillus subtilis* in Zimbabwe. *Eur. J. Plant Pathol.*, 108, 317–325 (2002).

28. Collins, D. P., and Jacobsen, B. J. Optimizing a *Bacillus subtilis* isolate for biological control of sugar beet cercospora leaf spot. *Biol. Control*, 26, 153–161 (2003).

29. Obagwu, J., and Korsten, L. Integrated control of citrus green and blue molds using *Bacillus subtilis* in combination with sodium bicarbonate or hot water. *Postharv. Biol. Technol.*, 28, 187–194 (2003).

30. Larkin, R. P., and Fravel, D. R. Efficacy of various fungal and bacterial biocontrol organisms for control of *Fusarium* wilt of tomato. *Plant Dis.*, 82, 1022–1028 (1998).

31. Raaijmakers, J. M., van der Sluis, I., Koster, M., Bakker, P. A. H. M., Weisbeek, P. J., and Schippers, B. Utilization of heterologous siderophores and rhizosphere competence of fluorescent *Pseudomonas* spp. *Can. J. Microbiol.*, 41, 126–135 (1995).

32. Holmes, A., Govan, J., and Goldstein, R. Agricultural use of *Burkholderia* (Pseudomonas) *cepacia*: a threat to human health? *Emerg. Infect. Dis.*, 4, 221–227 (1998).

33. Hayman, D. S. The influence of cotton seed exudate on seedling infection by *Rhizoctonia solani*, pp. 99–102. In Toussoun, T. A., Bega, R. V., and Nelson, P. E. (eds.), *Root Diseases and Soilborne Pathogens*. University of California Press, Berkeley, CA (1970).

34. Asaka, O., and Shoda, M. Biocontrol of *Rhizoctonia solani* damping-off of tomato with *Bacillus subtilis* RB14-C. *Appl. Environ. Microbiol.*, 62, 408–4085 (1996).

11 Practical Application of *B. subtilis* in Fruit Gardens for Biocontrol of Diseases

11.1 MASS PRODUCTION OF *B. SUBTILIS* CULTURE BROTH

One pharmaceutical company has produced approximately 500 liters of *Bacillus subtilis* RB14 culture, in which the cell concentration was between 10^{10} and 10^{11} cells/ml. The company guarantees this high concentration of *B. subtilis* for use in fruit gardens. Details for the medium composition and operational conditions to produce the bacterial culture at a reasonable price are not available mainly because the methods and original technique are proprietary information for the company. The bacterial culture was purchased and stored in a large chilled facility until use in the fruit garden.

11.2 APPLICATION OF *B. SUBTILIS* TO CHERRY

The fruit garden is managed as a tourist fruit garden, where customers can harvest fruits by themselves in the garden.

The area of the garden for cherry cultivation is 9000 m². The original cell suspension was diluted by 500 times to equal the approximate cell concentration of 10^8 cells/ml. The diluted cell suspension was loaded in a sprayer tank (as shown in Figure 11.1) that was originally developed for spraying chemical pesticides.

In spring, when the leaves began to emerge, the cell suspension was applied once every 1 or 2 weeks. When the possibility of disease occurrence was high, the spraying frequency was increased (Figure 11.1).

In June when the cherries are ripe and ready for harvest, the garden is open to customers who come in groups by cars or buses and enjoy harvesting the cherries (Figure 11.2). This practical prevention of diseases has continued for more than 15 years without any serious problems.

The use of this bacterium in this garden is intensively focused on the prevention of diseases before cherry is infested, which has been effective for maintaining its reputation among customers over such a long period of time.

FIGURE 11.1   Photos of a sprayer used for chemical pesticides and *B. subtilis* culture application using a sprayer.

11.3   CHEMICAL PESTICIDE TEST

Other fruit gardens use excessive amounts of chemical pesticides, and airborne contamination of the fruit garden cannot be avoided. Therefore, detection of chemical pesticides has been conducted by an authorized institution once every 3 years. The list of chemical pesticides tested is shown in Table 11.1. No significant amount of chemical pesticides has been detected in the fruit produced in the fruit garden.

11.4   APPLICATION TO GRAPES

The changes in grape leaves infested by plant pathogens before and after treatment with the bacterial culture are shown in Figure 11.3.

FIGURE 11.2   A photo of children enjoying chemical pesticide-free cherries.

Grapes needed treatment with chemical pesticides several times throughout normal cultivation. As a preliminary test, the application of *B. subtilis* suspension was conducted at the same time as the chemical pesticides were to be applied. Table 11.2 shows the spraying time of *B. subtilis* suspension for grapes at the same time when the application of chemical pesticides was practically performed. When phytohormone gibberellin was applied to grapes to produce seedless grapes, the application of *B. subtilis* did not influence the subsequent growth of grapes. After 3 weeks of treatment, the quality and quantity of the grapes harvested were the same as those dosed with chemical pesticides.

11.5   ADVANTAGES OF *B. SUBTILIS* USE

1. Heavy protective equipment (masks, hat, gloves, and clothes) is needed when chemical pesticides are applied as shown in Figure 11.4. However, no such equipment is necessary when *B. subtilis* is used (Figure 11.1).
2. Many children suffering from chemical pesticide allergies enjoy the garden's fruit without allergic reactions (Figure 11.2).
3. Customers who visit the garden come repeatedly, and the number of visitors increases every year.

TABLE 11.1

Chemical Pesticides List Applied for Detection Test

Chemical Pesticide	Detected Concentrations (mg/l)	Chemical Pesticide	Detected Concentrations (mg/l)
Acrinathrin	ND	Triadimefon	ND
Acetamiprid	ND	Tolclofos-methyl	ND
Isoxathrion	ND	Tolufenpyrad	ND
Isoprothiolane	ND	Bitertanol	ND
Imibenconazole	ND	Bifenthrin	ND
Edifenphos	ND	Pyridaben	ND
Ethofenprox	ND	Fipronil	ND
Endosulfan	ND	Fenarimol	ND
Kresoxim-methyl	ND	Fenitrothion	ND
Chlorpyrifos	ND	Fenthion	ND
Chlorpyrifos-methyl	ND	Phenthoate	ND
Chlorfenapyr	ND	Fenbuconazole	ND
Diethophen-carb	ND	Fenpropathrin	ND
Diclocymet	ND	Fthalide	ND
Dichlofenthion	ND	Buprofezin	ND
Difenoconazole	ND	Fludioxonil	ND
Cyfluthrin	ND	Flucythrinate	ND
Cyproconazole	ND	Fluvalinate	ND
Cypermethrin	ND	Procymidone	ND
Dimethoate	ND	Prothiofos	ND
Diazinon	ND	Propargite	ND
Tebuconazole	ND	Hexaconazole	ND
Tebufenpyrad	ND	Permethrin	ND
Delatamethrin	ND	Phosalone	ND
Triadimefon	ND	Malathion	ND
		Metalaxyl	ND
		Methidathion	ND
		Mepronil	**ND**

Note: ND, not detected.

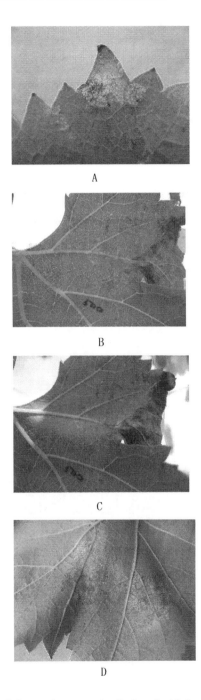

FIGURE 11.3 Photos of changes in a grape leaf infested with downy mildew. (a) Grape leaf infested with downy mildew. (b) The leaf treated with *B. subtilis* culture. (c) The leaf 2 days after treatment with *B. subtilis* culture. (d) The leaf 4 days after treatment with *B. subtilis* culture.

TABLE 11.2

***B. subtilis* Spraying Schedule to Grapes**

Time	Date	Dilution Ratio[a]	Volume Sprayed (l)	Comments
1	March 18	500	250	Spraying to leaves
2	23	500	250	
3	28	500	250	No damage observed
4	30	300	15	
5	April 1	500	300	
	5	Gibberellin treatment		Co-use of *Bacillus* and gibberellin is possible
6	10	500	250	
7	16	500	250	
	18	Gibberellin treatment		
8	May 15	500	250	
9	June 1	500	250	
	13	Detection test of chemical pesticides		
	20			Harvesting, no damage

[a] Crude *Bacillus* suspension was diluted 500 times.

FIGURE 11.4 An example of a farmer spraying chemical pesticides wearing heavy protective equipment.

12 Conclusions

In this book, several kinds of isolated *Bacillus subtilis* were introduced, and the conclusions are as follows based on their characteristics and the results obtained.

1. *B. subtilis* is a safe microorganism and is related to *B. natto*.
2. *B. subtilis* showed suppressive effects against many plant pathogens *in vitro*.
3. In a pot test, *B. subtilis* exhibits suppressive effects against damping-off, *Fusarium* wilt, and crown and root rot caused by *Fusarium oxysporum*, as well as bacterial wilt caused by *Pseudomonas solanacearum*, and root rot caused by *Phomopsis* sp.
4. *B. subtilis* suppression of plant pathogens is based on a multifunctional mechanism.
   Three lipopeptide antibiotics are identified, and one siderophore is involved. Some enzyme production is speculated.
   The scheme of this result is shown in Figure 12.1.
   The siderophore captures Fe^{3+} in soil as shown in Figure 12.2.
5. To obtain *B. subtilis* cells, both submerged fermentation (SmF) and solid-state fermentation (SSF) can be used. The productivity of lipopeptide antibiotics is significantly higher in SSF than in SmF.
6. Products obtained from SSF using soybean curd residue as a solid material can be reutilized in agricultural fields as a dual-functional product of an organic fertilizer and a biocontrol agent. The recycling system is shown in Figure 12.3. The product obtained from SSF was used for plant tests for the suppression of the damping-off caused by *Rhizoctonia solani*.
7. Metallic ions are important factors for the mass production of spores of *B. subtilis* in SmF.
8. Products mainly containing *B. subtilis* spores can be preserved for months without any significant changes in their suppressive activity.
9. Once soil is treated with *B. subtilis* spores, the bacterial cells can survive in soil for a long period of time as spores.
10. The direct treatment of diluted *B. subtilis* culture to fruits in a fruit garden was effective to suppress the growth of plant pathogens without use of chemical pesticides.
11. The use of *B. subtilis* can be effective not only for the direct treatment to plants or soil but also for seed coating or by dipping plant roots in a bacterial solution as shown in Figure 12.4.

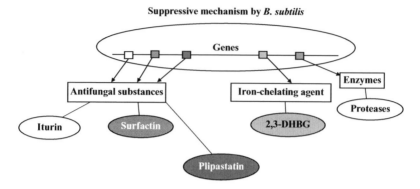

FIGURE 12.1 The scheme of isolated multifunctional *B. subtilis*, which suppresses various plant pathogens. The structures of iturin and surfactin, plipastatin, and 2,3-DHBG are shown in Figure 3.7, Figure 5.15, and Figure 5.11, respectively.

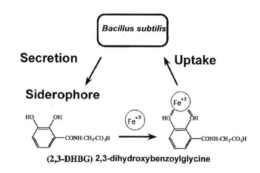

FIGURE 12.2 The role of the siderophore produced by *B. subtilis* in soil.

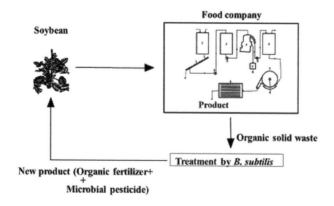

FIGURE 12.3 Recycling system in which solid waste from a food company is treated with *B. subtilis*, and the product containing a bifunction of biocontrol agent composed of *B. subtilis* and an organic fertilizer can be fed back to the agricultural field.

Development of new multifunctional *Bacillus subtilis*

A. Application of *B. subtilis* to seeds or roots of plants

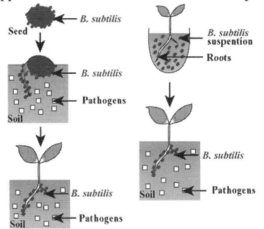

B. Introduction of new genes to *B. subtilis*

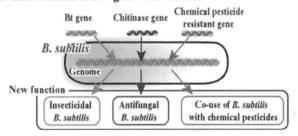

FIGURE 12.4 Future applications of *B. subtilis*. (A) Seed coating with *B. subtilis* cells and dipping plant roots in a *B. subtilis* solution. (B) Enhancement of the abilities of *B. subtilis* by genetic engineering.

12. Isolated *B. subtilis* is resistant to some chemical pesticides and, thus, co-utilization with chemical pesticides can reduce the required amount of chemical pesticides.

13. Inoculation of *B. subtilis* in a composting process is possible to produce effective compost.

14. Genetic manipulation is possible for enhancing activity, as shown in Figure 12.4.

Index